Singularities, Mirror Symmetry, and the Gauged Linear Sigma Model

CONTEMPORARY MATHEMATICS

763

Singularities, Mirror Symmetry, and the Gauged Linear Sigma Model

Summer School
Crossing the Walls in Enumerative Geometry
May 21 - June 1, 2018
Snowbird, Utah

Tyler J. Jarvis
Nathan Priddis
Editors

AMS AMERICAN MATHEMATICAL SOCIETY

2020 *Mathematics Subject Classification*. Primary 14J33, 14N35, 14J32, 14B05, 14D23, 14H10; Secondary 53D37, 53D45, 32Q25, 32S05, 32S30.

Library of Congress Cataloging-in-Publication Data

Library of Congress Cataloging-in-Publication Data

Names: Jarvis, Tyler Jamison, editor. | Priddis, Nathan C., 1983- editor. | Summer School on Crossing the Walls in Enumerative Geometry (2018 : Snowbird, Utah)

Title: Singularities, mirror symmetry, and the gauged linear sigma model : Summer School on Crossing the Walls in Enumerative Geometry, May 21-June 1, 2018, Snowbird, Utah / Tyler J. Jarvis, Nathan Priddis, editor.

Description: Providence, Rhode Island : American Mathematical Society, [2021] | Series: Contemporary mathematics, 0271-4132 ; volume 763 | "This volume contains lecture notes from several [Summer School] short courses as well as some additional papers on current topics"–Preface. | Includes bibliographical references.

Identifiers: LCCN 2020042993 | ISBN 9781470457006 (paperback) | ISBN 9781470464196 (ebook)

Subjects: LCSH: Gromov-Witten invariants–Congresses. | Singularities (Mathematics)–Congresses. | Mirror symmetry–Congresses. | Intersection theory (Mathematics)–Congresses. | Calabi-Yau manifolds–Congresses. | Moduli theory–Congresses. | AMS: Algebraic geometry – Surfaces and higher-dimensional varieties – Mirror symmetry. | Algebraic geometry – Projective and enumerative geometry – Gromov-Witten invariants, quantum cohomology, Gopakumar-Vafa invariants, Donaldson-Thomas invariants. | Algebraic geometry – Surfaces and higher-dimensional varieties – Calabi-Yau manifolds. | Algebraic geometry – Local theory – Singularities. | Algebraic geometry – Families, fibrations – Stacks and moduli problems. | Algebraic geometry – Curves – Families, moduli (algebraic). | Differential geometry – Symplectic geometry, contact geometry | Differential geometry – Symplectic geometry, contact geometry | Several complex variables and analytic spaces – Complex manifolds – Calabi-Yau theory. | Several complex variables and analytic spaces – Singularities – Local singularities.

Classification: LCC QA665 .S56 2021 | DDC 516.3/6–dc23

LC record available at https://lccn.loc.gov/2020042993

Contemporary Mathematics ISSN: 0271-4132 (print); ISSN: 1098-3627 (online)

DOI: https://doi.org/10.1090/conm/763

∞ The paper used in this book is acid-free and falls within the guidelines
established to ensure permanence and durability.
Visit the AMS home page at `https://www.ams.org/`

10 9 8 7 6 5 4 3 2 1 26 25 24 23 22 21

Contents

Preface

As part of activities supported by a National Science Foundation FRG grant (DMS-1564502), a summer school was held at Snowbird, Utah, in 2018, offering an array of short topics courses for graduate students in the general area of Gromov–Witten theory and enumerative geometry. We felt that it would help graduate students to access a broad area of advanced topics which they may not have through the normal graduate course offerings of their math departments. This volume contains lecture notes from several of these short courses as well as some additional papers on current topics.

Several years earlier another summer school was held in Colorado, providing an extensive treatment of the more basic topics in Gromov–Witten theory. Building on the success of Colorado summer school, the focus of current summer school was to capture the more recent developments of the subject, such as quasimaps and the gauged linear sigma model.

The fundamental object of Gromov–Witten theory is the notion of stable map $f : C \to X$ from a nodal curve C to a target X which is a Kähler or symplectic manifold. But it is very difficult to do any computation for a general target. During last thirty years, nearly all of the computations that have been done are for when X is a complete intersection in a toric variety. Recently, there has been a strong push to extend the calculations to complete intersections of nonabelian targets such as Grassmannians or flag varieties. In all these cases, the target X is embedded in an ambient space of the form of a GIT quotient $V/\!/_\theta G$, where V is a representation of a reductive group G and θ is a polarization.

It was known classically that when we vary the polarization θ, the quotient $V/\!/_\theta G$ undergoes a certain birational transformation called a *wall crossing*. In this setting, Gromov–Witten theory can be described as what is known as a *gauged theory*. In 1992, Witten proposed a physical model exactly for this situation. He named his model the *gauged linear sigma model* or *GLSM*. During the next twenty years, the GLSM became popular in physics, and was studied by many physicists. In retrospect, it is surprising that is has taken so long to get into mathematics.

Let me explain the basic set-up for a GLSM. Take the famous quintic 3-fold defined by

$$X_5 = \left\{ [x_0, x_1, x_2, x_3, x_4] \in \mathbb{P}^4 | \sum_i x_i^5 = 0 \right\}.$$

Instead of $\mathbb{P}^4$, the GLSM target is the total space of the GIT-quotient

$$(\mathbb{C}^5 - (0,0,0,0,0,0)) \times \mathbb{C}/\!/_\theta \mathbb{C}^* = \mathcal{O}(-5)$$

over $\mathbb{P}^4$, together with a *superpotential*

$$W = p \sum_i x_i^5.$$

In this setting X_5 is interpreted as the critical locus

$$\mathrm{Crit}_W = \{x \mid dW_x = 0\}.$$

In general, it is always a good idea to express the object of interest as the critical locus of a holomorphic function. An additional key ingredient is an **R**-*charge*—a $\mathbb{C}^*$-action such that W has degree one with respect to this action. The GLSM can be thought as the quantization of the preceding picture by a $2d$ quantum field theory.

The collection of lecture notes in this book explains many ingredients of the GLSM as well as some related theories. Following is a short description for each article that the reader can take as a guide.

(1) **Rachel Webb. Quasimaps and some examples of stacks for everyone.**
The mathematical GLSM is built upon quasimap theory. This article is an introduction to quasimaps, beginning with the motivation of allowing basepoints in maps. It also includes a discussion on stack quotients of a reductive group acting on an affine variety and GIT quotients (as orbifolds). The main goal is to introduce the moduli space of stable quasimaps together with its virtual cycle. The article ends with a discussion of nonabelian targets.

(2) **Emily Clader. Introduction to the gauged linear sigma model.**
This article begins with a discussion of orbifolds, orbifold line bundles, and orbifold cohomology. It introduces p-fields and variation of GIT and shows how Gromov–Witten theory and certain other curve-counting theories can be realized as a GLSM, ending with a discussion of wall-crossing formulas.

(3) **Dustin Ross. Localization and mirror symmetry.**
The GLSM contains both Gromov-Witten theory and FJRW theory as special cases. The relation between them is described by the variation of polarization of the ambient GIT quotient. This article gives a proof of mirror symmetry statements for the quintic, using localization. It introduces FJRW theory and Gromov–Witten theory, shows how to construct the J-functions and I-functions for these theories and then proves the mirror theorems for these two theories.

(4) **Wei-Ping Li. A brief introduction to cosection localization and p-fields.**
There are two motivating examples for the GLSM: Chang-Li's theory of p-fields and Clader's theory of the hybrid model. In particular, Chang-Li introduced a fundamental technique called *cosection localization*, originating from Kiem and Li, to construct a virtual cycle in p-field theory. This technique is also very important for the GLSM. This article gives a good introduction to this topic as well as the topic of p-fields, using basic examples as motivation and for understanding the construction.

The main goal of this article is to show how the virtual fundamental class can be constructed using these techniques.

(5) **Mark Shoemaker. Virtual classes via matrix factorizations.**
The cosection localization technique only constructs the virtual class in the case of narrow or compact-type insertions. This article describes the construction of the virtual class by 2-periodic complexes using the example of the zero section of a vector bundle. The main goal of this article is to present the ideas of the construction of the virtual class for the case of broad insertions using an algebraic method (an analytic approach was carried out by a different group). It should be read as a preparatory article to the more advanced article by Shoemaker and his coauthors detailing the algebraic construction of the virtual class.

(6) **Jeongseok Oh. Localized Chern characters for 2-periodic complexes and virtual cycles.**
This article reviews the construction of a virtual fundamental class via a perfect obstruction theory. It describes the virtual fundamental class via localized Chern characters and compares the classical virtual class for the quintic to the virtual class constructed this way.

(7) **Todor Milanov. Singularity theory and mirror symmetry.**
This article gives an introduction to Frobenius manifolds and Givental's R-matrix. These topics are less directly connected to the GLSM, but are essential in mirror symmetry. It details the construction of the Frobenius manifold associated to the Fermat hypersurfaces, including the I-function, the mirror map, and analytic continuation on the Frobenius manifold.

The remaining articles are more research oriented and have a looser connection to the GLSM. They should give readers a broader perspective of the subject.

(8) **Ursula Whitcher. Counting points with Berglund-Hübsch-Krawitz mirror symmetry.**
This article gives a different perspective to mirror symmetry, namely it considers Calabi-Yau varieties over finite fields and BHK mirror symmetry in this setting. The main result is that if two polynomials are both dual (under BHK mirror symmetry) to two polynomials with the same weights, then the zeta functions share a common factor. Furthermore, in this case, the number of points on the hypersurfaces over a finite field of order q on each variety is equivalent modulo q.

(9) **Honglong Fan and Y.P. Lee. Variations on the theme of quantum Lefschetz.**
This paper discusses several recursions that allow one to compute Gromov–Witten invariants via twisted invariants of some master space, reframing the computations in terms of Givental's formalism in order to extend from the case of Calabi–Yau hypersurfaces to more general targets. It explores the recursions one obtains by varying the master space. Finally, it considers the case where the Gromov–Witten theory of the hypersurfaces is trivial (such as for a K3 surface), and thereby obtains interesting relations on the Gromov–Witten theory of the ambient space.

(10) **Rongxiao Mi. Extremal transitions and quantum D-modules.**

This article presents a conjecture about the relationship between quantum D-modules for two smooth projective varieties that are related by extremal transitions. It also presents several examples of Type II extremal transitions that support the conjecture.

(11) **Chiu-Chu Melissa Liu. A Lecture on Holomorphic Anomaly Equations and Extended Holomorphic Anomaly Equations.**

This article presents a review of the holomorphic anomaly equations and introduces extended holomorphic anomaly equations. It includes a discussion of Gromov–Witten invariants for closed strings and for open strings, Teichmüller space, the Hodge bundle and Gauss-Manin connection, and various other constructions. It concludes with mirror symmetry conjectures about the potential functions for Gromov–Witten invariants in various genera and for various numbers of marked points.

Yongbin Ruan
Nathan Priddis

Contemporary Mathematics
Volume **763**, 2021
https://doi.org/10.1090/conm/763/15340

Quasimaps and some examples of stacks for everybody

Rachel Webb

ABSTRACT. This note introduces the theory of quasimaps to GIT quotients with intuition and concrete examples, with the goal of explaining a closed formula for the quasimap I-function. Along the way, it emphasizes aspects of this story that illustrate general stacky concepts.

1. Introduction

In modern algebraic geometry, stacks are pervasive. To name just a few appearances, they arise as quotients of varieties by group actions, as the algebraic analog of orbifold curves, and as moduli spaces of objects with nontrivial automorphism groups. Unfortunately the theory is laden with technicalities, making it difficult for newcomers. This note provides concrete examples of stacks that arise in the theory of *quasimaps to GIT quotients.* Quasimaps are themselves a topic in Gromov-Witten theory with a growing list of applications.

Primarily a friendly introduction to quasimaps, this note is also a "second course" to Fantechi's brief introduction to stacks [**16**], a tour of some examples of stacks "in the wild." As an introduction to quasimaps, the goal of this note is to present a closed formula for the quasimap I-function. Along the way, we'll highlight examples that illustrate general stacky concepts with the heading STACK EXAMPLE. We will always work over $\mathbb{C}$.

2. Motivation for the study of quasimaps

The theory of ϵ-stable quasimaps to GIT quotients, as presented in this note, was introduced by Ciocane-Fontanine, Kim, and Maulik in a series of papers: [**9**], [**8**], [**7**], [**12**], and [**13**]. For the sake of novel and concrete exposition, this note omits much of the historical and mathematical context of the theory, which can be found in the original papers and in the survey articles [**11**] and [**22**].

Gromov-Witten theory begins with the study of maps from a smooth genus-g curve with n marks to a projective target, for example to $\mathbb{P}^n$. The set of such maps having a fixed degree d forms a moduli space, denoted (in this example) $\mathcal{M}_{g,n}(\mathbb{P}^n, d)$. Unfortunately, $\mathcal{M}_{g,n}(\mathbb{P}^n, d)$ is not compact; in order to define Gromov-Witten invariants of $\mathbb{P}^n$ as integrals over a moduli space of maps, we must

2010 *Mathematics Subject Classification.* Primary 14-02.
Key words and phrases. Quasimaps, abelian/nonabelian correspondence, algebraic stacks.
The author was supported in part by NSF RTG grant 1045119.

replace $\mathcal{M}_{g,n}(\mathbb{P}^n, d)$ with a compactification. One such compactification is the moduli space of Kontsevich-stable maps $\overline{\mathcal{M}}_{g,n}(\mathbb{P}^n, d)$. These stable map moduli spaces are the central objects of Gromov-Witten theory.

The Kontsevich moduli space compactifies $\mathcal{M}_{g,n}(\mathbb{P}^n, d)$ by allowing the source curve to be nodal, but there are other ways to compactify. One may see hints of these other ways by looking at possible limits of families of maps in $\mathcal{M}_{g,n}(\mathbb{P}^n, d)$. Let C be equal to $\mathbb{P}^1$ with homogeneous coordinates $[x : y]$ and markings at $[1 : 1]$ and $[2 : 1]$. Define

$$\phi : \mathbb{C}^* \times \mathbb{P}^1 \to \mathbb{P}^2 \qquad \phi_a(x, y) = [ax^2 : xy : y^2] \quad \text{for} \quad a \in \mathbb{C}^*. \tag{2.1}$$

This is a family in $\mathcal{M}_{0,2}(\mathbb{P}^2, 2)$ with base $\mathbb{C}^*$. To extend it over the origin to a flat family over $\mathbb{C}$, we need to define a map ϕ_0. The natural choice seems to be

$$\phi_0(s, t) = [0 : xy : y^2], \tag{2.2}$$

but this has a basepoint (is undefined) at $y = 0$. To recover the limit in $\overline{\mathcal{M}}_{0,2}(\mathbb{P}^3, 2)$, we resolve the rational map $\phi : \mathbb{A}^1 \times \mathbb{P}^1 \dashrightarrow \mathbb{P}^2$ (given by (2.1) and (2.2)) by blowing up this basepoint, adding an extra rational curve in the fiber over 0. This produces a morphism $\widetilde{\phi}$ from the blowup to $\mathbb{P}^2$, and $\widetilde{\phi}_0$ has a source curve that is two copies of $\mathbb{P}^1$, glued at a node. The limiting map has degree 1 on each copy. This is depicted in Figure 1.

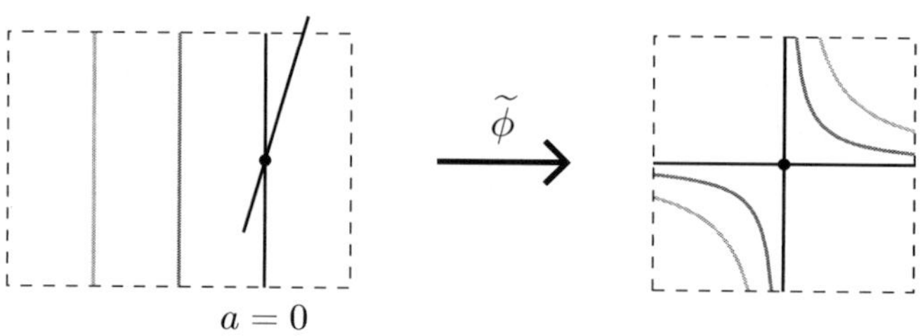

FIGURE 1. The morphism $\widetilde{\phi}$ maps from $\mathrm{Bl}_{a=y=0}\mathbb{A}^1 \times \mathbb{P}^1$ on the left to $\mathbb{P}^2$ on the right, depicted here in the chart of $\mathbb{P}^2$ where the middle coordinate is nonzero. The varying shades of gray show the fibers of this family of stable maps. The fiber over $a = 0$ is a map from a nodal curve.

However, what happens if we compactify $\mathcal{M}_{0,2}(\mathbb{P}^2, 2)$ by allowing basepoints? That is, what if we take $\phi_0(x, y) = [0 : xy : y^2]$ as a rational map from $\mathbb{P}^1$ to $\mathbb{P}^2$ to be the limit of (2.1)? Indeed, this rational map is a *stable quasimap*, depicted in Figure 2.

We will see that with the right definitions, stable quasimaps form a moduli space that is just as well-behaved as the Kontsevich moduli spaces. In fact, stability for quasimaps depends on a postive rational parameter ϵ, giving us a whole collection of moduli spaces! This collection has the following advantages:

(1) When $\epsilon > 2$, the quasimap moduli space is equal to the familiar Kontsevich moduli space, and its invariants are Gromov-Witten invariants.
(2) When ϵ is sufficiently small, certain quasimap invariants (the genus-0 invariants) are easier to compute. A generating function for these explicit invariants is called the *quasimap I-function.*

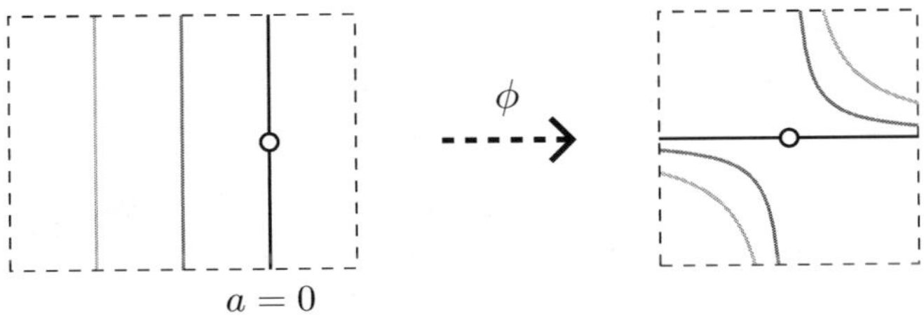

FIGURE 2. The rational map ϕ takes $\mathbb{A}^1 \times \mathbb{P}^1$ on the left to $\mathbb{P}^2$ on the right. The varying shades of gray show the fibers of this family of quasimaps. In particular, the fiber over $a = 0$ is a rational map of degree 1 with a basepoint of length 1 (see Definition 5.3).

(3) One can "cross the wall," relating quasimap invariants for differing values of ϵ, thereby (roughly) expressing Gromov-Witten invariants in terms of the quasimap I-function.

These statements are a heuristic only; for careful statements and their proofs, see [**8**] and [**13**]. For certain targets, invariants with small ϵ and $g = 0$ were first studied by Givental [**18**], while the strategy in (3) above for computing Gromov-Witten invariants was first employed by Bertram [**3**].

3. Quotients in algebraic geometry

When a reductive algebraic group G acts on an affine variety X, we'd like to take the quotient, producing an algebraic object X/G. Unfortunately, there may not be a scheme X/G satisfying the universal property of a quotient—if it exists, this scheme is called a *categorical quotient*. Even if a categorical quotient scheme exists, it may not have the expected topology (e.g., be a *geometric quotient*). Therefore, defining a quotient either requires us to modify the original data X and G, or to leave the category of schemes. Both strategies arise in quasimap theory. The resulting quotients are respectively called *geometric invariant theory (GIT) quotients* and *stack quotients*.

We briefly summarize the key definitions of GIT for affine X when G acts with no kernel, as found in [**23**]. Fix a character θ of G. Given a nonnegative integer n, a function $f \in \Gamma(X, \mathscr{O}_X)$ is a *relative invariant of weight* θ^n if for every $x \in X$ we have $f(g \cdot x) = \theta(g)^n f(x)$.

DEFINITION 3.1. A point $x \in X$ is *θ-semistable* if there exists an integer $n \geq 1$ and a relative invariant f of weight θ^n such that $f(x) \neq 0$. If moreover the dimension of the orbit $G \cdot x$ is equal to the dimension of G and the G-action on $\{y \in X \mid f(y) \neq 0\}$ is closed, then x is *θ-stable*.

The upshot of these definitions is that we get an open locus $X^{ss}_\theta(G) \subset X$ of semistable points with respect to θ, and a smaller locus $X^s_\theta(G) \subset X^{ss}_\theta(G)$ of stable points. We define the locus of *unstable points* to be $X^{us}_\theta(G) = X \setminus X^{ss}_\theta(G)$. The sets $X^s_\theta(G)$ and $X^{ss}_\theta(G)$ can be computed with [**23**, Prop 2.5]. We include the group G in our notation here because later we will vary it; when there is no risk of misunderstanding, we will omit both the character and the group from the notation. In this note, we will assume

- $X^{ss}_\theta(G) = X^s_\theta(G)$ and this set is nonempty and smooth

- G acts freely on $X^{ss}_\theta(G)$

In this case, the GIT quotient is the smooth variety

$$X/\!\!/_\theta G := X^s_\theta(G)/G$$

where the right hand side is defined to be the categorical quotient (it is a theorem that this exists as a scheme). As a topological space, $X/\!\!/_\theta G$ is the usual topological quotient $X^s_\theta(G)/G$. Again, when θ is understood, we will omit it from the notation. In this note, we will also assume that $X/\!\!/G$ is a projective variety. In general it is projective over the affine quotient $\operatorname{Spec}(\Gamma(X, \mathscr{O}_X)^G)$.

Whereas the GIT quotient "forgets" the unstable locus X^{us}, this information is retained in the stack quotient. The objects of the stack quotient $[X/G]$ over a scheme S are diagrams

$$\begin{array}{ccc} \mathscr{P} & \longrightarrow & X \\ \downarrow & & \\ S & & \end{array}$$

where $\mathscr{P}$ is a principal G-bundle (locally trivial in the étale topology) and $\mathscr{P} \to X$ is a G-equivariant map. Morphisms in this category are given by fiber diagrams.

One should picture the GIT quotient $X/\!\!/G$ as an open subset of the stack quotient $[X/G]$. In a moment we will define quasimaps to $X/\!\!/G$ to be certain "stable" maps to the stack quotient $[X/G]$. Motivation for this definition comes from the previous section and the next example.

STACK EXAMPLE 3.2 (GIT vs stack quotient). Let $\mathbb{C}^*$ act on $\mathbb{C}^{n+1}$ by

$$\lambda \cdot (x_1, \dots, x_n) = (\lambda x_1, \dots, \lambda x_n). \tag{3.1}$$

Then if $\theta : \mathbb{C}^* \to \mathbb{C}^*$ is the identity, we have $(\mathbb{C}^{n+1})^{ss} = \mathbb{C}^{n+1} \backslash \{0\}$ and $\mathbb{C}^{n+1}/\!\!/_\theta \mathbb{C}^* = \mathbb{P}^n$. A map from a scheme S to $\mathbb{P}^n$ is given by a line bundle $\mathcal{L}$ on S and $n+1$ sections of $\mathcal{L}$ that do not simultaneously vanish.

On the other hand, the stack $[\mathbb{C}^{n+1}/\mathbb{C}^*]$ "remembers the origin." By definition, a map from a scheme S to $[\mathbb{C}^{n+1}/\mathbb{C}^*]$ is a principal $\mathbb{C}^*$-bundle $\mathscr{P}$ on S and an equivariant map $\mathscr{P} \to \mathbb{C}^{n+1}$. As we will see in Example 4.2, this equivariant map is equivalent to a section of $\mathscr{P} \times_{\mathbb{C}^*} \mathbb{C}^{n+1} = \mathcal{L}^{\oplus n+1} \to S$ where $\mathcal{L}$ is the line bundle associated to $\mathscr{P}$. Comparing this to the data of a map to $\mathbb{C}^{n+1}/\!\!/\mathbb{C}^*$, we see that the only difference is that now, the sections of $\mathcal{L}$ are allowed to vanish simultaneously—i.e., we allow the map to "hit the origin."

We close this section with an important example in which a desirable quotient does exist as a scheme, namely certain *mixing spaces*. Let $\mathscr{P} \to S$ be a principal G-bundle with a left G-action and let X be an affine G-variety. Then $\mathscr{P} \times X$ carries a natural G-action defined on closed points by

$$g \cdot (p, x) = (gp, gx).$$

We define the *mixing space* to be the categorical quotient

$$\mathscr{P} \times_G X := (\mathscr{P} \times X)/G.$$

One can show that the categorical quotient exists as a scheme, and in fact it is a geometric quotient (see for example the discussion in [**5**, Sec 3]). It has a natural map to $\mathscr{P}/G = S$, with closed fibers isomorphic to X. In particular, when X is a vector space and G acts linearly on X, the mixing space $\mathscr{P} \times_G V$ is the total space of a vector bundle on S.

In fact, there is a bijection between principal GL_r-bundles and rank-r vector bundles on S given by sending $\mathscr{P}$ to $\mathscr{P} \times_G \mathbb{C}^r$, where $\mathbb{C}^r$ is a GL_r-module via left multiplication. We call $\mathscr{P} \times_G \mathbb{C}^r$ the *vector bundle associated to* $\mathscr{P}$, and $\mathscr{P}$ the *underlying principal bundle of* $\mathscr{P} \times_G \mathbb{C}^r$.

4. Quasimaps

This note will discuss the theory of quasimaps to $V/\!\!/G$ when V is a vector space. (The original papers present this theory in more generality, allowing V to be an affine variety with l.c.i. singularities contained in V^{us}.) Let V be a vector space and G a reductive algebraic group that acts linearly from the left on V. We've seen that a quasimap to $\mathbb{P}^2$ is, loosely speaking, a rational map to $\mathbb{P}^2$, and that such a map can be realized as a map to the stack quotient $[\mathbb{C}^3/\mathbb{C}^*]$. This motivates the following definition, which essentially says that a prestable quasimap to $V/\!\!/G$ is a map to the stack quotient $[V/G]$ that recovers a rational map.

DEFINITION 4.1. A prestable *quasimap* to $V/\!\!/G$ is data $(C, x_1, \ldots, x_n, \mathscr{P}, \tilde{u})$ where

- C is a curve with at worst nodal singularities and marked points $x_1, \ldots, x_n$
- $\varrho : \mathscr{P} \to C$ is a principal G-bundle with a left G-action
- $\tilde{u} : \mathscr{P} \to V$ is a G-equivariant map with $\varrho(\tilde{u}^{-1}(V^{us}))$ a finite set that is disjoint from nodes and markings.

The *genus* of the quasimap is the genus of C, and the *degree* of the quasimap is the degree of $\mathscr{P}$, i.e., a group homomorphism $\beta \in Hom(\chi(G), \mathbb{Z})$ defined by

$$\beta(\omega) = \deg_C(\mathscr{P} \times_G \mathbb{C}_\omega)$$

where $\omega \in \chi(G)$ is a character of G and $\mathbb{C}_\omega$ is the corresponding 1-dimensional representation

$$g \cdot z = \omega(g)z \qquad \text{for } g \in G \text{ and } z \in \mathbb{C}_\omega.$$

To write down examples of quasimaps, it is helpful to replace $\tilde{u}$ with an associated section of $\mathscr{P} \times_G V \to C$, which we will typically denote u. In fact this is equivalent data, as explained in the example below.

STACK EXAMPLE 4.2 (Morphisms of stacks). If $\varrho : \mathscr{P} \to C$ is a principal G-bundle, then equivariant maps $\mathscr{P} \to V$ are in bijection with sections of the associated vector bundle $\mathscr{P} \times_G V \to C$. I'll sketch this bijection.

From the universal property of fiber products, we have a natural bijection between $\mathrm{Hom}(\mathscr{P}, V)$ and $\mathrm{Sec}(\mathscr{P}, \mathscr{P} \times V)$ where Sec denotes the space of sections. Letting G act on morphisms by conjugation, this is a G-equivariant bijection, so we have

$$\mathrm{Hom}_G(\mathscr{P}, V) \cong \mathrm{Sec}_G(\mathscr{P}, \mathscr{P} \times V).$$

where the subscript G indicates the G-invariant part (in particular, $\mathrm{Hom}_G(\mathscr{P}, V)$ is the set of G-equivariant maps). On the other hand, I'll sketch a bijection

$$\mathrm{Sec}(C, \mathscr{P} \times_G V) \cong \mathrm{Sec}_G(\mathscr{P}, \mathscr{P} \times V)$$

by interpreting $\mathscr{P} \times_G V$ as the stack quotient $[(\mathscr{P} \times V)/G]$. A map from C to $\mathscr{P} \times_G V$ is by definition a principal G-bundle $\mathscr{Q}$ on C and an equivariant map to

$\mathscr{P} \times V$; to say that it is a section means that composition with projection to C is the identity:

$$\begin{array}{ccccc} \mathscr{Q} & \longrightarrow & \mathscr{P} \times V & \longrightarrow & \mathscr{P} \\ \downarrow & & & & \downarrow \\ C & = & = & = & C \end{array}$$

Then the induced map $\mathscr{Q} \to \mathscr{P}$ is a morphism of principal bundles, hence an isomorphism. After identifying $\mathscr{P}$ with $\mathscr{Q}$ this way, the map $\mathscr{Q} \to \mathscr{P} \times V$ becomes a G-equivariant section $\mathscr{P} \to \mathscr{P} \times V$. The reader is invited to find the inverse to this correspondence.

EXAMPLE 4.3. For two GIT quotients $V/\!\!/G$, we'll write down all quasimaps from $\mathbb{P}^1$ to $V/\!\!/G$ as vectors of homogeneous polynomials. This example is generalized in Section 8.

(1) If $V = \mathbb{C}^{n+1}$ and $G = \mathbb{C}^*$ with action (3.1), and if θ is the identity character of $\mathbb{C}^*$, then V^{ss} is $V \setminus \{0\}$ and $V/\!\!/_\theta G$ is $\mathbb{P}^n$. A quasimap to $\mathbb{P}^n$ of degree $d \in \mathrm{Hom}(\mathbb{Z}, \mathbb{Z}) = \mathbb{Z}$ with source curve $\mathbb{P}^1$ is
   - A principal $\mathbb{C}^*$-bundle $\mathscr{P}$ such that $deg_{\mathbb{P}^1}(\mathscr{P} \times_{\mathbb{C}^*} \mathbb{C}_\theta) = d$. Therefore $\mathscr{P}$ is the underlying principal bundle of $\mathscr{O}_{\mathbb{P}^1}(d)$.
   - A section of $\mathscr{P} \times_{\mathbb{C}^*} \mathbb{C}^{n+1}$. Here the action of $\mathbb{C}^*$ on $\mathbb{C}^{n+1}$ is given by (3.1), so that this vector bundle is $n+1$ copies of the associated bundle to $\mathscr{P}$, i.e., $\mathscr{P} \times_{\mathbb{C}^*} \mathbb{C}^{n+1}$ equals $\mathscr{O}(d)^{\oplus n+1}$.

   Therefore, a prestable quasimap to $\mathbb{P}^n$ of degree $d \in \mathbb{Z}$ is given by a vector
$$(\ p_1(x, y),\ p_2(x, y),\ \ldots,\ p_{n+1}(x, y)\ )$$
   of homogeneous polynomials of degree d that are not all zero. In particular (2.2) is a prestable quasimap of degree 2.

(2) If $V = M_{k\times n}$ is $k \times n$ matrices over $\mathbb{C}$ and $G = GL_k$ acts on V by left multiplication and θ is the determinant character, then V^{ss} is full-rank matrices and $V/\!\!/_\theta G$ is the Grassmannian $Gr(k, n)$. A quasimap to $Gr(k, n)$ of degree $d \in \mathrm{Hom}(\mathbb{Z}, \mathbb{Z}) = \mathbb{Z}$ with source curve $\mathbb{P}^1$ is
   - A principal GL_k-bundle $\mathscr{P}$ on $\mathbb{P}^1$ such that $\deg_{\mathbb{P}^1}(\mathscr{P} \times_{GL_k} \mathbb{C}_\theta) = d$. Let E denote the vector bundle associated to $\mathscr{P}$. From Grothendieck's classification of principal bundles [**20**], we have $E = \oplus_{i=1}^k \mathscr{O}(d_i)$ for some d_i with $\sum_{i=1}^k d_i = d$.
   - A section of $\mathscr{P} \times_{GL_k} M_{k\times n}$, i.e., of $E^{\oplus n}$.

   So a prestable quasimap to $Gr(k, n)$ of degree d is given by a matrix of polynomials
$$[p_{ij}(x, y)]_{1\le i\le k,\ 1\le j\le n} \tag{4.1}$$
   where $p_{ij}(x, y)$ is homogeneous of degree d_i. Because $M^{us}_{k\times n}$ is matrices of low rank, to define a prestable quasimap the matrix (4.1) must have low rank on a finite set.

5. Stability and moduli spaces of quasimaps

In analogy with Gromov-Witten invariants, we want to define quasimap invariants of a target $V/\!\!/G$ to be integrals on certain moduli spaces. This cues the entrance of our second example of a stack: the moduli space of quasimaps. We will see that this is a "master" moduli space containing $\mathcal{M}_{g,n}(V/\!\!/G,\beta)$ and many compactifications of it. Then we will define some of these compactifications (substacks of the master space) to be the ϵ-stable moduli spaces of quasimaps.

STACK EXAMPLE 5.1 (Moduli stack of prestable quasimaps). Let $\mathfrak{M}_{g,n}(V/\!\!/G,\beta)$ be the moduli stack of prestable quasimaps of genus g, degree β, and n marks. Objects in this category over a base S are families of prestable quasimaps on S; i.e., they are triples $(\mathcal{C},\mathscr{P},\tilde{u})$ where $\mathcal{C}\to S$ is a flat family of genus-g nodal curves (not necessarily stable) on S, $\mathscr{P}\to\mathcal{C}$ is a principal G-bundle, $\tilde{u}:\mathscr{P}\to V$ is G-equivariant, and geometric fibers over S are prestable quasimaps of degree β. An isomorphism between objects in $\mathfrak{M}_{g,n}(V/\!\!/G,\beta)(S)$ is a commuting diagram

$$\begin{array}{ccccc} & & \tilde{u}' & & \\ \mathscr{P}' & \xrightarrow{\sim} & \mathscr{P} & \xrightarrow{\tilde{u}} & V \\ \downarrow & & \downarrow & & \\ \mathcal{C}' & \xrightarrow{\sim} & \mathcal{C} & & \end{array}$$

where $\mathcal{C}'\xrightarrow{\sim}\mathcal{C}$ commutes with the maps to S and the square is fibered (such a diagram is an isomorphisms of quasimap families). The stack $\mathfrak{M}_{g,n}(V/\!\!/G,\beta)$ is not Deligne-Mumford, as some prestable quasimaps have non-finite automorphism groups. We see the familiar offenders from stable map theory: for example, a degree-0 map sending $\mathbb{P}^1$ to a point in $\mathbb{P}^2$ is invariant under the entire automorphism group of $\mathbb{P}^1$. However, there are new examples as well. For instance, $[x^2:3x^2:x^2]$ defines a prestable quasimap of degree 2 from $\mathbb{P}^1$ to $\mathbb{P}^2$ which is invariant under $[x:y]\mapsto[x:ty]$.

EXAMPLE 5.2. Let's find isomorphisms between the quasimaps in Example 4.3.

(1) Let V,G, and θ be as in Example 4.3 part 1. From that example, a prestable quasimap from $\mathbb{P}^1$ to $\mathbb{P}^n$ of degree d is given by a vector of homogeneous degree-d polynomials $(p_i(x,y))_{i=1}^{n+1}$. Then a quasimap isomorphism is an element $\alpha\in\operatorname{Aut}(\mathscr{O}(d))=\mathbb{C}^*$, and it sends $(p_i(x,y))$ to $(\alpha p_i(x,y))$.

(2) Let V,G, and θ be as in Example 4.3 part 2. From that example, a prestable quasimap from $\mathbb{P}^1$ to $Gr(k,n)$ of degree d is n sections of a vector bundle $\oplus_{i=1}^k\mathscr{O}(d_i)$ of degree d, which may be denoted by a $k\times n$ matrix $(p_{ij}(x,y))$ of polynomials where $p_{ij}(x,y)$ is homogeneous of degree d_i. Then a quasimap isomorphism is an element A of $\operatorname{Aut}(E)=\operatorname{Hom}(\oplus\mathscr{O}(d_i),\oplus\mathscr{O}(d_i))^\times$ which we may identify with a $k\times k$ matrix $(a_{\ell i}(x,y))$ of polynomials where $a_{\ell i}(x,y)$ is homogeneous of degree $d_\ell-d_i$. Such an isomorphism acts on a the quasimap $(p_{ij}(x,y))$ by matrix multiplication. (Notice that if $d_1\geq d_2\geq\ldots\geq d_k$, then $(a_{\ell i}(x,y))$ will be block upper triangular.)

For many reasons, the stack $\mathfrak{M}_{g,n}(V/\!\!/G,\beta)$ is not the right one for defining invariants. Instead, we impose a *stability condition* that cuts out a proper separated Deligne-Mumford substack of $\mathfrak{M}_{g,n}(V/\!\!/G,\beta)$ which contains $\mathcal{M}_{g,n}(V/\!\!/G,\beta)$ as an open subset.

DEFINITION 5.3. Let $(C,x_1,\ldots,x_n,\mathscr{P},\tilde{u})$ be a prestable quasimap to $V/\!\!/_\theta G$ and let $L=\mathscr{P}\times_G\mathbb{C}_\theta$. This quasimap is ϵ-stable if

(1) On every component C' of C we have

$$2g_{C'} - 2 + n_{C'} + \epsilon \deg(L|_{C'}) > 0$$

where $g_{C'}$ is the genus of C' and $n_{C'}$ is the number of marked points and nodes on C', and

(2) For every $x \in C$ we have

$$\ell(x) \leq 1/\epsilon$$

where $\ell(x)$, called the *length* of x, is the order of contact of $u(C)$ with $\mathscr{P} \times_G V^{us}$, and u is the section of $\mathscr{P} \times_G V \to C$ determined by $\tilde{u}$. See [**9**, Def 7.1.1] for more on the definition of length. For $x \in C$, the length $\ell(x)$ is nonzero if and only if $u(x)$ is in $\mathscr{P} \times_G V^{us}$, and in this case we say x is a *basepoint* of the quasimap.

The first condition (1) is sometimes stated as "$\omega_C(\sum x_i) \otimes L^\epsilon$ is ample." A family of prestable quasimaps is ϵ-stable if every geometric fiber is ϵ-stable. The moduli space of ϵ-stable quasimaps is denoted $\overline{\mathcal{M}}^\epsilon_{g,n}(V /\!\!/ G, \beta)$. An ϵ-stable quasimap only has finitely many automorphisms. Hence the moduli spaces $\overline{\mathcal{M}}^\epsilon_{g,n}(V /\!\!/ G, \beta)$ are Deligne-Mumford, not just Artin stacks. They also have other good geometric properties, as stated in the following theorem.

THEOREM 5.4. [**9**, Thm 7.1.6] *The moduli space $\overline{\mathcal{M}}^\epsilon_{g,n}(V /\!\!/ G, \beta)$ is a proper separated Deligne-Mumford stack of finite type.*

As explained in the introduction, the benefit of these spaces, especially in genus 0, is that when ϵ is greater than 2, the space $\overline{\mathcal{M}}^\epsilon_{0,n}(V /\!\!/ G, \beta)$ is the familiar space $\overline{\mathcal{M}}_{0,n}(V /\!\!/ G, \beta)$. On the other hand, when ϵ is sufficiently small, the moduli spaces $\overline{\mathcal{M}}^\epsilon_{0,n}(V /\!\!/ G, \beta)$ do not depend on ϵ; this is called the 0+-stable quasimap moduli space. The space $\overline{\mathcal{M}}^{0+}_{0,n}(V /\!\!/ G, \beta)$ is more "computable" (see Example 5.7). Hence the wall-crossing theorem of [**8**], which translates between invariants of $\overline{\mathcal{M}}^\epsilon_{0,n}(V /\!\!/ G, \beta)$ for differing values of ϵ, gives a way to relate the Gromov-Witten invariants of $V /\!\!/ G$ to invariants that are more computable.

EXAMPLE 5.5. In Section 2 we described two possible limits of the family (2.1) of quasimaps to $\mathbb{P}^2$. One limit was the stable map $\widetilde{\phi}$. Because $\widetilde{\phi}$ has no basepoints, it satisfies condition (2) in Definition 5.3 for each ϵ. However, on the component with no marks—call it C'—we have

$$2g_{C'} - 2 + n_{C'} + \epsilon \deg(L|_{C'}) = -1 + \epsilon,$$

which is possible only for $\epsilon > 1$. So this map is ϵ stable for $\epsilon > 1$.

The second limit was the rational map $\phi_0(x, y) = [0 : xy : y^2]$. This quasimap satisfies condition (1) of Definition 5.3 for every $\epsilon > 0$. However, we have $\ell([1 : 0]) = 1$, so that this map is ϵ-stable only when $\epsilon \leq 1$.

Hence, the family (2.1) in $\overline{\mathcal{M}}^\epsilon_{g,n}(V /\!\!/ G, \beta)$ has the stable map $\widetilde{\phi}$ for a limit when $\epsilon > 1$, and the rational map ϕ_0 for a limit when $\epsilon \leq 1$.

A variant of $\overline{\mathcal{M}}^\epsilon_{g,n}(V /\!\!/ G, \beta)$ are the ϵ-stable *quasimap graph spaces.* Graph spaces are used in many contexts to prove wall crossing or mirror theorems. In particular, we will use the graph moduli space with $g = n = 0$ and $\epsilon = 0^+$ in Section 7 to define the I-function, so we'll define the quasimap graph space with these parameters now.

DEFINITION 5.6. Fix a "reference copy" of $\mathbb{P}^1$ and denote it $\mathbb{P}^{\mathbf{1}}$. The quasimap graph space $QG(V/\!\!/G,\beta)$ is the stack whose objects over a scheme S are prestable quasimaps $(\mathcal{C},\mathscr{P},\tilde{u})$, with an additional datum

$$\phi : \mathcal{C} \to \mathbb{P}^{\mathbf{1}}$$

which restricts to an isomorphism on every geometric fiber of $\mathcal{C} \to S$. An isomorphism between two objects $(\mathcal{C},\mathscr{P},\tilde{u},\phi)$ and $(\mathcal{C}',\mathscr{P}',\tilde{u}',\phi')$ in $QG(V/\!\!/G,\beta)(S)$ is a commuting diagram

$$\begin{tikzcd} & \mathscr{P}' \arrow[r, "\sim"] \arrow[d] \arrow[rr, bend left, "\tilde{u}'"] & \mathscr{P} \arrow[r, "\tilde{u}"] \arrow[d] & V \\ \mathbb{P}^{\mathbf{1}} & \mathcal{C}' \arrow[l, "\phi'"'] \arrow[r, "\sim"] & \mathcal{C} \arrow[ll, bend left, "\phi"] & \end{tikzcd} \tag{5.1}$$

such that $\mathcal{C}' \xrightarrow{\sim} \mathcal{C}$ commutes with the maps to S and the square is fibered.

When S is Spec($\mathbb{C}$), The additional datum ϕ realizes $\mathcal{C}$ as $\mathbb{P}^{\mathbf{1}}$. So closed points of $QG(V/\!\!/G,\beta)$ are in bijection with quasimaps from $\mathbb{P}^{\mathbf{1}}$ to $V/\!\!/G$. In addition, notice that ϕ prevents the existence of automorphisms, so we do not need marked points to achieve stability. According to [**9**, Thm 7.2.2], the space $QG(V/\!\!/G,d)$ is a proper Deligne-Mumford stack of finite type. In the next example, we identify $QG(\mathbb{P}^n,d)$ as a smooth scheme.

STACK EXAMPLE 5.7 (A moduli stack represented by a scheme). We have seen (Example 4.3) that a quasimap from $\mathbb{P}^1$ to $\mathbb{P}^n$ of degree d is a nonzero element of $\Gamma(\mathbb{P}^1,\mathscr{O}(d))^{\oplus n+1}$. However, two such sections define isomorphic quasimaps exactly when they differ by a complex scalar (Example 5.2). Hence, we naively expect

$$QG(\mathbb{P}^n,d) = \left(\, \Gamma(\mathbb{P}^1,\mathscr{O}(d))^{\oplus n+1} \;\setminus\; \{0\}\,\right) / \;\mathbb{C}^* \cong \mathbb{P}^N,$$

where $N = dn+d+n$. Indeed, this space carries a tautological family of quasimaps as follows. On $\mathbb{P}^N$ we have the trivial family of curves $\mathbb{P}^N \times \mathbb{P}^1$, and on this family the vector bundle $V = \mathscr{O}_{\mathbb{P}^N}(1)^{\oplus n+1} \otimes \mathscr{O}_{\mathbb{P}^1}(d)$. An element of $\mathbb{P}^N \times \mathbb{P}^1$ may be written $(\sigma,\mathbf{x})$ where $\sigma \in \Gamma(\mathbb{P}^1,\mathscr{O}(d))^{\oplus n+1}$ is a vector of $n+1$ degree-d homogeneous polynomials in two variables and $\mathbf{x} = (x,y)$. The tautological quasimap is given by the section of V sending $(\sigma,\mathbf{x})$ to $\sigma(\mathbf{x})$.

The tautological family on $\mathbb{P}^N$ defines a map $F : \mathbb{P}^N \to QG(\mathbb{P}^n,d)$ which is a bijection on closed points by Examples 4.3 and 5.2. It can be shown that $QG(\mathbb{P}^n,d)$ is a smooth algebraic space (see Examples 6.1 and 7.1). Since $\mathbb{P}^N$ is smooth as well, it follows that F must be an isomorphism (see [**26**, Lem 3.6.2]).

6. Virtual cycles via perfect obstruction theories

We want to define invariants of $X/\!\!/G$ which are integrals over the moduli spaces $\overline{\mathcal{M}}^{\epsilon}_{g,n}(X/\!\!/G,\beta)$. But these spaces may not be smooth or even equidimensional, so the usual definition of integration doesn't make sense. We need to choose some class to be the (virtual) fundamental class with which to define integration. One way to get such a class is via a perfect obstruction theory and the intrinsic normal cone construction of Behrend-Fantechi [**2**]. We will summarize their construction; for a more thorough but still elementary introduction see [**1**].

Let X be a Deligne-Mumford stack with a map to an Artin stack Y. We'll assume X and Y are separated and of locally finite type, and that the map $X \to Y$ is of *relative Deligne-Mumford type*, a technical condition that is always satisfied in our examples. In this case, the *relative cotangent complex* of X over Y is a certain object $\mathcal{L}^\bullet_{X/Y}$ in the derived category $D^{\leq 0}(X)$ of quasi-coherent sheaves. One property of $\mathcal{L}^\bullet_{X/Y}$ is that $H^0(\mathcal{L}^\bullet_{X/Y})$ is isomorphic to the sheaf of Kahler differentials $\Omega^1_{X/Y}$.

A *relative perfect obstruction theory* on X is a perfect complex $E^\bullet \in D^{[-1,0]}(X)$ and a morphism $\phi : E^\bullet \to \mathcal{L}^\bullet_{X/Y}$ such that $h^0(\phi)$ is an isomorphism and $h^{-1}(\phi)$ is surjective. From this data, Behrend-Fantechi construct

- A *relative intrinsic normal cone* $\mathfrak{C}_{X/Y}$ that is an Artin stack over X
- A cone stack $\mathfrak{E} \to X$ that contains the intrinsic normal cone via a closed embedding $\mathfrak{C}_{X/Y} \hookrightarrow \mathfrak{E}$

By the second bullet, the intrinsic normal cone is a closed substack of $\mathfrak{E}$; intuitively, the virtual class induced by $E^\bullet$ is the intersection of the intrinsic normal cone with the 0-section of $\mathfrak{E}$. Symbolically,

$$[X]^{\mathrm{vir}} := 0^!_{\mathfrak{E}}[\mathfrak{C}_{X/Y}] \in A_{\dim Y + \mathrm{rk}\, E^0 - \mathrm{rk}\, E^{-1}}(X).$$

The symbol $0^!_{\mathfrak{E}}$ is defined in [**2**]. One incidental use of perfect obstruction theories is to detect when a stack is smooth, using the following example (see [**2**, Prop 7.3]).

EXAMPLE 6.1. Let X and Y be as above and let $\phi : E^\bullet \to \mathcal{L}^\bullet_{X/Y}$ be a relative perfect obstruction theory. If $E^{-1} = 0$ then we say *the obstructions vanish.* If this is the case and moreover E^0 is locally free, then $h^0(E^\bullet) = E^0$ is locally free, but $h^0(\phi)$ is an isomorphism, so $h^0(\mathcal{L}^\bullet_{X/Y}) = \Omega^1_{X/Y}$ is locally free and $X \to Y$ is smooth. In this case, the construction of Behrend-Fantechi returns the usual fundamental class of X; i.e., $[X]^{\mathrm{vir}} = [X]$.

There is a natural relative perfect obstruction theory on $\overline{\mathcal{M}}^\epsilon_{g,n}(V/\!\!/G, \beta)$. This stack has a forgetful map to the moduli stack of principal G-bundles on genus g curves

$$\mu : \overline{\mathcal{M}}^\epsilon_{g,n}(V/\!\!/G, \beta) \to \mathcal{B}un_G$$

given by forgetting the section from the quasimap data.

THEOREM 6.2. [**9**, Thm 7.1.6] *On $\overline{\mathcal{M}}^\epsilon_{g,n}(V/\!\!/G, \beta)$ let $\pi : \mathcal{C} \to \overline{\mathcal{M}}^\epsilon_{g,n}(V/\!\!/G, \beta)$ be the universal curve, let $\mathfrak{P}$ denote the universal principal bundle, and let $\varrho : \mathfrak{P} \times_G V \to \mathcal{C}$ be the associated vector bundle with T_ϱ the relative tangent bundle of this map. Then the complex*

$$\left(R^\bullet \pi_*(u^* T_\varrho)\right)^\vee$$

is a μ-relative perfect obstruction theory for $\overline{\mathcal{M}}^\epsilon_{g,n}(V/\!\!/G, \beta)$.

According to [**9**], the analogous result holds for the quasimap graph space $QG(V/\!\!/G, \beta)$.

STACK EXAMPLE 6.3 (Perfect obstruction theory). We compute the relative perfect obstruction theory on $QG(\mathbb{P}^n, d)$ and show that it recovers the usual fundamental class of $\mathbb{P}^N$. From Stack Example 5.7, the relative perfect obstruction theory is the complex

$$\left(R^\bullet \pi_*(u^* T_\varrho)\right)^\vee = \left(R^\bullet \pi_*(\mathscr{O}_{\mathbb{P}^1}(d) \otimes \mathscr{O}_{\mathbb{P}^N}(1)^{\oplus n+1})\right)^\vee$$

where $\pi : \mathbb{P}^N \times \mathbb{P}^1 \to \mathbb{P}^N$ is projection to the first factor. By the projection formula, this is the dual of the complex

$$R^0\pi_*(\mathscr{O}_{\mathbb{P}^1}(d)) \otimes \mathscr{O}_{\mathbb{P}^N}(1)^{\oplus n+1} \to R^1\pi_*(\mathscr{O}_{\mathbb{P}^1}(d)) \otimes \mathscr{O}_{\mathbb{P}^N}(1)^{\oplus n+1}.$$

The second term vanishes, so the μ-relative perfect obstruction theory is the vector bundle $\mathscr{O}_{\mathbb{P}^N}^{\oplus d+1} \otimes \mathscr{O}_{\mathbb{P}^N}(1)^{\oplus n+1} = \mathscr{O}_{\mathbb{P}^N}(1)^{\oplus N+1}$ in degree 0. By Example 6.1 the virtual class defined by this theory is the usual fundamental class on $\mathbb{P}^N$.

7. Quasimap I-functions

For any positive rational ϵ, we have a proper separated Deligne-Mumford stack $\overline{\mathcal{M}}^{\epsilon}_{0,n}(V/\!\!/G, \beta)$ with a perfect obstruction theory. This is exactly what we need to define invariants of $V/\!\!/G$ as integrals on these spaces. As discussed at the end of Section 2, when ϵ is at least 2, these invariants are equal to Gromov-Witten invariants. The mirror theorem of [**8**] relates these invariants to localization residues on the *graph moduli space* $QG(V/\!\!/G, \beta)$ for the $\epsilon = 0^+$ stability parameter (see Definition 5.6). The residues on $QG(V/\!\!/G, \beta)$ are indexed by the *quasimap I-function.* In contrast to other Gromov-Witten generating functions, the advantage of the I-function is that it can be written down with a closed formula. The goal of this section is to define the quasimap I-function as a formal power series indexing certain localization residues on $QG(V/\!\!/G, \beta)$.

We already know from [**9**, Thm 7.2.2] that $QG(V/\!\!/G, \beta)$ is a Deligne-Mumford stack, but in fact more is true: it is an *algebraic space.*

Stack Example 7.1 ($QG(V/\!\!/G, \beta)$ is an algebraic space). By [**14**, Thm 2.2.5] it suffices to show that a quasimap in this space has a trivial automorphism group. Essentially, this is because quasimaps are required to map into V^s generically, and G acts on V^s with trivial stabilizers by our assumptions in Section 2. Let $\sigma : \mathbb{P}^1 \to \mathscr{P} \times_G V$ be a quasimap, and suppose ϕ is an automorphism of $\mathscr{P}$ commuting with σ. In a local chart U where $\mathscr{P}$ is trivial, let $\sigma_U : U \to V$ and $\phi_U : U \to G$ be coordinate representations of σ and ϕ. Then

$$\phi_U(u)\sigma_U(u) = \sigma_U(u).$$

For every $u \in U$ that is not a basepoint, we know $\sigma_U(u) \in V^s$, so this equation implies $\phi_U(u) = 1$ in G. Hence ϕ_U is the identity on a dense subset of U, hence on all of U.

To define the coefficients of the I-function as localization residues, we need a $\mathbb{C}^*$-action on $QG(V/\!\!/G, \beta)$. This space carries a $\mathbb{C}^*$ action as follows. Let x_0, x_1 be homogeneous coordinates on $\mathbb{P}^1$ and let $\mathbb{C}^*$ act on $\mathbb{P}^1$ by

$$\lambda \cdot [x_0 : x_1] = [\lambda x_0 : x_1], \qquad \lambda \in \mathbb{C}^*.$$

This induces an action on $QG(V/\!\!/G, \beta)$ given by

$$\lambda(\mathcal{C}, \mathscr{P}, \tilde{u}, \phi) = (\mathcal{C}, \mathscr{P}, \tilde{u}, \lambda \circ \phi).$$

The fixed locus of $\mathbb{C}^*$ is the closed subspace of $QG(V/\!\!/G, \beta)$ whose objects over a scheme S are families fixed by the action.

Stack Example 7.2 (Fixed locus of a group action). What are the basepoints of a fixed graph quasimap? If a graph quasimap $(\mathcal{C}, \mathscr{P}, \tilde{u}, \phi)$ over Spec($\mathbb{C}$) is $\mathbb{C}^*$-fixed, then for every $\lambda \in \mathbb{C}^*$ we have a diagram (5.1) with $(\mathcal{C}', \mathscr{P}', \tilde{u}', \phi') = (\mathcal{C}, \mathscr{P}, \tilde{u}, \lambda \circ \phi)$ and ϕ an isomorphism. Then the map $\mathcal{C}' \to \mathcal{C}$ in (5.1) must be

$\phi^{-1} \circ \lambda \circ \phi$. But $\phi^{-1} \circ \lambda \circ \phi$ must fix basepoints of $\tilde{u}$, for every λ. This means that basepoints of $\tilde{u}$ have to be $\phi^{-1}([0:1])$ or $\phi^{-1}([1:0])$.

We can use this information to identify components of the fixed locus: the lengths of these basepoints $\ell([0:1])$ and $\ell([1:0])$ are constant in families, so specifying these integers specifies a component of the fixed locus.

As explained in the previous example, the fixed locus of $\mathbb{C}^*$ acting on $QG(V/\!\!/G, \beta)$ has components determined by the length of the basepoints at $[0:1]$ and $[1:0]$. Let F_β be the component where all the basepoints are at $[0:1] \in \mathbb{P}^1$. This component has a natural map $ev_\bullet$ to $V/\!\!/G$.

Stack Example 7.3 (Morphism of stacks). Define a map $ev_\bullet : F_\beta \to V/\!\!/G$ as follows. Let $(\mathcal{C}, \mathscr{P}, \tilde{u}, \phi)$ be an object of F_β lying over S. Recall from Example 4.2 that $\tilde{u}$ defines a section u of $\mathscr{P} \times_G V \to \mathcal{C}$; since $[1:0]$ is not a basepoint, $u([1:0])$ is in $\mathscr{P} \times_G V^s$. So define $ev_\bullet$ to send $(\mathcal{C}, \mathscr{P}, \tilde{u}, \phi)$ to the morphism $S \to V/\!\!/G$ apparent in the following diagram:

$$\begin{array}{ccccc} & & \mathscr{P} \times_G V & \hookleftarrow & \mathscr{P} \times_G V^s \longrightarrow V^s/G = V/\!\!/G \\ & & {\scriptstyle u}\uparrow\downarrow & & \\ S \times \{[1:0]\} & \longrightarrow & \mathcal{C} & & \end{array}$$

Finally we can define the I-function as follows. The μ-relative perfect obstruction theory on $QG(V/\!\!/G, \beta)$ defines an *absolute perfect obstruction theory* which is $\mathbb{C}^*$-equivariant. The moving part of this equivariant theory may be used to define the euler class of the normal bundle to the fixed locus, denoted $e_{\mathbb{C}^*}(N^{\mathrm{vir}}_{F_\beta})$. The fixed part, when restricted to F_β, is a perfect obstruction theory for F_β, and hence defines a virtual class $[F_\beta]^{\mathrm{vir}}$ (see [**6**, Sec 3]). With these definitions, we may write the I-function of $V/\!\!/G$ as a formal sum: we use formal symbols q^β to index coefficients that are localization residues on the stacks F_β. Precisely,

(7.1)

$$I^{V/\!\!/G}(z) = 1 + \sum_{\beta \neq 0} q^\beta I_\beta^{V/\!\!/G}(z) \qquad \text{where} \qquad I_\beta^{V/\!\!/G}(z) = (ev_\bullet)_* \left(\frac{[F_\beta]^{vir}}{e_{\mathbb{C}^*}(N^{vir}_{F_\beta})} \right).$$

The promise of 0^+-stable quasimaps was that their invariants would be computable. Indeed, the moduli spaces F_β in the formula (7.1) are represented by smooth schemes, as stated in the following theorem:

Theorem 7.4. *The moduli space F_β is a disjoint union of flag bundles on smooth subvarieties of $V/\!\!/G$.*

In [**25**] this identification is explicit. The strategy of the proof is the same as that used in Example 5.7: show that F_β is a smooth algebraic space, and then find a tautological family on the desired smooth variety (a closed subscheme of a flag bundle on $V/\!\!/G$) whose geometric fibers are in bijection with objects parametrized by (a component of) F_β. This theorem can be used to derive an explicit formula for the quasimap I-function, a formula we'll write down in the next section.

Example 7.5. Let us describe F_d when the target is the Grassmannian $Gr(k,n)$. Recall from Example 4.3 that a degree-d graph quasimap to $Gr(k,n)$ is n sections of a vector bundle $\oplus_{\mathbb{P}^1}(\mathscr{O}(d_i))$ with $\sum d_i = d$, which may be represented by a $k \times n$ matrix of homogeneous polynomials. For convenience assume

$d_1 \geq d_2 \geq \ldots \geq d_k$. Clearly, full-rank matrices $(c_{ij}x^i)$ for complex numbers c_{ij} define elements of F_β. Referring to Example 5.2 we see that invertible matrices of the form $(a_{ij}x^{d_i-d_j})$ define isomorphisms of this collection of elements of F_β, where a_{ij} are complex numbers with $a_{ij} = 0$ if $d_j > d_i$. From this, one may naively guess that there is a component of F_β given by

$$F_{d_1,\ldots,d_k} = (M_{k\times n} \setminus \Delta)/U$$

where $\Delta \subset M_{k\times n}$ is matrices of rank less than k and U is the subgroup of GL_k equal to block upper triangular matrices with block sizes given by the multiplicities of the d_i. In fact, this identification is correct (see [**25**]).

8. A formula for the I-function

The quasimap I-function was advertised as a more tractable member of the Gromov-Witten family, a generating function that recovers Gromov-Witten invariants after a mirror map, but which (supposedly) has a closed formula. But how, in practice, can one write down the quasimap I-function of $V/\!\!/G$?

When G is abelian, the answer is well-known (see [**17**] or [**7**]). To handle nonabelian groups, the strategy is to prove an *abelian/nonabelian correspondence* for I-functions. "Abelian/nonabelian correspondences" can be found throughout representation theory and geometry; the basic idea of these results, if T is a maximal torus of G, is to relate some data determined by G to the corresponding data determined by T. In our situation, a character θ of G restricts to a character of T, so one may attempt to recover the various invariants (including the I-function) of $V/\!\!/_\theta G$ from the more accessible invariants of $V/\!\!/_\theta T$. Most abelian/nonabelian correspondences use the *Weyl group* W of T in G, which equals the quotient of the normalizer of T by the torus T itself: $W = N_G(T)/T$.

EXAMPLE 8.1. We describe an "abelian/nonabelian correspondence" for principal bundles on $\mathbb{P}^1$. Let $\mathscr{T}$ be a principal T-bundle; then if we let T act on G by left multiplication, the associated bundle $\mathscr{T} \times_T G$ is a principal G-bundle. There is a group homomorphism $\tau : \mathrm{Hom}(\chi(T), \mathbb{Z}) \to \mathrm{Hom}(\chi(G), \mathbb{Z})$ induced by restriction of characters, and if $\mathscr{T}$ has degree $\tilde{\beta} \in \mathrm{Hom}(\chi(T), \mathbb{Z})$, then the degree of $\mathscr{T} \times_T G$ is $\tau(\tilde{\beta})$. Moreover, the Weyl group of T in G acts on $\mathrm{Hom}(\chi(T), \mathbb{Z})$. Grothendieck's classification theorem [**20**] says that every principal G-bundle may be written $\mathscr{T} \times_T G$ for some $\mathscr{T}$, and that the isomorphism class of $\mathscr{T} \times_T G$ is determined by the Weyl orbit of $\tilde{\beta}$ in $\mathrm{Hom}(\chi(T), \mathbb{Z})$.

We may understand the cohomology of $V/\!\!/G$ as follows. There is a rational map $p : V/\!\!/T \dashrightarrow V/\!\!/G$ defined on $A = V^s(G)/T \subset V^s(T)/T$. Let W denote the Weyl group of T in G; then W acts on A and hence on $H^*(A, \mathbb{Q})$. A classical argument shows that p^* defines an isomorphism from $H^*(V/\!\!/G, \mathbb{Q})$ to $H^*(A, \mathbb{Q})^W$ (this may be read as an abelian/nonabelian correspondence for cohomology [1]). This is useful because we can write down explicit classes in $H^*(A, \mathbb{Q})$ from characters of T: If $\xi \in \chi(T)$, let L_ξ denote the line bundle $V^s(G) \times_{\mathbb{C}^*} \mathbb{C}_\xi$ on A. This gives a class $c_1(L_\xi)$, and with the right definitions, the map $\xi \mapsto c_1(L_\xi)$ is W-equivariant.

[1]The full relationship of $H^*(V/\!\!/G, \mathbb{Q})$ and $H^*(V/\!\!/T, \mathbb{Q})$ (which requires understanding the restriction $H^*(V/\!\!/T, \mathbb{Q}) \to H^*(A, \mathbb{Q})$) is worked out in [**24**] by Martin; the analogous result in Chow is proved in [**15**] by Ellingsrud-Stromme.

With this language, we can now write down a formula for the quasimap I-function. Because the formula expresses the I-function for $V/\!/G$ in terms of the I-function for $V/\!/T$ and the Weyl group, we call it an abelian/nonabelian correspondence for I-functions.

THEOREM 8.2. [**25**] *Let G be a reductive algebraic group and let T be a maximal torus of G. Suppose G acts on a vector space V, and that the weights of the restriction of this action to T are $\xi_1, \ldots, \xi_n \in \chi(T)$. Then the coefficients of the I-function of $V/\!/G$ satisfy*

$$I_\beta^{V/\!/G}(z) = \sum_{\tilde{\beta}\to\beta} \left(\prod_\alpha \frac{\prod_{k=-\infty}^{\tilde{\beta}\cdot c_1(L_\alpha)}(c_1(L_\alpha)+kz)}{\prod_{k=-\infty}^{0}(c_1(L_\alpha)+kz)} \right) I_{\tilde{\beta}}^{V/\!/T}(z) \tag{8.1}$$

where

$$I_{\tilde{\beta}}^{V/\!/T}(z) = \prod_{i=1}^{n} \frac{\prod_{k=-\infty}^{0}(c_1(L_{\xi_i})+kz)}{\prod_{k=-\infty}^{\tilde{\beta}(\xi_i)}(c_1(L_{\xi_i})+kz)}. \tag{8.2}$$

The sum is over all effective $\tilde{\beta}$ in $\tau^{-1}(\beta)$, the variable α runs over the roots of G, and the equality in (8.1) *is equality after pulling both sides back to $H^*(A, \mathbb{Q})$.*

The formula (8.2) for the coefficients of a toric I-function was originally given by Givental in [**17**] (note that the quasimap I-function differs from Givental's I-function by an exponential factor). Theorem 8.2 has natural generalizations to twisted and equivariant I-functions.

One application of this theorem is an abelian/nonabelian correspondence in quantum cohomology for a large class of targets. Indeed, for these targets, the correspondence was reduced to a correspondence of small J-functions in [**10**], and the wall-crossing result of [**8**] translates this to the relationship of small I-functions in Theorem 8.2. (A more general abelian/nonabelian correspondence theorem in quantum cohomology was proved using symplectic geometry by Gonzalez-Woodward in [**19**].) A second application, due to Kalashnikov [**21**], is to identify nonisomorphic Fano 4-folds by comparing their quantum periods, which are determined by their I-functions (this reference also provides an independent proof of Theorem 8.2 when the target is a quiver flag variety). Thirdly, the proof of Theorem 8.2 is very geometric, and so translates easily to the setting of K-theory, providing an analogous formula for the K-theoretic I-function. By providing explicit formulas for I-functions, Theorem 8.2 should facilitate the computation of examples in Gromov-Witten theory, and may lead to the discovery of new relationships.

We close this paper with a sketch of the proof of Theorem 8.2.

SKETCH OF PROOF OF THEOREM 8.2. To compute the I-function coefficients $I_\beta^{V/\!/G}(z)$ in (7.1), the key step is to identify the fixed locus F_β. In fact, connected components of F_β are indexed by the isomorphism type of the principal bundle $\mathscr{P}$ of the quasimap, or equivalently by the Weyl orbits in $\tau^{-1}(\beta)$ (see Example 8.1). Choose representatives $\tilde{\beta}_i$ for these orbits and let $F_{\tilde{\beta}_i}$ represent the corresponding component of the fixed locus. As we did in Example 7.5 for Grassmannians, we can naïvely guess what variety $F_{\tilde{\beta}_i}$ should be.

Generalizing Example 4.3, a quasimap from $\mathbb{P}^1$ to $V/\!/G$ may be written as a vector of homogeneous polynomials. If it is in F_β, then it has a unique basepoint at $[0:1]$, so we hope that we can choose the polynomials to be polynomials in x

only. Hence, if T acts on V with weights $\xi_1, \ldots, \xi_n$, define $V_{\tilde{\beta}_i}$ to be the vector space with elements

$$(u_1(x), u_2(x), \ldots, u_n(x)), \tag{8.3}$$

where $u_j(x)$ is a constant multiple of $x^{\tilde{\beta}_i(\xi_j)}$ if $\tilde{\beta}_i(\xi_j) \geq 0$, and $u_j(x)$ is zero otherwise. There is an evaluation map $ev_\bullet : V_{\tilde{\beta}_i} \to V$ sending (8.3) to $(u_1(1), \ldots, u_n(1))$. This map embeds $V_{\tilde{\beta}}$ as the subspace of V where $\tilde{\beta}_i(\xi_j) \geq 0$, which allows us to intersect $V_{\tilde{\beta}_i}$ with $V^s(G)$ and obtain a space $V^s_{\tilde{\beta}_i}$. Our guess is that every closed point of $F_{\tilde{\beta}_i}$ may be represented by a point of $V^s_{\tilde{\beta}_i}$.

However, distinct points of $V^s_{\tilde{\beta}_i}$ may not represent distinct quasimaps. Generalizing Example 5.2, an isomorphism of quasimaps in $V^s_{\tilde{\beta}_i}$ is an invertible $n \times n$ matrix of polynomials $(a_{\ell j}(x))_{\ell j}$ where $a_{\ell j}(x)$ is homogeneous of degree $\tilde{\beta}_i(\xi_\ell) - \tilde{\beta}_i(\xi_j)$ if this quantity is nonnegative, and $a_{\ell j}(x) = 0$ otherwise. Let $P_{\tilde{\beta}_i}$ denote this group of matrices; we also have an evaluation map $ev_\bullet : P_{\tilde{\beta}_i} \to \mathrm{Aut}(V)$ by evaluating a matrix at 1. In fact, the image of $ev_\bullet$ is in G, and $ev_\bullet$ identifies $P_{\tilde{\beta}_i}$ with a parabolic subgroup of G.

Finally we notice that $V^s_{\tilde{\beta}_i}/P_{\tilde{\beta}_i}$ carries a tautological family of quasimaps with the following section:

$$U : \frac{V^s_{\tilde{\beta}_i} \times (\mathbb{C}^2 \setminus \{0\})}{(\mathbf{u}, \mathbf{x}) \sim (A\mathbf{u}, t\mathbf{x})} \longrightarrow \frac{V^s_{\tilde{\beta}_i} \times (\mathbb{C}^2 \setminus \{0\}) \times V}{(\mathbf{u}, \mathbf{x}, \mathbf{v}) \sim (A\mathbf{u}, t\mathbf{x}, [t^{\tilde{\beta}(\boldsymbol{\xi})}]A(\mathbf{x})\mathbf{v})} \tag{8.4}$$
$$(\mathbf{u}, \mathbf{x}) \mapsto (\mathbf{u}, \mathbf{x}, \mathbf{u}(\mathbf{x}))$$

where $(\mathbf{u}, \mathbf{x}, \mathbf{v})$ is in $V^s_{\tilde{\beta}_i} \times (\mathbb{C}^2 \setminus \{0\}) \times V$ and $(A, t) \in P_{\tilde{\beta}_i} \times \mathbb{C}^*$, and $[t^{\tilde{\beta}_i(\boldsymbol{\xi})}]$ denotes the matrix with diagonal $(t^{\tilde{\beta}_i(\xi_1)}, \ldots, t^{\tilde{\beta}_i(\xi_n)})$. This matrix factor guarantees that every geometric fiber of this family of vector bundles has isomorphism type $\tilde{\beta}_i$. This tautological family defines a map from $V^s_{\tilde{\beta}_i}/P_{\tilde{\beta}_i}$ to $F_{\tilde{\beta}_i}$, and the hardest part of Theorem 8.2 is showing that this morphism is an isomorphism. The proof uses that obstructions on $F_{\tilde{\beta}_i}$ vanish, and moreover this stack is smooth.

From here, the computation of the formula in Theorem 8.2 is relatively straightforward. The evaluation map $V^s_{\tilde{\beta}_i}/P_{\tilde{\beta}_i} \to V /\!\!/ G$ (which sends x to 1) factors as

$$V^s_{\tilde{\beta}_i}/P_{\tilde{\beta}_i} \xrightarrow{i} V^s/P_{\tilde{\beta}_i} \xrightarrow{h} V^s/G,$$

where the first map is a closed embedding (recall $V_{\tilde{\beta}_i}$ is a subspace of V) and h is a flag bundle ($P_{\tilde{\beta}_i}$ is a parabolic subgroup of G). Recall that the equality (8.1) holds only after pulling back both sides to $V^s(G)/T$, so let $\Psi : V^s(G)/T \to V^s(G)/G$ be the projection. We compute

$$\Psi^* I_\beta^{V/\!\!/G}(z) = \Psi^* \sum_{\tilde{\beta}_i} (ev_\bullet)_* \frac{[F_{\tilde{\beta}_i}]^{\mathrm{vir}}}{e_{\mathbb{C}^*}(N^{\mathrm{vir}}_{F_{\tilde{\beta}_i}})} = \sum_{\tilde{\beta}_i} \Psi^* h_* i_* \frac{1}{e_{\mathbb{C}^*}(N^{\mathrm{vir}}_{F_{\tilde{\beta}_i}})}$$

where we have used that $[F_{\tilde{\beta}_i}]^{\mathrm{vir}} = 1$ because $F_{\tilde{\beta}_i}$ is smooth and the obstructions vanish (see Example 6.1).

To compute $e_{\mathbb{C}^*}(N^{\mathrm{vir}}_{F_{\tilde{\beta}_i}})$, we use that the restriction of the absolute obstruction theory to $F_{\tilde{\beta}_i}$ is $R^\bullet \pi_* \mathcal{F}$, where $\mathcal{F}$ is the sheaf on $(V^s_{\tilde{\beta}_i}/P_{\tilde{\beta}_i}) \times \mathbb{P}^1$ is defined by the exact sequence

$$0 \to \mathcal{A} \to \mathcal{B} \to \mathcal{F} \to 0$$

where

$$\mathcal{A} = \frac{V^s_{\tilde{\beta}_i} \times (\mathbb{C}^2 \setminus \{0\}) \times \mathfrak{g}}{(\mathbf{u}, \mathbf{x}, X) \sim (A\mathbf{u}, t\mathbf{x}, [t^{\tilde{\beta}_i(\boldsymbol{\xi})}]A(\mathbf{x}) \cdot X)} \quad \text{and} \quad \mathcal{B} = \frac{V^s_{\tilde{\beta}_i} \times (\mathbb{C}^2 \setminus \{0\}) \times V}{(\mathbf{u}, \mathbf{x}, v) \sim (A\mathbf{u}, t\mathbf{x}, [t^{\tilde{\beta}_i(\boldsymbol{\xi})}]A(\mathbf{x}) \cdot v)}.$$

So $e_{\mathbb{C}^*}(N^{\mathrm{vir}}_{F_{\tilde{\beta}_i}})$ is a certain ratio of the moving parts of the pushforwards of $\mathcal{A}$ and $\mathcal{B}$ (see (8.5)). We compute i_* with the projection formula, noting that $V^s_{\tilde{\beta}_i}$ is the zero locus of the tautological section of the bundle $V^s \times (V/V_{\tilde{\beta}_i})$ on V^s. We compute $\Psi^* h_*$ with a lemma of Brion [4], which introduces an additional sum over the Weyl group. After these steps we have

(8.5)

$$\Psi^* I^{V/\!/G}_{\beta}(z) = \sum_{\tilde{\beta}_i} \sum_{\omega \in W/W_{L_i}} \omega \left[\frac{e(V^s \times_T (V/V_{\tilde{\beta}_i}))\ \ e_{\mathbb{C}^*}(R^0\pi_*\mathcal{A})^{\mathrm{mov}} e_{\mathbb{C}^*}(R^1\pi_*\mathcal{B})^{\mathrm{mov}}}{\prod_{\alpha \in R^+ \setminus R_{L_i}} c_1(L_\alpha)\ \ e_{\mathbb{C}^*}(R^0\pi_*\mathcal{B})^{\mathrm{mov}} e_{\mathbb{C}^*}(R^1\pi_*\mathcal{A})^{\mathrm{mov}}} \right]$$

where $\mathcal{A}$ and $\mathcal{B}$ now denote the analogous bundles on $V^s(G)/T$, and L_i is the Levi subgroup of $P_{\tilde{\beta}_i}$ containting T with roots R_{L_i} and Weyl group W_{L_i}, and R^+ are the opposite roots of a Borel subgroup of G contained in $P_{\tilde{\beta}_i}$.

One can check that W_{L_i} is precisely the stabilizer of $\tilde{\beta}_i$ in $\mathrm{Hom}(\chi(T), \mathbb{Z})$ and that the double sum and group action may be simplified to a single sum over all $\tilde{\beta}$ in $\tau^{-1}(\beta)$, giving

$$\Psi^* I^{V/\!/G}_{\beta}(z) = \sum_{\tilde{\beta} \to \beta} \frac{e(V^s \times_T (V/V_{\tilde{\beta}}))\ \ e_{\mathbb{C}^*}(R^0\pi_*\mathcal{A})^{\mathrm{mov}} e_{\mathbb{C}^*}(R^1\pi_*\mathcal{B})^{\mathrm{mov}}}{\prod_{\alpha \in R^+ \setminus R_{L_{\tilde{\beta}}}} c_1(L_\alpha)\ \ e_{\mathbb{C}^*}(R^0\pi_*\mathcal{B})^{\mathrm{mov}} e_{\mathbb{C}^*}(R^1\pi_*\mathcal{A})^{\mathrm{mov}}}.$$

From here it remains to compute the euler classes. One finds that the classes coming from $\mathcal{A}$ and the flag pushforward h_* yield the twist factor in (8.1):

$$\frac{e_{\mathbb{C}^*}(R^0\pi_*\mathcal{A})^{\mathrm{mov}}}{\prod_{\alpha \in R^+ \setminus R_{L_{\tilde{\beta}}}} c_1(L_\alpha)\ e_{\mathbb{C}^*}(R^1\pi_*\mathcal{A})^{\mathrm{mov}}} = \left(\prod_{\alpha} \frac{\prod_{k=-\infty}^{\tilde{\beta}\cdot c_1(L_\alpha)} (c_1(L_\alpha) + kz)}{\prod_{k=-\infty}^{0} (c_1(L_\alpha) + kz)} \right)$$

while the classes coming from $\mathcal{B}$ and the inclusion pushforward i_* yield the toric I-function:

$$\frac{e(V^s \times_T (V/V_{\tilde{\beta}}))\ e_{\mathbb{C}^*}(R^1\pi_*\mathcal{B})^{\mathrm{mov}}}{e_{\mathbb{C}^*}(R^0\pi_*\mathcal{B})^{\mathrm{mov}}} = I^{V/\!/T}_{\tilde{\beta}}(z).$$

□

References

[1] Luca Battistella, Francesca Carocci, and Cristina Manolache, *Virtual classes for the working mathematician*, SIGMA Symmetry Integrability Geom. Methods Appl. **16** (2020), Paper No. 026, 38 pages, DOI 10.3842/SIGMA.2020.026. MR4082488

[2] K. Behrend and B. Fantechi, *The intrinsic normal cone*, Invent. Math. **128** (1997), no. 1, 45–88, DOI 10.1007/s002220050136. MR1437495

[3] Aaron Bertram, *Another way to enumerate rational curves with torus actions*, Invent. Math. **142** (2000), no. 3, 487–512, DOI 10.1007/s002220000094. MR1804158

[4] Michel Brion, *The push-forward and Todd class of flag bundles*, Parameter spaces (Warsaw, 1994), Banach Center Publ., vol. 36, Polish Acad. Sci. Inst. Math., Warsaw, 1996, pp. 45–50. MR1481478

[5] Michel Brion, *On automorphism groups of fiber bundles*, Publ. Mat. Urug. **12** (2011), 39–66. MR3012239

[6] Huai-Liang Chang, Young-Hoon Kiem, and Jun Li, *Torus localization and wall crossing for cosection localized virtual cycles*, Adv. Math. **308** (2017), 964–986, DOI 10.1016/j.aim.2016.12.019. MR3600080

[7] Ionuţ Ciocan-Fontanine and Bumsig Kim, *Moduli stacks of stable toric quasimaps*, Adv. Math. **225** (2010), no. 6, 3022–3051, DOI 10.1016/j.aim.2010.05.023. MR2729000

[8] Ionuţ Ciocan-Fontanine and Bumsig Kim, *Wall-crossing in genus zero quasimap theory and mirror maps*, Algebr. Geom. **1** (2014), no. 4, 400–448, DOI 10.14231/AG-2014-019. MR3272909

[9] Ionuţ Ciocan-Fontanine, Bumsig Kim, and Davesh Maulik, *Stable quasimaps to GIT quotients*, J. Geom. Phys. **75** (2014), 17–47, DOI 10.1016/j.geomphys.2013.08.019. MR3126932

[10] Ionuţ Ciocan-Fontanine, Bumsig Kim, and Claude Sabbah, *The abelian/nonabelian correspondence and Frobenius manifolds*, Invent. Math. **171** (2008), no. 2, 301–343, DOI 10.1007/s00222-007-0082-x. MR2367022

[11] Ionuţ Ciocan-Fontanine and Bumsig Kim, *Quasimap theory*, Proceedings of the International Congress of Mathematicians—Seoul 2014. Vol. II, Kyung Moon Sa, Seoul, 2014, pp. 615–633. MR3728630

[12] Ionuţ Ciocan-Fontanine and Bumsig Kim, *Big I-functions*, Development of moduli theory—Kyoto 2013, Adv. Stud. Pure Math., vol. 69, Math. Soc. Japan, [Tokyo], 2016, pp. 323–347, DOI 10.2969/aspm/06910323. MR3586512

[13] Ionuţ Ciocan-Fontanine and Bumsig Kim, *Higher genus quasimap wall-crossing for semipositive targets*, J. Eur. Math. Soc. (JEMS) **19** (2017), no. 7, 2051–2102, DOI 10.4171/JEMS/713. MR3656479

[14] Brian Conrad, *Arithmetic moduli of generalized elliptic curves*, J. Inst. Math. Jussieu **6** (2007), no. 2, 209–278, DOI 10.1017/S1474748006000089. MR2311664

[15] Geir Ellingsrud and Stein Arild Strømme, *On the Chow ring of a geometric quotient*, Ann. of Math. (2) **130** (1989), no. 1, 159–187, DOI 10.2307/1971479. MR1005610

[16] Barbara Fantechi, *Stacks for everybody*, European Congress of Mathematics, Vol. I (Barcelona, 2000), Progr. Math., vol. 201, Birkhäuser, Basel, 2001, pp. 349–359. MR1905329

[17] Alexander Givental, *A mirror theorem for toric complete intersections*, Topological field theory, primitive forms and related topics (Kyoto, 1996), Progr. Math., vol. 160, Birkhäuser Boston, Boston, MA, 1998, pp. 141–175. MR1653024

[18] Alexander B. Givental, *Equivariant Gromov-Witten invariants*, Internat. Math. Res. Notices **13** (1996), 613–663, DOI 10.1155/S1073792896000414. MR1408320

[19] Eduardo González and Chris Woodward, *Quantum Witten localization and abelianization for qde solutions*, `arXiv:0811.3358`, 2016.

[20] A. Grothendieck, *Sur la classification des fibrés holomorphes sur la sphère de Riemann* (French), Amer. J. Math. **79** (1957), 121–138, DOI 10.2307/2372388. MR87176

[21] Elana Kalashnikov, *Four-dimensional Fano quiver flag zero loci*, Proc. A. **475** (2019), no. 2225, 20180791, 23. MR3964122

[22] Bumsig Kim, *Stable quasimaps*, Commun. Korean Math. Soc. **27** (2012), no. 3, 571–581, DOI 10.4134/CKMS.2012.27.3.571. MR2978724

[23] A. D. King, *Moduli of representations of finite-dimensional algebras*, Quart. J. Math. Oxford Ser. (2) **45** (1994), no. 180, 515–530, DOI 10.1093/qmath/45.4.515. MR1315461

[24] Shaun Martin, *Symplectic quotients by a nonabelian group and by its maximal torus*, `arXiv:math/0001002v1`, Jan 2000.

[25] Rachel Webb, *The abelian-nonabelian correspondence for I-functions*, `arXiv:1804.07786v2`, 2018.

[26] Rachel Webb, *Functoriality and the moduli of sections, with applications to quasimaps*, Ph.D. thesis, University of Michigan, 5 2020.

Department of Mathematics, University of Michigan, Ann Arbor, Michigan 48109

Current address: Department of Mathematics, University of California, Berkeley, Berkeley, California, 94720-3840

Email address: `webbra@umich.edu`

Contemporary Mathematics
Volume **763**, 2021
https://doi.org/10.1090/conm/763/15321

Introduction to the gauged linear sigma model

Emily Clader

ABSTRACT. We describe the definition of the gauged linear sigma model (GLSM), focusing specifically on Fan–Jarvis–Ruan–Witten theory and its generalization, the hybrid model. These theories are related, via the celebrated Landau–Ginzburg/Calabi–Yau (LG/CY) correspondence, to the Gromov–Witten theory of complete intersections in weighted projective space. We discuss how the LG/CY correspondence motivates the meaning of the GLSM, and conversely, how a GLSM perspective can be used to prove the correspondence in many cases.

The goal of these lectures is to motivate the definition of the *gauged linear sigma model*, or GLSM, a generalization of Gromov–Witten theory as well as other (perhaps less familiar) objects such as Fan–Jarvis–Ruan–Witten (FJRW) theory. In particular, we show that, from a GLSM viewpoint, the definition of FJRW theory is an entirely natural analogue of the definition of Gromov–Witten theory. The celebrated Landau–Ginzburg/Calabi–Yau correspondence relates these two theories to one another, and placing both in the more general context of the GLSM hints at a method by which this correspondence could be proven. In the last lecture, we discuss the proof of the Landau–Ginzburg/Calabi–Yau correspondence in genus zero, and we mention ongoing work that points toward a proof in higher genus.

1. Preliminaries on orbifolds and Gromov–Witten theory

Before beginning our discussion of the GLSM, we must recall some basics about orbifolds and their cohomology, as well as the definition of Gromov–Witten theory and quantum cohomology.

1.1. What is an orbifold? Roughly speaking, an orbifold is a space locally homeomorphic to the quotient of an open subset of Euclidean space by the action of a finite group. More precisely, paralleling the definition of a manifold, the definition of an orbifold uses the language of an atlas of compatible orbifold charts.

DEFINITION 1.1. Let X be a topological space. An *orbifold chart* on X consists of

(1) an open set $U \subseteq X$;
(2) an open set $\widetilde{U} \subseteq \mathbb{R}^n$ for some n;

2010 *Mathematics Subject Classification.* Primary 14H81.
This work was partially supported by NSF DMS Grant 1810969.

(3) a smooth action of a finite group G on $\widetilde{U}$;
(4) a continuous, G-invariant map $\widetilde{\phi}: \widetilde{U} \to U$ such that the induced map
$$\phi: \widetilde{U}/G \to U$$
is a homeomorphism.

From here, one must prescribe what it means for two overlapping orbifold charts to be compatible. This is surprisingly subtle, and we encourage the careful reader to attempt to write down a reasonable definition for herself before reading further.

DEFINITION 1.2. We say that an orbifold chart
$$\widetilde{V}/H \xrightarrow{\psi} V \subseteq X$$
is a *subchart* of an orbifold chart
$$\widetilde{U}/G \xrightarrow{\phi} U \subseteq X$$
if $V \subseteq U$ and there is a group homomorphism $\tau: H \to G$ and a smooth embedding $\iota: \widetilde{V} \hookrightarrow \widetilde{U}$ such that

(1) $\iota(h \cdot x) = \tau(h) \cdot \iota(x)$ for all $x \in \widetilde{V}$ and $h \in H$, where $\cdot$ denotes the group actions of H on $\widetilde{V}$ and G on $\widetilde{Y}$;
(2) $\widetilde{\psi}(x) = \widetilde{\phi}(\iota(x))$ for all $x \in \widetilde{V}$;
(3) for each $x \in \widetilde{V}$, the stabilizer H_x is isomorphic to the stabilizer $G_{\iota(x)}$.

Given two orbifold charts
$$\widetilde{U}_1/G_1 \xrightarrow{\phi_1} U_1 \subseteq X \quad \text{and} \quad \widetilde{U}_2/G_2 \xrightarrow{\phi_2} U_2 \subseteq X,$$
we say that these charts are *compatible* if every point of $U_1 \cap U_2$ is contained in an open set U_3 for which
$$\widetilde{U}_3/G_3 \xrightarrow{\phi_3} U_3 \subseteq X$$
is a subchart of both of the two original charts.

Having defined compatibility of charts, we can finally give a precise definition of orbifolds.

DEFINITION 1.3. A (smooth) *orbifold* $\mathcal{X}$ is a topological space X together with a family of compatible charts covering X.

The key point is that, although U is homeomorphic to $\widetilde{U}/G$, the orbifold chart remembers more than just the topological quotient $\widetilde{U}/G$: it remembers the action of G, and specifically, where that action has isotropy. In particular, the definition of compatibility of orbifold charts implies that each point $x \in X$ has an isotropy group that is well-defined up to isomorphism.

DEFINITION 1.4. Let $\mathcal{X}$ be an orbifold and let $x \in X$. Choose any orbifold chart $\phi: \widetilde{U}/G \to U$ with $x \in U$, and any $\widetilde{x} \in \widetilde{U}$ such that $\widetilde{\phi}(\widetilde{x}) = x$. Then the *isotropy group* (or *stabilizer*) of x, denoted G_x, is the stabilizer of $\widetilde{x}$ under the action of G.

Although the definition of G_x appears to depend on the choice of both the orbifold chart and the preimage $\widetilde{x}$, choosing a different $\widetilde{x}$ in the same orbifold chart yields an isomorphic stabilizer because $\widetilde{\phi}$ is G-invariant, while choosing a different orbifold chart yields an isomorphic stabilizer by condition (3) of Definition 1.2.

Informally, we often think of orbifolds as though they were manifolds, in which each point with nontrivial stabilizer is labeled with that stabilizer. With this perspective in mind, let us turn to some of the key examples of orbifolds.

EXAMPLE 1.5. Let the cyclic group $\mathbb{Z}_n$ act on $\mathbb{C}$ via multiplication by nth roots of unity. As a topological space, the quotient $\mathbb{C}/\mathbb{Z}_n$ is homeomorphic to $\mathbb{C}$, via the map

$$\phi : \mathbb{C}/\mathbb{Z}_n \to \mathbb{C}$$
$$\phi([z]) = z^n$$

with inverse $z \mapsto [z^{1/n}]$. (Here, we use $[\cdot]$ to denote an equivalence class in the quotient $\mathbb{C}/\mathbb{Z}_n$.) Thus, if one sets $X = \mathbb{C}$, then there is a global orbifold chart on X given by $\widetilde{U} = \mathbb{C}$, $G = \mathbb{Z}_n$, and

$$\widetilde{\phi} : \widetilde{U} \to X$$
$$\widetilde{\phi}(z) = z^n.$$

We denote the resulting orbifold by $\mathcal{X} = [\mathbb{C}/\mathbb{Z}_n]$.

In $[\mathbb{C}/\mathbb{Z}_n]$, the isotropy group of any nonzero element $x \in X$ is trivial, whereas the isotropy group of $0 \in X$ is $\mathbb{Z}_n$. As a result, one typically represents the orbifold $[\mathbb{C}/\mathbb{Z}_n]$ graphically as the space $X = \mathbb{C}$ with a copy of the group $\mathbb{Z}_n$ attached to the origin.

EXAMPLE 1.6. Generalizing the above example, let M be any smooth manifold and let G be a finite group acting smoothly on M. Then there is an orbifold denoted $[M/G]$ with underlying topological space $X = M/G$. To construct the orbifold charts, one uses the fact that any $x \in M$ for which the G-action has isotropy group G_x is contained in a (manifold) chart $U_x \cong \widetilde{U}_x \subseteq \mathbb{R}^n$ for which U_x is invariant under the action of G_x. The orbifold charts then consist of the subsets $U_x/G_x \subseteq M/G$ with the quotient maps $\widetilde{\phi} : \widetilde{U}_x \to U_x/G_x$ and the action of G_x on $\widetilde{U}_x$ induced by the homeomorphism $\widetilde{U}_x \cong U_x$.

Any orbifold of the form $[M/G]$ is referred to as a **global quotient**. A special case, which is surprisingly rich, is the following.

EXAMPLE 1.7. If M is a single point and G is any finite group (acting in the only possible way on a point), we denote the resulting global quotient orbifold by $\mathcal{B}G$.

The following example, on the other hand, is provably not a global quotient for some values of the parameters $w_0, \ldots, w_n$:

EXAMPLE 1.8. Let $\mathbb{C}^*$ act on $\mathbb{C}^{n+1}$ by

$$\lambda \cdot (z_0, \ldots, z_n) = (\lambda^{w_0} z_0, \ldots, \lambda^{w_n} z_n),$$

where $w_0, \ldots, w_n$ are positive integers. Then there is an orbifold

$$\mathcal{X} = \mathbb{P}(w_0, \ldots, w_n),$$

referred to as a *weighted projective space*, in which X is the topological quotient

$$X = (\mathbb{C}^{n+1} \setminus \{0\})/\mathbb{C}^*.$$

To form the orbifold charts, first note that, as in the case of ordinary projective space, each $x \in X$ is contained in one of the subsets

$$U_i = \{[z_0, \ldots, z_n] \in (\mathbb{C}^{n+1} \setminus \{0\})/\mathbb{C}^* \mid z_i \neq 0\} \subseteq X.$$

Then $\widetilde{U}_i \cong \mathbb{C}^n$, on which the coordinates are denoted $(z_0, \ldots, z_{i-1}, z_{i+1}, \ldots, z_n)$ and the map $\widetilde{\phi}_i : \widetilde{U}_i \to U_i$ is

$$\widetilde{\phi}_i(z_0, \ldots, z_{i-1}, z_{i+1}, \ldots, z_n) = [z_0 : \cdots : z_{i-1} : 1 : z_{i+1} : \cdots : z_n].$$

The group $\mathbb{Z}_{w_i}$ of w_ith roots of unity acts on $\widetilde{U}_i$ by

$$\zeta \cdot (z_0, \ldots, z_{i-1}, z_{i+1}, \ldots, z_n) = (\zeta^{w_0} z_0, \ldots, \zeta^{w_{i-1}} z_{i-1}, \zeta^{w_{i+1}} z_{i+1}, \ldots, \zeta^{w_n} z_n),$$

and it is straightforward to check that this action makes the above into an orbifold chart.

EXERCISE 1.9. Consider the action of $\mathbb{Z}_2$ on the torus $\mathbb{T} = S^1 \times S^1 \subseteq \mathbb{C}^* \times \mathbb{C}^*$, where the nontrivial element of $\mathbb{Z}_2$ acts by

$$(e^{it_1}, e^{it_2}) \mapsto (e^{-it_1}, e^{-it_2}).$$

Find an explicit atlas of orbifold charts on $X = \mathbb{T}/\mathbb{Z}_2$. What, topologically, is X?

EXERCISE 1.10. Consider the action of the symmetric group S_3 on $\mathbb{P}^1 \times \mathbb{P}^1 \times \mathbb{P}^1$ given by permuting the three coordinates; the resulting topological quotient is the *symmetric product* $X = S^3(\mathbb{P}^1)$. At which points does the action of S_3 have nontrivial isotropy, and what is the isotropy group? Find an explicit atlas of orbifold charts on X, and convince yourself that, topologically, $X \cong \mathbb{P}^3$.

1.2. Orbifold vector bundles and a first pass at orbifold cohomology. In general, all of the geometric constructions one might associate to a manifold can be extended appropriately to orbifolds. The philosophy, when defining the orbifold analogue of a manifold construction, is to apply the manifold construction on the charts $\widetilde{U}$ and insist that it be equivariant with respect to the actions of the local groups G.

This idea is easiest to make precise in the case when the orbifold is a global quotient. For example, in that case, the orbifold analogue of a vector bundle can be defined as follows.

DEFINITION 1.11. Let $\mathcal{X} = [M/G]$ be a global quotient orbifold. Then an *orbifold vector bundle* on $\mathcal{X}$ consists of a vector bundle $\pi : E \to M$ together with an action of G on E that takes the fiber over each $x \in M$ linearly to the fiber over gx. A *section* of an orbifold vector bundle on $\mathcal{X}$ is a section s of π that is equivariant with respect to the G-actions on E and M.

By carrying out the above construction in orbifold charts, this definition can be extended to arbitrary orbifolds; we refer the reader to [**2**, Section 2.3] for a careful definition.

Orbifolds also have their own version of the tangent bundle. In the global quotient case, the definition is as follows.

DEFINITION 1.12. Let $\mathcal{X} = [M/G]$ be a global quotient orbifold. Then the *tangent bundle* to $\mathcal{X}$, denoted $T\mathcal{X}$, is the orbifold vector bundle defined by the ordinary tangent bundle $\pi : TM \to M$, on which G acts by the derivative of its action on M.

The typical operations on vector bundles extend to the orbifold setting, so, having defined the tangent bundle $T\mathcal{X}$, one can define the cotangent bundle, and its wedge powers $\wedge^k T^*\mathcal{X}$, and sections of $\wedge^k T^*\mathcal{X}$—that is, differential k-forms. This leads to a definition of orbifold de Rham cohomology $H^*_{\mathrm{dR}}(\mathcal{X})$.

Unfortunately, though, this definition of orbifold de Rham cohomology captures none of the information of the orbifold structure. More precisely, one has the following theorem.

THEOREM 1.13 (Satake). *For any orbifold $\mathcal{X}$ with underlying topological space X, there is an isomorphism*

$$H^*_{dR}(\mathcal{X}) \cong H^*(X;\mathbb{R}).$$

So, for example, the orbifold de Rham cohomology of $\mathcal{B}G$ is simply the cohomology of a point; it sees nothing of the group G.

The way out of this sad state of affairs was proposed by Chen and Ruan [**5**]: one should first define orbifold *quantum* cohomology, and then restrict to its degree-zero part to obtain a definition of orbifold cohomology that truly takes the orbifold structure into account.

Before discussing Chen–Ruan's definition of orbifold cohomology, we digress to recall the basics of quantum cohomology in the non-orbifold setting.

1.3. Review of Gromov–Witten theory and quantum cohomology. Let X be a smooth projective variety, and for simplicity, fix an embedding $i : X \hookrightarrow \mathbb{P}^r$. For any integers $n, \beta \geq 0$, the *moduli space of genus-zero stable maps to X* is the parameter space $\overline{\mathcal{M}}_{0,n}(X,\beta)$ of tuples $(C; q_1, \ldots, q_n; f)$, where

- C is an algebraic curve of (arithmetic) genus zero with at worst nodal singularities;
- $q_1, \ldots, q_n \in C$ are a collection of distinct smooth points of C;
- $f : C \to X$ is a morphism for which $i_* f_*[C] = \beta$ as an element of $H_2(\mathbb{P}^r;\mathbb{Z}) \cong \mathbb{Z}$;
- the data $(C; q_1, \ldots, q_n; f)$ is *stable* in the sense that it has finitely many automorphisms, or equivalently, any irreducible component of C on which f is constant has at least three "special points," which are either marked points q_i or nodes.

This moduli space is equipped with evaluation maps

$$\mathrm{ev}_i : \overline{\mathcal{M}}_{0,n}(X,\beta) \to X$$
$$\mathrm{ev}_i(C; q_1, \ldots, q_n; f) = f(q_i).$$

From here, the idea of Gromov–Witten invariants is to choose n cohomology classes $\alpha_1, \ldots, \alpha_n \in H^*(X)$ and study

$$\int_{\overline{\mathcal{M}}_{0,n}(X,\beta)} \mathrm{ev}_1^*(\alpha_1) \cdots \mathrm{ev}_n^*(\alpha_n). \tag{1.1}$$

In particular, if $\alpha_1, \ldots, \alpha_n$ were Poincaré dual to subvarieties $Y_1, \ldots, Y_n$ of X, then (1.1) would, naïvely, capture the number of points in $\mathrm{ev}_1^{-1}(Y_1) \cap \cdots \mathrm{ev}_n^{-1}(Y_n)$—in other words, the number of degree-β, genus-zero curves passing through these n subvarieties.[1]

However, $\overline{\mathcal{M}}_{0,n}(X,\beta)$ can be a very singular space, and it can have different components of different dimensions, so it is in general unclear how to make sense of the integral (1.1) in a meaningful way. It is a deep and difficult foundational fact that, despite its singularity, $\overline{\mathcal{M}}_{0,n}(X,\beta)$ admits a pure-dimensional homology class

[1]This naïve hope is useful motivation for the definition of Gromov–Witten invariants, but the reader should be warned that it is merely motivation; only in rare special cases are Gromov–Witten invariants actually equal to such enumerations of curves.

$[\overline{\mathcal{M}}_{0,n}(X,\beta)]^{\mathrm{vir}}$ that enjoys some of the properties that a fundamental class on a smooth space would have. Integrating against this *virtual fundamental class* gives the precise definition of Gromov–Witten invariants.

DEFINITION 1.14. Let $n, \beta \in \mathbb{Z}^{\geq 0}$, and let $\alpha_1, \ldots, \alpha_n \in H^*(X)$. Then the associated genus-zero *Gromov–Witten invariant* is

$$\langle \alpha_1 \cdots \alpha_n \rangle^X_{0,n,\beta} := \int_{[\overline{\mathcal{M}}_{0,n}(X,\beta)]^{\mathrm{vir}}} \mathrm{ev}_1^*(\alpha_1) \cdots \mathrm{ev}_n^*(\alpha_n).$$

We are now prepared to define the quantum cohomology of a variety.

DEFINITION 1.15. Let q denote a formal parameter, and, for $\omega, \nu \in H^*(X)$, define the Poincaré pairing

$$(\omega, \nu) := \int_X \omega \cup \nu.$$

The *quantum cohomology* of X is the vector space $H^*(X)[[q]]$ equipped with the product $*$ defined by

$$\Big(\alpha * \beta, \gamma\Big) := \sum_{\beta \geq 0} q^\beta \langle \alpha\ \beta\ \gamma \rangle^X_{0,3,\beta},$$

for $\alpha, \beta \in H^*(X)$, extended by linearity in q.

The fundamental feature of this definition is that its degree-zero part recovers the usual cohomology.

EXERCISE 1.16. Verify that setting $q = 0$ in the quantum product recovers $H^*(X)$ with its usual cup product.

1.4. Orbifold cohomology. We would like to generalize the definition of Gromov–Witten invariants and quantum cohomology to orbifolds, but in order to do so carefully, one must define what is meant by a *morphism* of orbifolds. This definition is remarkably difficult to make precise; it can be defined via equivariant morphisms in local orbifold charts, but the atlas of orbifold charts on the domain orbifold may need to be refined before doing so.

Rather than covering the careful definition of an orbifold morphism (which the reader can find in [**2**, Section 2.4]), we will content ourselves with one key example and one key fact.

EXAMPLE 1.17. Let $\mathcal{X} = [M/G]$ and $\mathcal{Y} = [N/H]$ be global quotient orbifolds. Then one example of an orbifold morphism is a pair (f_0, f_1) consisting of a continuous map

$$f_0 : M \to N$$

and a homomorphism

$$f_1 : G \to H$$

such that f_0 is equivariant with respect to f_1—that is,

$$f_0(g \cdot x) = f_1(g) \cdot f_0(x)$$

for all $g \in G$ and $x \in M$.

FACT 1.18. Let $\mathcal{X}$ and $\mathcal{Y}$ be any orbifolds, and let $f : \mathcal{X} \to \mathcal{Y}$ be an orbifold morphism. Then f induces a map $\overline{f} : X \to Y$ between the underlying topological spaces of $\mathcal{X}$ and $\mathcal{Y}$, as well as a homomorphism of isotropy groups

$$\lambda_x : G_x \to G_{\overline{f}(x)}$$

for each $x \in X$.

EXERCISE 1.19. Verify that, in the special case of Example 1.17, one indeed obtains maps $\overline{f}$ and λ_x as in Fact 1.18.

The key upshot of Fact 1.18 is the following. Equipped with the notion of an orbifold morphism, one can define a moduli space $\overline{\mathcal{M}}_{0,n}(\mathcal{X}, \beta)$ of stable maps $f : C \to \mathcal{X}$ from genus-zero orbifold curves C to $\mathcal{X}$. (Roughly, an *orbifold curve* is an orbifold of complex dimension 1; see [**1**] for the precise definition.) In the orbifold setting, though, there are two pieces of local data around each marked point q_i:

- the image $\overline{f}(q_i) \in X$, and
- the homomorphism λ_{q_i}.

Moreover, the isotropy groups of orbifold curves are cyclic with a canonical generator ζ_i induced by the orientation, so the data of λ_{q_i} is encoded in a single element $\lambda_{q_i}(\zeta_i)$ of the isotropy group of $\mathcal{X}$ at $f(q_i)$.

As a result of these considerations, the natural evaluation maps on $\overline{\mathcal{M}}_{0,n}(\mathcal{X}, \beta)$ land not in $\mathcal{X}$ itself but in a richer object known as the inertia orbifold.

DEFINITION 1.20. The *inertia orbifold* of $\mathcal{X}$ is

$$I\mathcal{X} := \{(x, g) \mid x \in X,\ g \in G_x\}. \tag{1.2}$$

(To be more precise, equation (1.2) describes only the underlying topological space of $I\mathcal{X}$, but this topological space can be given the structure of an orbifold. We illustrate the orbifold structure only in a single example, which is the most important example for what follows. The reader is referred to [**2**, Section 2.5] for a more careful general definition.)

EXAMPLE 1.21. Let $\mathcal{X} = [\mathbb{C}/\mathbb{Z}_r]$. Then $X = \mathbb{C}$, and any nonzero $x \in \mathbb{C}$ has trivial isotropy whereas the isotropy group of $0 \in \mathbb{C}$ is $\mathbb{Z}_r$. It follows that, on the level of the underlying space,

$$\begin{aligned} I[\mathbb{C}/\mathbb{Z}_r] &= \{(x,1) \mid x \in \mathbb{C}\} \sqcup \{(0,\zeta)\} \sqcup \{(0,\zeta^2)\} \sqcup \cdots \sqcup \{(0,\zeta^{r-1})\} \\ &\cong \mathbb{C} \sqcup \{0\} \sqcup \{0\} \sqcup \cdots \sqcup \{0\} \end{aligned} \tag{1.3}$$

in which $\zeta \in \mathbb{Z}_r$ is a primitive rth root of unity. To give $I[\mathbb{C}/\mathbb{Z}_r]$ the structure of an orbifold, let $\mathbb{Z}_r$ act on the above by sending

$$(x, g \in G_x) \mapsto (hx, hgh^{-1} \in G_{hx})$$

for each $h \in \mathbb{Z}_r$. Because $\mathbb{Z}_r$ is abelian, this action amounts to sending (x,g) to (hx, g), so it preserves the r components in (1.3). As an orbifold, then,

$$I[\mathbb{C}/\mathbb{Z}_r] = [\mathbb{C}/\mathbb{Z}_r] \sqcup \mathcal{B}\mathbb{Z}_r \sqcup \cdots \sqcup \mathcal{B}\mathbb{Z}_r.$$

Having defined the inertia orbifold, the evaluation maps are

$$\mathrm{ev}_i : \overline{\mathcal{M}}_{0,n}(\mathcal{X}, \beta) \to I\mathcal{X}$$

$$\mathrm{ev}_i(C; q_1, \ldots, q_n; f) = (\overline{f}(q_i), \lambda_{q_i}(\zeta_i)).$$

From here, the definition of Gromov–Witten invariants carries over verbatim, but the insertions $\alpha_1, \ldots, \alpha_n$ to these invariants are chosen from $H^*(I\mathcal{X})$ as opposed to $H^*(\mathcal{X})$. In particular, quantum cohomology in the orbifold setting is a product structure on $H^*(I\mathcal{X})[[q]]$.

In light of Exercise 1.16, one should expect the "ordinary" cohomology of orbifolds to be obtained from their quantum cohomology by setting $q = 0$. Rather than proving this as a theorem, one takes this result as the definition of orbifold cohomology.

DEFINITION 1.22. Let $\mathcal{X}$ be an orbifold. Then the *orbifold* (or *Chen–Ruan*) *cohomology* of $\mathcal{X}$ is the vector space

$$H^*_{\mathrm{CR}}(\mathcal{X}) := H^*(I\mathcal{X}),$$

where the right-hand side denotes the de Rham (or singular) cohomology of $I\mathcal{X}$, which is the same as the cohomology of its underlying topological space. The product structure on $H^*_{\mathrm{CR}}(\mathcal{X})$ is the $q = 0$ limit of the quantum product on $H^*(I\mathcal{X})[[q]]$.

EXERCISE 1.23. What is $H^*_{\mathrm{CR}}([\mathbb{C}/\mathbb{Z}_r])$, as a vector space?

EXERCISE 1.24. Consider a weighted projective space $\mathcal{X} = \mathbb{P}(w_0, w_1)$ with coprime weights w_0 and w_1. Calculate the inertia orbifold and the orbifold cohomology (as a vector space) of $\mathcal{X}$.

EXERCISE 1.25. Now consider a weighted projective space $\mathcal{X} = \mathbb{P}(d, \ldots, d)$ in which every weight is d. Verify that the underlying topological space is $X = \mathbb{P}^n$, and that every element $x \in \mathbb{P}^n$ has isotropy group $G_x = \mathbb{Z}_d$. Deduce that the coarse underlying space of $I\mathbb{P}(d, \ldots, d)$ is a disjoint union of d copies of $\mathbb{P}^n$, and hence, as a vector space, $H^*_{\mathrm{CR}}(\mathbb{P}(d, \ldots, d))$ is a direct sum of d copies of $H^*(\mathbb{P}^n)$.

2. Fan–Jarvis–Ruan–Witten theory

Equipped with the background developed in the previous section, we are prepared to define one of our primary objects of study in these lectures: Fan–Jarvis–Ruan–Witten (FJRW) invariants.

2.1. The structure of FJRW theory. Structurally, FJRW theory is analogous to the Gromov–Witten theory of a smooth projective variety X. In particular, it consists of

(1) a vector space $\mathcal{H}$ (the "state space"), analogous to $H^*(X)$ in Gromov–Witten theory;
(2) a moduli space $\overline{\mathcal{M}}$ equipped with a virtual fundamental cycle, analogous to $\overline{\mathcal{M}}_{g,n}(X, \beta)$ in Gromov–Witten theory;
(3) a notion of "correlators", analogous to Gromov–Witten invariants, which are integrals against the virtual cycle of $\overline{\mathcal{M}}$ associated to any tuple of elements in $\mathcal{H}$.

The input data for FJRW theory, though, is not a variety X but a polynomial $W \in \mathbb{C}[x_1, \ldots, x_N]$. This polynomial is required to be of a particular type.

DEFINITION 2.1. A polynomial $W \in \mathbb{C}[x_1, \ldots, x_N]$ is *quasihomogeneous* if there exist positive integers $w_1, \ldots, w_N$ (the *weights*) and d (the *degree*) such that

$$W(\lambda^{w_1} x_1, \ldots, \lambda^{w_N} x_N) = \lambda^d W(x_1, \ldots, x_N)$$

for all $\lambda \in \mathbb{C}$.

Note, in particular, that quasihomogeneity implies that the vanishing locus of W defines a hypersurface X_W in the weighted projective space $\mathbb{P}(w_1, \ldots, w_N)$.

In addition to quasihomogeneity, in what follows, we require our polynomials W to satisfy two further technical conditions:

(1) **Nondegeneracy**: The hypersurface $X_W \subseteq \mathbb{P}(w_1, \ldots, w_N)$ is nonsingular as an orbifold;

(2) **Invertibility**: The number of monomials of W is equal to the number of variables.

A particularly simple example of a nondegenerate, invertible, quasihomogeneous polynomial—to which we return repeatedly in what follows—is the Fermat quintic.

EXAMPLE 2.2. Let $W(x_1, \ldots, x_5) = x_1^5 + \cdots + x_5^5$. This is a quasihomogeneous polynomial with $w_1 = \cdots = w_5 = 1$ and $d = 5$.

In the following subsections, we define the ingredients of the FJRW theory of W precisely. Before doing so, though, it is worth grounding the discussion by looking ahead to the ultimate goal: in Section 3, we present the Landau–Ginzburg/Calabi–Yau correspondence, which asserts an equivalence between the FJRW theory of W and the Gromov–Witten theory of X_W in certain cases. We encourage the reader to keep this correspondence in mind as motivation for the definitions that follow.

2.2. The state space of FJRW theory. Let $W \in \mathbb{C}[x_1, \ldots, x_N]$ be a nondegenerate, invertible, quasihomogeneous polynomial with weights $w_1, \ldots, w_N$ and degree d.

DEFINITION 2.3. The *FJRW state space* associated to W is the $\mathbb{Q}$-vector space

$$\mathcal{H}_W := H^*_{\mathrm{CR}}([\mathbb{C}^N/J], W^{+\infty}; \mathbb{Q}),$$

where

$$J := \left\langle \left(e^{2\pi i \frac{w_1}{d}}, \ldots, e^{2\pi i \frac{w_N}{d}} \right) \right\rangle \subseteq (\mathbb{C}^*)^N$$

acts diagonally on $\mathbb{C}^N$ and

$$W^{+\infty} := W^{-1}(\rho)$$

for any sufficiently large real number ρ. (The resulting vector space is independent of ρ for $\rho \gg 0$, because topologically, $W^{-1}(\rho)$ is eventually independent of ρ.)

We note that the action of J on $\mathbb{C}^N$ is cooked up so that $W(g\vec{x}) = W(\vec{x})$ for all $g \in J$, so W gives a well-defined map out of the quotient $\mathbb{C}^N/J$. To express $\mathcal{H}_W$ more explicitly, we recall that $H^*_{\mathrm{CR}}(\mathcal{X}) := H^*(I\mathcal{X})$ and

$$I[\mathbb{C}^N/J] := \{(x,g) \mid x \in \mathbb{C}^N/J,\ g \in G_x \subseteq J\} \cong \bigsqcup_{g \in J} \mathbb{C}^N_g/J,$$

where $\mathbb{C}^N_g \subseteq \mathbb{C}^N$ is the fixed locus of g. Therefore, we have

$$\mathcal{H}_W = \bigoplus_{g \in J} H^*(\mathbb{C}^N_g/J, W^{+\infty}; \mathbb{Q}).$$

The cohomology groups appearing in the above are ordinary (for example, de Rham) cohomology groups, for which the cohomology of a quotient Y/G can be calculated as the G-invariant part of the cohomology of Y. Thus,

$$\mathcal{H}_W = \bigoplus_{g \in J} H^*(\mathbb{C}^N_g, W^{+\infty}; \mathbb{Q})^J, \tag{2.1}$$

in which the superscript denotes the J-invariant part.

EXAMPLE 2.4. Let $W(x_1,\ldots,x_5) = x_1^5 + \cdots + x_5^5$. Then

$$J = \left\langle \left(e^{2\pi i \frac{1}{5}}, \ldots, e^{2\pi i \frac{1}{5}} \right) \right\rangle \cong \mathbb{Z}_5,$$

acting diagonally on $\mathbb{C}^5$ by multiplication by fifth roots of unity. Given that the action of $1 \in J$ fixes all of $\mathbb{C}^5$ but the action of any nontrivial $g \in J$ fixes only the origin, we have

$$I[\mathbb{C}^5/J] = [\mathbb{C}^5/\mathbb{Z}_5] \sqcup \mathcal{B}\mathbb{Z}_5 \sqcup \mathcal{B}\mathbb{Z}_5 \sqcup \mathcal{B}\mathbb{Z}_5 \sqcup \mathcal{B}\mathbb{Z}_5.$$

It follows that

$$\mathcal{H}_W = H^*(\mathbb{C}^5, W^{+\infty}; \mathbb{Q})^{\mathbb{Z}_5} \oplus \bigoplus_{g \neq 1 \in \mathbb{Z}_5} H^*(\{0\}, \emptyset; \mathbb{Q}) = H^*(\mathbb{C}^5, W^{+\infty}; \mathbb{Q})^{\mathbb{Z}_5} \oplus \mathbb{Q}^4.$$

As the above example illustrates, the state space decomposes into two types of elements.

DEFINITION 2.5. An element $g \in J$ is called *narrow* if it fixes only $0 \in \mathbb{C}^N$, meaning that the corresponding summand of $\mathcal{H}_W$ is a copy of $\mathbb{Q}$. Elements that are not narrow are called *broad*.

For instance, in Example 2.4, the element $g = 1$ is broad and all other $g \in J$ are narrow.

EXERCISE 2.6. Find an example of a quasihomogeneous polynomial for which there exists a nontrivial broad element $g \in J$.

2.3. The moduli space of FJRW theory. Let W be as in the previous subsection. Then the moduli space of FJRW theory is the parameter space $\overline{\mathcal{M}}_{g,n}^{W}$ of tuples $(C; q_1, \ldots, q_n; L; \phi)$, where

- $(C; q_1, \ldots, q_n)$ is a genus-g, n-pointed orbifold stable curve;
- L is an orbifold line bundle on C;
- ϕ is an isomorphism

$$\phi : L^{\otimes d} \xrightarrow{\sim} \omega_{\log},$$

where $\omega_{\log} := \omega_C([q_1] + \cdots + [q_n])$,

subject to a certain stability condition on the actions of the isotropy groups on the fibers of L. The term "orbifold stable curve" here has a precise meaning that we will not specify; in particular, though, it means that C has nontrivial isotropy only at the marked points q_i and the nodes.

EXAMPLE 2.7. When $W(x_1,\ldots,x_5) = x_1^5 + \cdots + x_5^5$, the moduli space $\overline{\mathcal{M}}_{g,n}^{W}$ is otherwise known as the moduli space of 5-spin structures and is often denoted $\overline{\mathcal{M}}_{g,n}^{1/5}$.

2.4. The correlators of FJRW theory. To define the correlators of FJRW theory, we must refine the definition of the moduli space by decomposing it into components depending on the "multiplicities" of the bundle L.

DEFINITION 2.8. Let C be an orbifold curve, let L be an orbifold line bundle on C, and let $q \in C$ be a point with isotropy group $G_q = \mathbb{Z}_r$. Then q is contained in an orbifold chart isomorphic as an orbifold to the global quotient $[\mathbb{C}/\mathbb{Z}_r]$, on which the data of L consists of a bundle $\widetilde{L}$ on $\mathbb{C}$ together with an action of $\mathbb{Z}_r$ according to which the projection map $\widetilde{L} \to \mathbb{C}$ is equivariant. In particular, $\mathbb{Z}_r$ acts on the fiber of $\widetilde{L}$ over q. The *multiplicity* of L at q is defined as the number

$$\mathrm{mult}_q(L) := m_q \in \{0, 1, \ldots, r-1\}$$

such that the action of $\mathbb{Z}_r$ on $\widetilde{L}$ is

$$\zeta \cdot (x, v) = \left(e^{2\pi i \frac{1}{r}} x, e^{2\pi i \frac{m_q}{r}} v\right)$$

in local coordinates around q.

In the case of $\overline{\mathcal{M}}^W_{g,n}$, the stability condition ensures that all of the isotropy groups have order dividing d. Thus, one can view the multiplicities of L at any special point as an element of $\{0, 1, \ldots, d-1\}$. For a tuple

$$m_1, \ldots, m_n \in \{0, 1, \ldots, d-1\},$$

we set

$$\overline{\mathcal{M}}^W_{g,(m_1,\ldots,m_n)} := \{(C; q_1, \ldots, q_n; L; \phi) \mid \mathrm{mult}_{q_i}(L) = m_i \text{ for all } i\} \subseteq \overline{\mathcal{M}}^W_{g,n}.$$

The key observation needed to define FJRW correlators, at least in the narrow case, is that for any $g \in J$,

$$g = \left(e^{2\pi i m \frac{w_1}{d}}, \ldots, e^{2\pi i m \frac{w_N}{d}}\right)$$

for some $m \in \{0, 1, \ldots, d-1\}$, and the set $\{0, 1, \ldots, d-1\}$ is the same set that indexes the summands of $\mathcal{H}_W$ in the decomposition (2.1).

More specifically, we define narrow correlators as follows.

DEFINITION 2.9. Decompose the FJRW state space as

$$\mathcal{H}_W = \bigoplus_{g \text{ broad}} H^*(\mathbb{C}^N_g, W^{+\infty}; \mathbb{Q})^J \oplus \bigoplus_{g \text{ narrow}} \mathbb{Q}\{e_g\},$$

where e_g denotes the fundamental class on the component of $I[\mathbb{C}^N/J]$ indexed by g. Choose n narrow elements

$$\alpha_1, \ldots, \alpha_n \in \mathcal{H}_W,$$

and write $\alpha_k = c_k \cdot e_{g_k}$ for $c_k \in \mathbb{Q}$ and

$$g_k = \left(e^{2\pi i m_k \frac{w_1}{d}}, \ldots, e^{2\pi i m_k \frac{w_N}{d}}\right),$$

where $m_k \in \{0, 1, \ldots, d-1\}$. Then we define the associated *FJRW correlator* as

$$\langle \alpha_1 \cdots \alpha_N \rangle^W_{g,n} = c_1 \cdots c_n \int_{[\overline{\mathcal{M}}^W_{g,(m_1,\ldots,m_n)}]^{\mathrm{vir}}} 1.$$

Two difficult fundamental facts are needed in order to make this definition complete. First, there indeed exists a virtual cycle $[\overline{\mathcal{M}}^W_{g,(m_1,\ldots,m_n)}]^{\mathrm{vir}}$ against which the above integral can be defined; we will say some more words about how this cycle is constructed in later sections of these notes. And second, the definition of correlators can be extended to broad insertions. This second fact is harder, and we will not address it here. For the cases of interest below, narrow correlators are sufficient; the reader is directed to **[19]** for the details of the broad case.

We conclude this section with several exercises to practice with the notion of multiplicity.

EXERCISE 2.10. Let $(C; q_1, \ldots, q_n)$ be a smooth orbifold curve at which each marked point q_i has isotropy group $\mathbb{Z}_d$, and let L be an orbifold line bundle on C. A fundamental fact about the multiplicities of L is that the bundle

$$|L| := L \otimes \mathcal{O}\left(-\sum_{i=1}^{n} \frac{m_i}{d}[q_i]\right)$$

is pulled back from the underlying topological space of C. Use this fact to explain why the multiplicities must satisfy

$$\sum_{i=1}^{n} m_i \equiv 0 \mod d. \tag{2.2}$$

EXERCISE 2.11. A node q of C is said to be *separating* if the partial normalization of C at q is disconnected, and a curve C is said to be of *compact type* if all of its nodes are separating. Convince yourself that, if L is an orbifold line bundle on an orbifold curve of compact type, then the multiplicity of L at each branch of each node is determined by (2.2) together with the "kissing condition" that $m' + m'' \equiv 0$ mod d for the multiplicities m', m'' at opposite branches of the same node. What happens if the curve is not of compact type?

EXERCISE 2.12. Work out the "fundamental fact" from Exercise 2.10 explicitly in local coordinates, by considering the diagram

$$\begin{array}{ccc} L = [(\mathbb{C} \times \mathbb{C})/\mathbb{Z}_d] & \xrightarrow{[x,v] \mapsto [x, x^{-m} v]} & [(\mathbb{C} \times \mathbb{C})/\mathbb{Z}_d] = |L| \\ & \searrow \quad \swarrow & \\ & \mathcal{C} = [\mathbb{C}/\mathbb{Z}_d] & \end{array},$$

where L is defined by the action $\zeta(x, v) = (\zeta x, \zeta^m v)$ and $|L|$ is defined by the action $\zeta(x, v) = (\zeta x, v)$ of $\mathbb{Z}_d$ on $\mathbb{C} \times \mathbb{C}$. (See [**23**, Section 2.1.4] for related discussion.)

3. The Landau–Ginzburg/Calabi–Yau correspondence

Two questions would be reasonable for the reader to ask at this point: where do all of these FJRW theory definitions come from, and what does all of this have to do with the Gromov–Witten theory of the hypersurface defined by W?

The answers to both of these questions come from putting FJRW theory in the context of the Landau–Ginzburg/Calabi–Yau (LG/CY) correspondence, which we describe in this section. In order to tell that story in a more readable form, we restrict our attention to the case where

$$W = W(x_1, \ldots, x_5) = x_1^5 + \cdots + x_5^5,$$

whose vanishing locus is the quintic threefold $Q \subseteq \mathbb{P}^4$. We describe, in this case, an isomorphism between the state spaces $H^*(Q)$ and $\mathcal{H}_W$ of Gromov–Witten and FJRW theory, respectively, and a "matching" between the generating functions of genus-zero Gromov–Witten and FJRW correlators under a particular identification of the state spaces.

3.1. The state space correspondence. The first step in the LG/CY correspondence is to explain why the state spaces $H^*(Q)$ and $\mathcal{H}_W$ are isomorphic. To do so, we introduce a larger space out of which both of these can be built.

Consider the space $\mathbb{C}^5 \times \mathbb{C}$ with coordinates $(x_1, \ldots, x_5, p)$, and let $\mathbb{C}^*$ act on this space by

$$\lambda(x_1, \ldots, x_5, p) = (\lambda x_1, \ldots, \lambda x_5, \lambda^{-5} p).$$

Then the polynomial

$$\overline{W}(x_1, \ldots, x_5, p) = p(x_1^5 + \cdots + x_5^5)$$

is $\mathbb{C}^*$-invariant, so it gives a well-defined map out of the quotient $(\mathbb{C}^5 \times \mathbb{C})/\mathbb{C}^*$.

This quotient is not separated, but it admits two maximal separated subquotients; in the language of geometric invariant theory, these are the GIT quotients $(\mathbb{C}^5 \times \mathbb{C}) /\!\!/_\theta \, \mathbb{C}^*$ associated to the positive and negative characters $\theta \in \chi(\mathbb{C}^*) \cong \mathbb{Z}$. Namely, they are

$$X_+ := \frac{(\mathbb{C}^5 \setminus \{0\}) \times \mathbb{C}}{\mathbb{C}^*} = \mathcal{O}_{\mathbb{P}^4}(-5)$$

and

$$X_- := \frac{\mathbb{C}^5 \times (\mathbb{C} \setminus \{0\})}{\mathbb{C}^*} = [\mathbb{C}^5/\mathbb{Z}_5].$$

The orbifold cohomology of X_- relative to a fiber $\overline{W}^{+\infty}$ of $\overline{W} : X_- \to \mathbb{C}$ is, by definition, the FJRW state space $\mathcal{H}_W$. On the other hand, the (orbifold) cohomology of X_+ relative to a fiber $\overline{W}^{+\infty}$ of $\overline{W} : X_+ \to \mathbb{C}$ is

$$H^*(\mathcal{O}_{\mathbb{P}^4}(-5), \overline{W}^{+\infty}) \cong H^*(\mathbb{P}^4, \mathbb{P}^4 \setminus Q) \cong H^*(Q),$$

where the first isomorphism follows from the deformation retraction of $\mathcal{O}_{\mathbb{P}^4}(-5)$ onto $\mathbb{P}^4$ (which carries $\overline{W}^{+\infty}$ bijectively onto $\mathbb{P}^4 \setminus Q$) and the second isomorphism is the Thom isomorphism.

Via this argument, the state spaces of Gromov–Witten and FJRW theory arise in completely analogous ways from the larger quotient $(\mathbb{C}^5 \times \mathbb{C})/\mathbb{C}^*$. In fact, this framework can be leveraged—with the help of a number of exact sequences—to prove that the two state spaces are isomorphic:

$$H^*(Q) \cong \mathcal{H}_W.$$

This is the first piece of the LG/CY correspondence; it was first proved by Chiodo–Ruan [**7**].

EXERCISE 3.1. Show, furthermore, that there is an isomorphism between the ambient part of the cohomology of $H^*(Q)$—that is, the cohomology classes pulled back from $\mathbb{P}^4$—and the narrow part of $\mathcal{H}_W$.

3.2. The moduli spaces of Gromov–Witten and FJRW theory. The second, more substantial, piece of the correspondence is a comparison of correlators, and in particular, of the moduli spaces and their virtual cycles.

The basic structure of this comparison is modeled on the argument for the state space correspondence explained above; namely, just as the state space correspondence arose by placing both $H^*(Q)$ and $\mathcal{H}_W$ in the context of a larger but badly-behaved quotient $(\mathbb{C}^5 \times \mathbb{C})/\mathbb{C}^*$, the moduli comparison begins by finding a larger but badly-behaved moduli space out of which $\overline{\mathcal{M}}_{g,n}(Q, \beta)$ and $\overline{\mathcal{M}}_{g,n}^W$ can both be built.

That larger moduli space, denoted $X_{g,n,\beta}$, parameterizes tuples

$$(C; q_1, \dots, q_n; L; x_1, \dots, x_5, p),$$

in which $(C; q_1, \dots, q_n)$ is again a genus-g, n-pointed orbifold stable curve, L an orbifold line bundle of degree β on C, and

$$(x_1, \dots, x_5, p) \in \Gamma\left(L^{\oplus 5} \oplus (L^{\otimes -5} \otimes \omega_{\log})\right),$$

subject to certain stability and compatibility conditions that we will not make explicit here. (See, for example, [**12**] for further details.)

Just as in the case of the quotient $(\mathbb{C}^5 \times \mathbb{C})/\mathbb{C}^*$, the moduli space $X_{g,n,\beta}$ is non-separated; furthermore, it is non-compact (as the sections can be scaled by any complex number) and not Deligne–Mumford (as scaling the line bundle leads to infinitely many automorphisms). Nevertheless, it contains two separated Deligne–Mumford substacks: the substack $X^+_{g,n,\beta}$ on which $\vec{x}$ is nowhere-vanishing, and the substack $X^-_{g,n,\beta}$ on which p is nowhere-vanishing.

When $\vec{x}$ is nowhere-vanishing, the tuple $(C; q_1, \dots, q_n; L; \vec{x})$ is equivalent to the data of an n-pointed map $f : C \to \mathbb{P}^4$, and the stability condition on $X_{g,n,\beta}$ implies that this is a stable map. Furthermore, from this perspective, we have $L = f^*\mathcal{O}(1)$, so

$$X^+_{g,n,\beta} = \{(C; q_1, \dots, q_n; f; p) \mid f : C \to \mathbb{P}^4 \text{ stable, } p \in \Gamma(f^*\mathcal{O}(-5) \otimes \omega_{\log})\}.$$

On the other hand, when p is nowhere-vanishing, it trivializes $L^{\otimes -5} \otimes \omega_{\log}$, so

$$X^-_{g,n,\beta} = \{(C; q_1, \dots, q_n; L; x_1, \dots, x_5; \varphi) \mid x_i \in \Gamma(L),\ \varphi : L^{\otimes -5} \cong \omega_{\log}\}.$$

These two moduli spaces, while separated and Deligne–Mumford, are still not compact, since each still parameterizes the choice of a section of a bundle. Analogously to the way in which we passed, in the state space correspondence, to cohomology relative to the fiber $\overline{W}^{+\infty}$, we restrict on the moduli level to the loci $Z^\pm_{g,n,\beta} \subseteq X^\pm_{g,n,\beta}$ on which the section $(x_1, \dots, x_5, p)$ satisfies the defining equations of the critical locus

$$\text{Crit}(\overline{W}) \subseteq \mathbb{C}^5 \times \mathbb{C}.$$

That is, we require that

$$\frac{\partial \overline{W}}{\partial x_i}(x_1, \dots, x_5, p) = 0 \in \Gamma\big(L^{\otimes 4} \otimes (L^{\otimes -5} \otimes \omega_{\log})\big)$$

and

$$\frac{\partial \overline{W}}{\partial p}(x_1, \dots, x_5, p) = 0 \in \Gamma\left(L^{\otimes 5}\right).$$

The following exercise makes these conditions explicit.

EXERCISE 3.2. Viewing $\overline{W}$ as a function on $X_+ = \mathcal{O}_{\mathbb{P}^4}(-5)$, verify that the critical locus of $\overline{W}$ is the quintic threefold Q inside the zero section $\mathbb{P}^4 \subseteq X_+$. Similarly, viewing $\overline{W}$ as a function on $X_- = [\mathbb{C}^5/\mathbb{Z}_5]$, verify that the critical locus of $\overline{W}$ is $[\{0\}/\mathbb{Z}_5]$.

In particular, then, we have

$$Z^+_{g,n,\beta} = \{(C; q_1, \dots, q_n; f; p) \mid f : C \to Q \text{ stable, } p = 0\} = \overline{\mathcal{M}}_{g,n}(Q, \beta),$$

which is the moduli space of Gromov–Witten theory, whereas

$$Z^-_{g,n,\beta} = \{(C; q_1, \dots, q_n; L; x_1, \dots, x_5; \varphi) \mid x_1 = \dots = x_5 = 0,\ \varphi : L^{\otimes -5} \cong \omega_{\log}\} = \overline{\mathcal{M}}^W_{g,n},$$

which is the moduli space of FJRW theory.

We have thus succeeded in constructing the moduli spaces of the two theories in entirely analogous ways. In fact, this framework can be upgraded to give a uniform construction of the virtual fundamental cycles on the two moduli spaces, via the *cosection construction* of Kiem–Li [**24**]. The basic idea, which was first applied to this setting by Chang–Li–Li [**3**, **4**], is that $X^{\pm}_{g,n,\beta}$ has a natural deformation theory—and hence, a natural virtual fundamental cycle—coming from the cohomology of $L^{\oplus 5} \oplus (L^{\otimes -5} \otimes \omega_{\log})$, and the cosection construction is designed to produce a virtual fundamental cycle supported on a critical locus inside a larger space.

One caution is in order here: unlike in the state space correspondence, no equivalence between virtual cycles or correlators falls out immediately from the fact that the two moduli spaces arise in analogous ways. Rather, the above argument should be viewed, at this point, as merely a heuristic framework for motivating a relationship between the two theories.

Nevertheless, a correspondence of correlators does hold in genus zero, as Chiodo and Ruan proved.

THEOREM 3.3 (Chiodo–Ruan). *The genus-zero FJRW theory of W can be encoded in a generating function $J^{FJRW}(t)$ taking values in $\mathcal{H}_W[[z^{-1}, z]$, and the genus-zero Gromov–Witten theory of Q can be encoded in a generating function $J^{GW}(q)$ taking values in $H^*(Q)[[z^{-1}, z]$. After choosing a specific isomorphism $\mathcal{H}_W \cong H^*(Q)$, these two generating functions are related by changes of variables in q and t, an identification $q = t^{-5}$, and analytic continuation between the $t = 0$ and $q = 0$ expansions.*

We return in the last section to discuss some ideas in the proof of the genus-zero LG/CY correspondence. Before doing so, however, we address another natural question: if FJRW corresponds to the Gromov–Witten theory of the hypersurface Q, what theory corresponds in this way to the Gromov–Witten theory of a complete intersection?

4. The hybrid model and the GLSM

Let

$$W_1, \dots, W_r \in \mathbb{C}[x_1, \dots, x_N]$$

be a collection of quasihomogeneous polynomials that all have the same weights $w_1, \dots, w_N$ and the same degree d, and that define a nonsingular complete intersection

$$Z = \{W_1 = \cdots = W_r = 0\} \subseteq \mathbb{P}(w_1, \dots, w_N).$$

The answer to the question posed at the end of the previous section is provided by a theory referred to as the "hybrid model," whose definition occupies this section. Just as in the FJRW case, the theory consists of a state space, a moduli space (equipped with a virtual fundamental cycle), and a notion of correlators.

4.1. The state space of the hybrid model. To understand what the appropriate FJRW-type state space associated to Z might be, we return to the variation of GIT perspective from the previous section. Namely, introduce r auxiliary variables $p_1, \dots, p_r$, and let $\mathbb{C}^*$ act on $\mathbb{C}^N \times \mathbb{C}^r$ by

$$\lambda(x_1, \dots, x_N, p_1, \dots, p_r) = (\lambda^{w_1} x_1, \dots, \lambda^{w_N} p_N, \lambda^{-d} p_1, \dots, \lambda^{-d} p_r).$$

Then the polynomial

$$\overline{W}(x_1, \dots, x_N, p_1, \dots, p_r) = p_1 W_1(\vec{x}) + \cdots + p_r W_r(\vec{x})$$

defines a function on the quotient $(\mathbb{C}^N \times \mathbb{C}^r)/\mathbb{C}^*$.

Again, while this original quotient is not separated, it admits two maximal separated subquotients,

$$X_+ = \{\vec{x} \neq 0\} = \bigoplus_{j=1}^{r} \mathcal{O}_{\mathbb{P}(w_1,\ldots,w_N)}(-d) \tag{4.1}$$

and

$$X_- = \{\vec{p} \neq 0\} = \bigoplus_{i=1}^{N} \mathcal{O}_{\mathbb{P}(d,\ldots,d)}(-w_i). \tag{4.2}$$

The cohomology of X_+ relative to a fiber $\overline{W}^{+\infty}$ of $\overline{W}$ is

$$H^*_{\mathrm{CR}}\left(\bigoplus_{j=1}^{r} \mathcal{O}_{\mathbb{P}(w_1,\ldots,w_N)}(-d), \overline{W}^{+\infty}\right) \cong H^*_{\mathrm{CR}}(\mathbb{P}(w_1,\ldots,w_N), \mathbb{P}(w_1,\ldots,w_N) \setminus Z),$$

where the isomorphism can be deduced from the fact that $\overline{W}^{+\infty}$ intersects the fibers of $\mathcal{O}_{\mathbb{P}(w_1,\ldots,w_N)}(-d)^{\oplus r}$ in an affine space (with trivial cohomology) over points in $\mathbb{P}(w_1,\ldots,w_N) \setminus Z$, while it does not intersect the fibers of $\mathcal{O}_{\mathbb{P}(w_1,\ldots,w_N)}(-d)^{\oplus r}$ over points in Z at all. An application of the Thom isomorphism then shows that

$$H^*_{\mathrm{CR}}(\mathbb{P}(w_1,\ldots,w_N), \mathbb{P}(w_1,\ldots,w_N) \setminus Z) \cong H^*_{\mathrm{CR}}(Z).$$

With this in mind, we define the hybrid-model state space as

$$\mathcal{H}_{\overline{W}} := H^*_{\mathrm{CR}}\left(\bigoplus_{i=1}^{N} \mathcal{O}_{\mathbb{P}(d,\ldots,d)}(-w_i), \overline{W}^{+\infty}\right).$$

Note that, by the definition of orbifold cohomology, this space can be decomposed as a direct sum over $g \in \mathbb{Z}_d$. Analogously to the FJRW case, we refer to g as *narrow* if its fixed locus is $\{x_1 = \cdots = x_N = 0\}$ or *broad* otherwise. The narrow elements of $\mathbb{Z}_d$ each contribute a copy of $H^*(\mathbb{P}(d,\ldots,d)) = H^*(\mathbb{P}^{r-1})$ to the hybrid state space.

4.2. The moduli space of the hybrid model. To define the hybrid moduli space, we repeat the analogous variation-of-GIT argument once again. Namely, let $X_{g,n,\beta}$ be the (non-compact, non-separated, Artin) stack parameterizing tuples

$$(C; q_1, \ldots, q_n; L; x_1, \ldots, x_N; p_1, \ldots, p_r),$$

in which $(C; q_1, \ldots, q_n)$ is a genus-g, n-pointed orbifold stable curve, L is an orbifold line bundle on C, and

$$(x_1, \ldots, x_N, p_1, \ldots, p_r) \in \Gamma\left(\bigoplus_{i=1}^{N} L^{\otimes w_i} \oplus \bigoplus_{j=1}^{r}(L^{\otimes -d} \otimes \omega_{\log})\right).$$

Within $X_{g,n,\beta}$, there are again two separated Deligne–Mumford substacks: the substack $X^+_{g,n,\beta}$ on which $\vec{x}$ is nowhere-vanishing, and the substack $X^-_{g,n,\beta}$ on which $\vec{p}$ is nowhere-vanishing.

When $\vec{x}$ is nowhere-vanishing, the situation is exactly as in the previous section. Namely,

$$X^+_{g,n,\beta} = \{(C; q_1, \ldots, q_n; f; p_1, \ldots, p_r) \mid f : C \to \mathbb{P}(\vec{w}),\ p_j \in \Gamma(f^*\mathcal{O}(-d) \otimes \omega_{\log})\},$$

where $\mathbb{P}(\vec{w}) := \mathbb{P}(w_1, \ldots, w_N)$. When $\vec{p}$ is nowhere-vanishing, on the other hand, whereas it led to a trivialization in the setting of FJRW theory, it now corresponds to a map $f : C \to \mathbb{P}^{r-1}$. Thus, we have

$$\begin{aligned} X^-_{g,n,\beta} = \{(C; q_1, \ldots, q_n; L; x_1, \ldots, x_N; f) \mid \\ x_i \in \Gamma(L^{\otimes w_i}),\ f : C \to \mathbb{P}^{r-1},\ f^*\mathcal{O}(1) \cong L^{\otimes -d} \otimes \omega_{\log}\}. \end{aligned}$$

REMARK 4.1. It is perhaps worth pausing to note at this point that $X^+_{g,n,\beta}$ and $X^-_{g,n,\beta}$ are nearly equal to the moduli spaces of stable maps to X_+ and X_-, respectively; the only difference is the appearance of the bundle $\omega_{\log}$ in various factors. For a much more detailed study of how the theories with and without $\omega_{\log}$ relate to one another, we refer the reader to [**26**] or [**14**].

Continuing on in the template laid out in the FJRW case, we observe that $X^\pm_{g,n,\beta}$ are not compact, and we restrict to the loci $Z^\pm_{g,n,\beta} \subseteq X^\pm_{g,n,\beta}$ on which the section $(x_1, \ldots, x_N, p_1, \ldots, p_r)$ satisfies the defining equations of the critical locus

$$\mathrm{Crit}(\overline{W}) \subseteq \mathbb{C}^N \times \mathbb{C}^r.$$

The following exercise makes this more explicit.

EXERCISE 4.2. Viewing $\overline{W}$ as a function on $X_+ = \bigoplus_{j=1}^r \mathcal{O}_{\mathbb{P}(\vec{w})}(-d)$, verify that the critical locus of $\overline{W}$ is the complete intersection Z in the zero section $\mathbb{P}(\vec{w}) \subseteq X_+$. Similarly, viewing $\overline{W}$ as a function on $X_- = \bigoplus_{i=1}^N \mathcal{O}_{\mathbb{P}(d,\ldots,d)}(-w_i)$, verify that the critical locus of $\overline{W}$ is the zero section.

As a result, then, we have

$$Z^+_{g,n,\beta} = \overline{\mathcal{M}}_{g,n}(Z, \beta),$$

whereas we define the hybrid model state space as

$$Z^-_{g,n,\beta} = \{(C; q_1, \ldots, q_n; L; f) \mid f : C \to \mathbb{P}^{r-1},\ f^*\mathcal{O}(1) \cong L^{\otimes -d} \otimes \omega_{\log}\}.$$

Perhaps, at this point, the name "hybrid" makes some sense; the above moduli space combines the Gromov–Witten theory of $\mathbb{P}^{r-1}$ with the FJRW theory of a degree-d polynomial.

4.3. The correlators of the hybrid model. Just as in the FJRW case, because the moduli spaces $Z^\pm_{g,n,\beta}$ arise via critical loci inside a larger moduli space with a natural deformation theory, they inherit virtual fundamental cycles through Kiem–Li's cosection construction [**3, 4, 24**]. The following exercise makes the procedure somewhat more explicit.

EXERCISE 4.3. Let

$$(C; L; x_1, \ldots, x_N, p_1, \ldots, p_r) \in X^-_{g,0,\beta}.$$

Show, using Serre duality, that there is a map

$$\alpha : \bigoplus_{i=1}^N H^1(L^{\otimes w_i}) \oplus \bigoplus_{j=1}^r H^1(L^{\otimes -d} \otimes \omega) \to \mathbb{C}$$

$$\alpha(u_1, \ldots, u_N, v_1, \ldots, v_r) = \sum_{i=1}^N \frac{\partial \overline{W}}{\partial x_i}(\vec{x}, \vec{p}) \cdot u_i + \sum_{j=1}^r \frac{\partial \overline{W}}{\partial p_j}(\vec{x}, \vec{p}) \cdot v_j,$$

and that this map fails to be surjective if and only if

$$(C; L; x_1, \ldots, x_N, p_1, \ldots, p_r) \in Z^-_{g,0,\beta}.$$

(These fiberwise maps are used to define the cosection on $X^-_{g,0,\beta}$ from which the virtual cycle on $Z^-_{g,0,\beta}$ is constructed, and they are the reason why the $\omega_{\log}$'s are needed in the sections p_j.)

Furthermore, both of these moduli spaces have evaluation maps

$$\mathrm{ev}_i : Z^\pm_{g,n,\beta} \to X_\pm$$

for each $i \in \{1, \ldots, n\}$. To see this, we need the following fundamental fact (see, for example, [**30**] for further explanation).

FACT 4.4. Let $\mathcal{C}$ be the universal curve over $\overline{\mathcal{M}}_{g,n}$, and let $\pi : \mathcal{C} \to \overline{\mathcal{M}}_{g,n}$ be the projection map. If $\Delta_i \subseteq \mathcal{C}$ denotes the divisor corresponding to the ith marked point, then $\omega_{\log}|_{\Delta_i}$ is trivial. Furthermore, the analogous statement is true for $X_{g,n,\beta}$, $X^{\pm}_{g,n,\beta}$, or $Z^{\pm}_{g,n,\beta}$.

With this fact in mind, consider the moduli space $X_{g,n,\beta}$, with its universal curve $\pi : \mathcal{C} \to X_{g,n,\beta}$, its universal line bundle $\mathcal{L}$ on $\mathcal{C}$, and its universal section

$$\sigma \in \Gamma\left(\bigoplus_{i=1}^{N} \mathcal{L}^{\otimes w_i} \oplus \bigoplus_{j=1}^{r} (\mathcal{L}^{\otimes -d} \otimes \omega_{\log})\right).$$

Then, for any $k \in \{1, \ldots, n\}$, we have

$$\sigma|_{\Delta_k} \in \Gamma\left(\bigoplus_{i=1}^{N} \mathcal{L}^{\otimes w_i} \oplus \bigoplus_{j=1}^{r} \mathcal{L}^{\otimes -d}\right).$$

EXERCISE 4.5. Use Fact 4.4 and the above discussion to explain how to construct evaluation maps

$$\mathrm{ev}_i : X^{\pm}_{g,n,\beta} \to X_{\pm},$$

and hence, by restriction, $\mathrm{ev}_i : Z^{\pm}_{g,n,\beta} \to X_{\pm}$.

The essential idea, at this point, is to define correlators in the hybrid model by pulling back elements of $H^*_{\mathrm{CR}}(X_-)$ under the evaluation maps and integrating them against the virtual fundamental cycle on $Z^-_{g,n,\beta}$. There is an important subtlety, though: we have defined the hybrid model state space to be not $H^*_{\mathrm{CR}}(X_-)$ but $H^*_{\mathrm{CR}}(X_-, \overline{W}^{+\infty})$. It would seem, then, that the insertions to our correlators are drawn from the wrong vector space.

The solution to this issue is to decompose the hybrid state space into broad and narrow components:

$$H^*_{\mathrm{CR}}(X_-, \overline{W}^{+\infty}) = \bigoplus_{g \text{ broad}} H^*\left((X_-)_g, \overline{W}^{+\infty}\right) \oplus \bigoplus_{g \text{ narrow}} H^*(\mathbb{P}^{r-1}).$$

In particular, the narrow part of the hybrid state space can be viewed as a subspace of

$$\bigoplus_{g \in \mathbb{Z}_d} H^*(\mathbb{P}^{r-1}) = H^*_{\mathrm{CR}}(\mathbb{P}(d, \ldots, d)) = H^*_{\mathrm{CR}}(X_-). \tag{4.3}$$

We can then define correlators with narrow insertions as follows.

DEFINITION 4.6. Let $\alpha_1, \ldots, \alpha_n \in \mathcal{H}_W$ be narrow, and, via (4.3), view each as an element of $H^*_{\mathrm{CR}}(X_-)$. Then the associated *hybrid model correlator* is

$$\langle \alpha_1 \cdots \alpha_n \rangle^{\mathrm{hyb}}_{g,n,\beta} = \int_{[Z^-_{g,n,\beta}]^{\mathrm{vir}}} \mathrm{ev}_1^*(\alpha_1) \cdots \mathrm{ev}_n^*(\alpha_n).$$

EXERCISE 4.7. Convince yourself that, in the case where $r = 1$, this recovers the definition of narrow FJRW correlators.

REMARK 4.8. The same construction can be applied to $Z^+_{g,n,\beta} = \overline{\mathcal{M}}_{g,n}(Z, \beta)$. Here, the state space is $H^*_{\mathrm{CR}}(Z)$ whereas the insertions are drawn from $H^*_{\mathrm{CR}}(X_+) \cong H^*(\mathbb{P}(\vec{w}))$; thus, in the same way that hybrid correlators are defined only for narrow insertions, these correlators are defined only for ambient insertions—those pulled back under the inclusion $Z \to \mathbb{P}(\vec{w})$. If we define the virtual cycle of $Z^+_{g,n,\beta}$ via the cosection construction, however, it is not at all obvious that the resulting correlators recover the usual ambient Gromov–Witten invariants of Z. Chang and Li proved, in the case where Z is the quintic threefold, that these two versions of the correlators indeed do agree, up to a sign [**3**]; see also the work of Kim and Oh [**25**].

Having developed the definition of the hybrid model correlators, we are now prepared to state the generalization of the LG/CY correspondence to the setting of complete intersections, at least under restrictive assumptions.

THEOREM 4.9 ([**11**]). *Suppose that Z is a Calabi–Yau threefold and $w_1 = \cdots = w_N = 1$. The genus-zero hybrid theory can be encoded in a generating function $J^{hyb}(t)$ taking values in $\mathcal{H}_W[[z^{-1}, z]$, and the genus-zero Gromov–Witten theory of Z can be encoded in a generating function $J^{GW}(q)$ taking values in $H^*(Q)[[z^{-1}, z]$. After choosing a specific isomorphism $\mathcal{H}_W \cong H^*(Q)$, these two generating functions are related by changes of variables in q and t, an identification between q and t, and analytic continuation between the $t = 0$ and $q = 0$ expansions.*

In fact, a version of the genus-zero LG/CY correspondence has been proven in much greater generality, for (almost) any Calabi–Yau Z in weighted projective space for which $w_i | d$ for all i; this was carried out by Ross and the author in [**14, 15**], based on the work of Lee–Priddis–Shoemaker [**26**] in the hypersurface case. However, a careful statement of the correspondence in that generality requires more technical machinery than we have developed here.

4.4. The general gauged linear sigma model. The story we have told for FJRW theory and the hybrid model can be vastly generalized; it is an instance of the *gauged linear sigma model* (GLSM), which was developed mathematically by Fan, Jarvis, and Ruan [**18**].

In full generality, the GLSM is a theory that depends on three pieces of input data:

(1) a GIT quotient $X = [V /\!\!/_\theta G]$, where V is a complex vector space and $G \subseteq \mathrm{GL}(V)$;
(2) a polynomial function $W : X \to \mathbb{C}$ known as the "superpotential";
(3) an action of $\mathbb{C}^*$ on V known as the "R-charge."

The rough idea, from here, is to form a moduli space parameterizing "Landau–Ginzburg maps" from curves into the critical locus of W; these look nearly the same as ordinary maps to the critical locus of W, but there are additional occurrences of $\omega_{\log}$ dictated by the R-charge.

In order for the resulting moduli space to be compact, it is sometimes necessary to weaken the notion of maps to "quasimaps", an idea first introduced by Ciocan-Fontanine and Kim [**9, 10**]. Quasimaps also play a key role in the proof of the genus-zero LG/CY correspondence, so we discuss them further in the final section of these notes.

5. Wall-crossing and the proof of the LG/CY correspondence

We turn, in this final section, to the proof of the LG/CY correspondence in genus zero. A number of proof strategies have been developed at this point, including Chiodo and Ruan's Lagrangian cone approach [**6**] (based on ideas of [**16**]) and Lee–Priddis–Shoemaker's approach via the crepant transformation conjecture [**14, 26**] (which leverages [**17**]). In what follows, we describe a proof via quasimaps and wall-crossing, which was first applied to the LG/CY correspondence by Ross and Ruan [**28**]; one advantage of this approach is that it points toward a strategy for higher genus.

5.1. Quasimaps. As in the previous section, let $W_1, \ldots, W_r \in \mathbb{C}[x_1, \ldots, x_N]$ be quasihomogeneous polynomials of the same weights $w_1, \ldots, w_N$ and the same degree d, defining a nonsingular complete intersection $Z \subseteq \mathbb{P}(w_1, \ldots, w_N)$. Recall that the moduli spaces of Gromov–Witten theory and the hybrid model were both constructed as substacks of the larger (Artin) stack $X_{g,n,\beta}$ parameterizing tuples

$$(C; q_1, \ldots, q_n; L; x_1, \ldots, x_N; p_1, \ldots, p_r)$$

with

$$(x_1, \ldots, x_n, p_1, \ldots, p_r) \in \Gamma\left(\bigoplus_{i=1}^{N} L^{\otimes w_i} \oplus \bigoplus_{j=1}^{r} (L^{\otimes -d} \otimes \omega_{\log})\right),$$

by first imposing that either $\vec{x}$ or $\vec{p}$ was nowhere-vanishing.

Suppose, now, that we weaken this requirement, allowing $\vec{x}$ (on the Gromov–Witten side) or $\vec{p}$ (on the hybrid side) to have zeroes of bounded order. Specifically, on the Gromov–Witten side, we consider the following definition (due to Ciocan-Fontanine and Kim [**9, 10**]).

DEFINITION 5.1. Fix a positive rational number ϵ, and define $X^{+,\epsilon}_{g,n,\beta} \subseteq X_{g,n,\beta}$ to be the locus where:

(1) $\vec{x}(q) = 0$ for at most finitely many points $q \in C$, all of which are smooth and unmarked, and for each such q,

$$\text{ord}_q(\vec{x}) \leq 1/\epsilon,$$

where $\text{ord}_q(\vec{x})$ denotes the order of vanishing of $\vec{x}$;

(2) the bundle $L^{\otimes \epsilon} \otimes \omega_{\log}$ is ample.

Elements of $X^{+,\epsilon}_{g,n,\beta}$ are referred to as *ϵ-stable quasimaps* to X_+, where X_+ is defined by (4.1).

The second condition in Definition 5.1 is equivalent to requiring that $\deg(L|_{C'}) > 1/\epsilon$ on any genus-zero components $C' \subseteq C$ with only one special point (such components are referred to as "rational tails") and $\deg(L|_{C'}) > 0$ on any genus-zero components $C' \subseteq C$ with only two special points (referred to as "rational bridges"). The intuition behind this condition is the following. Having imposed a bound on the order of vanishing of $\vec{x}$, properness of the moduli space forces one to specify what happens in the limit as a nowhere-vanishing $\vec{x}$ approaches a zero of order $\beta > 1/\epsilon$. Just as in the stable maps setting, the limit in this case is a curve with a new rational component on which L has degree β. The second condition in Definition 5.1 is concocted to allow precisely such rational components; allowing any more would lead to non-uniqueness of limits, preventing the moduli space from being separated.

Having constructed the moduli space $X^{+,\epsilon}_{g,n,\beta}$, we mimic the previous section and restrict further to the locus

$$Z^{+,\epsilon}_{g,n,\beta} \subseteq X^{+,\epsilon}_{g,n,\beta}$$

where $(\vec{x}, \vec{p})$ satisfies the defining equations of $\text{Crit}(\overline{W}) = Z \subseteq X$—that is, where

$$W_j(\vec{x}) \equiv 0 \in \Gamma(L^{\otimes d}) \text{ for all } j \in \{1, \ldots, r\}$$

and

$$\vec{p} \equiv 0.$$

This recovers Ciocan-Fontanine and Kim's moduli space of ϵ-stable quasimaps to Z.

When $\epsilon > 2$, condition (1) of Definition 5.1 implies that $\vec{x}$ is nowhere-vanishing and hence gives rise to an honest map $f : C \to \mathbb{P}(w_1, \ldots, w_N)$, and condition (2) implies that L has positive degree on all rational components with fewer than three special points, so f is stable. Thus, for such ϵ, the moduli space of ϵ-stable quasimaps to Z, which we denote by $Z^{+,\infty}_{g,n,\beta}$, is nothing but the usual moduli space of stable maps.

Taking $\epsilon \to 0$, on the other hand (that is, requiring conditions (1) and (2) of Definition 5.1 for all $\epsilon > 0$) corresponds to allowing arbitrary isolated zeroes of $\vec{x}$ and disallowing all rational tails. The resulting moduli space, which we denote by $Z^{+,0}_{g,n,\beta}$, is the moduli space of stable quotients studied by Marian, Oprea, and Pandharipande [**27**].

All of this can be carried out analogously on the hybrid side. Namely, inside $X_{g,n,\beta}$, we replace the requirement that $\vec{p}$ is nowhere zero by a bound on the order of its zeroes.

DEFINITION 5.2. Fix a positive rational number ϵ, and define $X^{-,\epsilon}_{g,n,\beta} \subseteq X_{g,n,\beta}$ to be the locus where:

(1) $\vec{p}(q) = 0$ for at most finitely many points $q \in C$, all of which are smooth and unmarked, and for each such q,

$$\mathrm{ord}_q(\vec{p}) \leq 1/\epsilon,$$

where $\mathrm{ord}_q(\vec{p})$ denotes the order of vanishing of $\vec{p}$;

(2) the bundle $(L^{\otimes -d} \otimes \omega_{\log})^{\otimes \epsilon} \otimes \omega_{\log}$ is ample.

Then, within $X^{-,\epsilon}_{g,n,\beta}$, we restrict further to the locus $Z^{-,\epsilon}_{g,n,\beta}$ on which $\vec{p}$ satisfies the defining equations of $\mathrm{Crit}(\overline{W}) = \mathbb{P}(d,\ldots,d) \subseteq X_-$, where X_- is defined by (4.2). In other words, $Z^{-,\epsilon}_{g,n,\beta}$ is the locus in $X^{-,\epsilon}_{g,n,\beta}$ where $\vec{x} \equiv 0$. As above, taking $\epsilon \gg 0$ in the definition of $Z^{-,\epsilon}_{g,n,\beta}$ recovers the hybrid moduli space $Z^{-}_{g,n,\beta}$.

5.2. Proof of genus-zero LG/CY. One of the most important reasons for our interest in quasimaps is their intimate connection to mirror symmetry, which leads ultimately to the LG/CY correspondence.

The idea behind (one manifestation of) genus-zero mirror symmetry is to calculate the genus-zero Gromov–Witten invariants of the complete intersection Z by relating their generating function to an explicit power series. In particular, Givental showed in [**20**] that the generating function $J^{\mathrm{GW}}(q)$ of genus-zero Gromov–Witten invariants of Z (mentioned in Theorem 4.9) is related by change of variables to the "I-function"

$$I^{\mathrm{GW}}(q) = z\sum_{\beta\geq 0} q^{\beta}\frac{\prod_{b=1}^{d\beta}(dH+bz)^r}{\prod_{b=1}^{\beta}(H+bz)^{N+1}},$$

a power series in q with coefficients in $H^*(Z)[[z^{-1},z]$, where $H \in H^*(Z)$ denotes the hyperplane class. This is a powerful result that allows the genus-zero Gromov–Witten invariants of Z to be explicitly computed.

The original appearance of the I-function in the literature was in terms of period integrals and Picard–Fuchs equations, but Ciocan-Fontanine and Kim gave an alternative explanation for its role. Namely, they showed that one can define a generating function $J^{\mathrm{GW},\epsilon}(q)$ of genus-zero ϵ-stable quasimap invariants for any $\epsilon \in \mathbb{Q}^+$, and $J^{\mathrm{GW},\epsilon}(q)$ coincides with $J^{\mathrm{GW}}(q)$ when $\epsilon \gg 0$ whereas $J^{\mathrm{GW},0}(q)$ is precisely the I-function. Furthermore, it is apparent from the definition of quasimaps that the invariants change only when $1/\epsilon$ crosses an integer value, so $J^{\mathrm{GW},\epsilon}(q)$ changes only when ϵ crosses one of a discrete set of "walls." From this perspective, genus-zero mirror symmetry becomes the result of a "wall-crossing" formula exhibiting how $J^{\mathrm{GW},\epsilon}(q)$ varies when ϵ crosses a wall; this re-proof of mirror symmetry was carried out by Ciocan-Fontanine and Kim in [**9**].

Precisely the same story can be told on the hybrid side, using wall-crossing to relate the hybrid J-function $J^{\mathrm{hyb}}(t) = J^{\mathrm{hyb},\infty}(t)$ to the function $J^{\mathrm{hyb},0}(t)$, which is the "hybrid I-function" $I^{\mathrm{hyb}}(t)$ and can be calculated explicitly as a hypergeometric series; this was done by Ross–Ruan for FJRW theory and by Ross and the author for the hybrid model [**15**, **28**].

From here, one proves the LG/CY correspondence by relating the two I-functions to one another. More specifically, if one expands $I^{\mathrm{GW}}(q)$ in the form

$$I^{\mathrm{GW}}(q) = I_0(q)\cdot \mathbf{1} + I_1(q)\cdot H + I_2(q)\cdot H^2 + \cdots + I_{N-1-r}(q)H^{N-1-r},$$

then the coefficients $I_i(q)$ form a basis of solutions to a differential equation in q known as the Picard–Fuchs equation. In the same way, the hybrid I-function assembles a basis of solutions to a differential equation in t, and these two differential equations coincide after a particular identification between q and t. (For example, in the case of the quintic threefold, the identification is $q = t^{-5}$.) Thus, by analytically continuing the solutions in the q-coordinate patch to solutions in the t-coordinate patch, and then applying a linear isomorphism of the state space to change from one basis of solutions to another, one shows that $I^{\mathrm{GW}}(q)$ and $I^{\mathrm{hyb}}(t)$ are related by analytic continuation and a state space isomorphism. This, combined with the wall-crossing changes of variables on both sides, proves the genus-zero LG/CY correspondence.

5.3. Towards higher-genus LG/CY. The proof of the genus-zero LG/CY correspondence outlined in the previous subsection suggests a path toward a proof in all genus: first, prove all-genus wall-crossing theorems exhibiting how the generating functions of Gromov–Witten or hybrid invariants depend on ϵ, and second, relate the $\epsilon \to 0$ Gromov–Witten generating function to the $\epsilon \to 0$ hybrid generating function.

The first step of this program has been successfully carried out, by Ciocan-Fontanine and Kim on the Gromov–Witten side [**8**] and by Janda, Ruan, and the author on the hybrid side [**12**].

THEOREM 5.3. *Let* $\square \in \{GW, hyb\}$*. For any complete intersection* $Z \subseteq \mathbb{P}(w_1, \ldots, w_N)$ *of hypersurfaces of the same degree, we have*

$$[Z^{\square,\epsilon}_{g,n,\beta}]^{vir} = \sum_{\substack{k \\ \beta_0+\beta_1+\ldots+\beta_k=\beta}} b_{\vec{\beta}*}\left(\prod_{i=1}^{k} ev^*_{n+i}(\mu^{\square,\epsilon}_{\beta_i}(-\psi_{n+i})) \cap [Z^{\square,\infty}_{g,n+k,\beta_0}]^{vir}\right).$$

Here, $b_{\vec{\beta}} : Z^{\square,\infty}_{g,n+k,\beta_0} \to Z^{\square,\epsilon}_{g,n,\beta}$ *is a certain comparison map that converts marked points to zeroes of* $\vec{x}$ *or* $\vec{p}$*, and* $\mu^{\square,\epsilon}_{\beta_i}$ *is a certain coefficient of the Gromov–Witten or hybrid I-function.*

We should also note that the proof of [**12**] yields an alternative proof of the wall-crossing on the Gromov–Witten side [**13**], and on the hybrid side, a different proof of Theorem 5.3 was independently given by Y. Zhou [**29**] in the case where $r = 1$.

Theorem 5.3 represents significant progress toward the higher-genus LG/CY correspondence, but the second step of the program—relating the $\epsilon \to 0$ theories on the two sides—remains almost entirely mysterious. It has been established in genus zero (as described in the previous subsection) and in genus one [**21, 22**]; the latter is the subject of another of the mini-courses at this workshop.

References

[1] Dan Abramovich and Angelo Vistoli, *Compactifying the space of stable maps*, J. Amer. Math. Soc. **15** (2002), no. 1, 27–75, DOI 10.1090/S0894-0347-01-00380-0. MR1862797

[2] Alejandro Adem, Johann Leida, and Yongbin Ruan, *Orbifolds and stringy topology*, Cambridge Tracts in Mathematics, vol. 171, Cambridge University Press, Cambridge, 2007. MR2359514

[3] Huai-Liang Chang and Jun Li, *Gromov-Witten invariants of stable maps with fields*, Int. Math. Res. Not. IMRN **18** (2012), 4163–4217, DOI 10.1093/imrn/rnr186. MR2975379

[4] Huai-Liang Chang, Jun Li, and Wei-Ping Li, *Witten's top Chern class via cosection localization*, Invent. Math. **200** (2015), no. 3, 1015–1063, DOI 10.1007/s00222-014-0549-5. MR3348143

[5] Weimin Chen and Yongbin Ruan, *A new cohomology theory of orbifold*, Comm. Math. Phys. **248** (2004), no. 1, 1–31, DOI 10.1007/s00220-004-1089-4. MR2104605

[6] Alessandro Chiodo and Yongbin Ruan, *Landau-Ginzburg/Calabi-Yau correspondence for quintic three-folds via symplectic transformations*, Invent. Math. **182** (2010), no. 1, 117–165, DOI 10.1007/s00222-010-0260-0. MR2672282

[7] Alessandro Chiodo and Yongbin Ruan, *LG/CY correspondence: the state space isomorphism*, Adv. Math. **227** (2011), no. 6, 2157–2188, DOI 10.1016/j.aim.2011.04.011. MR2807086

[8] Ionuţ Ciocan-Fontanine and Bumsig Kim, *Quasimap wall-crossings and mirror symmetry*, Publ. Math. Inst. Hautes Études Sci. **131** (2020), 201–260, DOI 10.1007/s10240-020-00114-0. MR4106795

[9] Ionuţ Ciocan-Fontanine and Bumsig Kim, *Wall-crossing in genus zero quasimap theory and mirror maps*, Algebr. Geom. **1** (2014), no. 4, 400–448, DOI 10.14231/AG-2014-019. MR3272909

[10] Ionuţ Ciocan-Fontanine, Bumsig Kim, and Davesh Maulik, *Stable quasimaps to GIT quotients*, J. Geom. Phys. **75** (2014), 17–47, DOI 10.1016/j.geomphys.2013.08.019. MR3126932

[11] Emily Clader, *Landau-Ginzburg/Calabi-Yau correspondence for the complete intersections $X_{3,3}$ and $X_{2,2,2,2}$*, Adv. Math. **307** (2017), 1–52, DOI 10.1016/j.aim.2016.11.010. MR3590512

[12] Emily Clader, Felix Janda, and Yongbin Ruan, *Higher-genus quasimap wall-crossing in the gauged linear sigma model*, `arXiv:1706.05038`, 2017.

[13] ———, *Higher-genus quasimap wall-crossing via localization*, `arXiv:1702.03427`, 2017.

[14] Emily Clader and Dustin Ross, *Sigma models and phase transitions for complete intersections*, Int. Math. Res. Not. IMRN **15** (2018), 4799–4851, DOI 10.1093/imrn/rnx029. MR3842378

[15] ———, *Wall-crossing in genus-zero hybrid theory*, `arXiv:1806.08442`, 2018.

[16] Tom Coates, Alessio Corti, Hiroshi Iritani, and Hsian-Hua Tseng, *Computing genus-zero twisted Gromov-Witten invariants*, Duke Math. J. **147** (2009), no. 3, 377–438, DOI 10.1215/00127094-2009-015. MR2510741

[17] Tom Coates, Hiroshi Iritani, and Yunfeng Jiang, *The crepant transformation conjecture for toric complete intersections*, Adv. Math. **329** (2018), 1002–1087, DOI 10.1016/j.aim.2017.11.017. MR3783433

[18] Yongbin Ruan, *Nonabelian gauged linear sigma model*, Chin. Ann. Math. Ser. B **38** (2017), no. 4, 963–984, DOI 10.1007/s11401-017-1106-5. MR3673177

[19] Huijun Fan, Tyler Jarvis, and Yongbin Ruan, *The Witten equation, mirror symmetry, and quantum singularity theory*, Ann. of Math. (2) **178** (2013), no. 1, 1–106, DOI 10.4007/annals.2013.178.1.1. MR3043578

[20] Alexander Givental, *A mirror theorem for toric complete intersections*, Topological field theory, primitive forms and related topics (Kyoto, 1996), Progr. Math., vol. 160, Birkhäuser Boston, Boston, MA, 1998, pp. 141–175. MR1653024

[21] Si Li, *A mirror theorem between Landau-Ginzburg models*, Nuclear Phys. B **898** (2015), 707–714, DOI 10.1016/j.nuclphysb.2015.04.002. MR3377499

[22] Shuai Guo and Dustin Ross, *The genus-one global mirror theorem for the quintic 3-fold*, Compos. Math. **155** (2019), no. 5, 995–1024, DOI 10.1112/s0010437x19007231. MR3946282

[23] Paul D. Johnson, *Equivariant Gromov-Witten theory of one dimensional stacks*, ProQuest LLC, Ann Arbor, MI, 2009. Thesis (Ph.D.)–University of Michigan. MR2713030

[24] Young-Hoon Kiem and Jun Li, *Localizing virtual cycles by cosections*, J. Amer. Math. Soc. **26** (2013), no. 4, 1025–1050, DOI 10.1090/S0894-0347-2013-00768-7. MR3073883

[25] Bumsig Kim and Jeongseouk Oh, *Localized Chern characters for 2-periodic complexes*, `arXiv:1804.03774`, 2018.

[26] Yuan-Pin Lee, Nathan Priddis, and Mark Shoemaker, *A proof of the Landau-Ginzburg/Calabi-Yau correspondence via the crepant transformation conjecture* (English, with English and French summaries), Ann. Sci. Éc. Norm. Supér. (4) **49** (2016), no. 6, 1403–1443, DOI 10.24033/asens.2312. MR3592361

[27] Alina Marian, Dragos Oprea, and Rahul Pandharipande, *The moduli space of stable quotients*, Geom. Topol. **15** (2011), no. 3, 1651–1706, DOI 10.2140/gt.2011.15.1651. MR2851074

[28] Dustin Ross and Yongbin Ruan, *Wall-crossing in genus zero Landau-Ginzburg theory*, J. Reine Angew. Math. **733** (2017), 183–201, DOI 10.1515/crelle-2015-0005. MR3731328

[29] Yang Zhou, *Higher-genus wall-crossing in Landau-Ginzburg theory*, Adv. Math. **361** (2020), 106914, 42, DOI 10.1016/j.aim.2019.106914. MR4041201

[30] Dimitri Zvonkine, *An introduction to moduli spaces of curves and their intersection theory*, Handbook of Teichmüller theory. Volume III, IRMA Lect. Math. Theor. Phys., vol. 17, Eur. Math. Soc., Zürich, 2012, pp. 667–716, DOI 10.4171/103-1/12. MR2952773

Department of Mathematics, San Francisco State University, 1600 Holloway Avenue, San Francisco, California 94132

Email address: `eclader@sfsu.edu`

Contemporary Mathematics
Volume **763**, 2021
https://doi.org/10.1090/conm/763/15327

Localization and mirror symmetry

Dustin Ross

ABSTRACT. These notes were born out of a five-hour lecture series for graduate students at the May 2018 Snowbird workshop *Crossing the Walls in Enumerative Geometry*. After a short primer on equivariant cohomology and localization, we provide proofs of the genus-zero mirror theorems for the quintic threefold, first in Fan-Jarvis-Ruan-Witten theory and then in Gromov-Witten theory. We make no claim to originality, except in exposition, where special emphasis is placed on peeling away the standard technical machinery and viewing the mirror theorems as closed-formula manifestations of elementary localization recursions.

0. Introduction

The mirror theorem for the quintic threefold $Q \subset \mathbb{P}^4$, first conjectured in the physics literature by Candelas, de la Ossa, Green, and Parkes [**4**], reveals deep and surprising structure in the enumerative geometry of rational curves in Q. In the two decades since its original proof by Givental [**11**], the mirror theorem has undergone countless reinterpretations and generalizations. The proof we present here is an adaptation of ideas that were developed more generally in [**8**] and [**15**], and they generalize naturally to prove genus-zero mirror theorems for complete intersections in toric Deligne-Mumford stacks. En route to proving the mirror theorem for the Gromov-Witten theory of Q, we also take a detour to prove the mirror theorem for the Fan-Jarvis-Ruan-Witten theory of Q, a result that was first proved by Chiodo and Ruan [**7**]. The combination of these two mirror theorems leads to the proof of Witten's genus-zero "Landau-Ginzburg/Calabi-Yau correspondence", which we do not discuss explicitly, but which is lurking in the background [**7**].

Torus localization is the primary tool used in the proofs of the genus-zero mirror theorems. Indeed, a direct application of the localization theorem of Atiyah and Bott [**2**] leads to an explicit algorithm that computes any genus-zero Gromov-Witten invariant of Q. However, as is typical with such localization computations, the combinatorial complexity of the algorithm grows at an unmanageable rate, and it quickly becomes apparent that any hope to pin down a closed formula for the Gromov-Witten invariants must result from the recursive structure of the localization contributions. The main content of the proof, then, is a clever packaging of localization recursions. Givental was the first person to successfully package the localization relations to prove the mirror theorem in Gromov-Witten theory [**11**]. Givental's ideas have matured over the two decades since his original proof, and the proofs in these notes may not immediately resemble his original paper; however, all of the main ideas in the proofs have their genesis in Givental's original work.

While the mirror theorems are beautiful and important results in their own right, the intent of these lectures is not simply to present proofs of these two results for historical enlightenment. Rather, our goal is to help the student familiarize herself with the methods of manipulating localization relations in two different but related settings. Localization is one of only a few computational methods in the Gromov-Witten theory toolkit, and nearly all of the current approaches to understanding higher-genus Gromov-Witten and Fan-Jarvis-Ruan-Witten theory utilize clever applications of localization on certain auxiliary moduli spaces. It is in this light that we view the genus-zero mirror theorems as a useful setting through which the student can begin honing her localization skills today in order to apply them in more creative settings that push the boundaries of the field tomorrow.

0.1. Target audience. These lectures are intended for advanced graduate students who have already taken a few courses in algebraic geometry. Some exposure to moduli spaces of stable curves and stable maps would certainly be useful; we suggest [**13**, Chapters 22 - 25] for the requisite background on $\overline{\mathcal{M}}_{g,n}$ and $\overline{\mathcal{M}}_{g,n}(X,d)$. A working knowledge of orbifold curves and line bundles on them would also be useful; a good starting place is Chapter 1 of [**1**].

0.2. Disclaimer. For the sake of brevity, we have chosen to make many omissions, and we have made no attempt to make this a comprehensive account of mirror symmetry. Two such accounts already exist, and we highly recommend them: [**9**] and [**13**]. Our hope for these notes is that they will serve as a more compact reference that only emphasizes the aspects of the story that are essential to the proofs.

1. Equivariant cohomology and localization

In this lecture, we introduce equivariant cohomology by way of several instructive examples that will be important throughout the subsequent lectures. We state the localization theorem of Atiyah and Bott in the form that will be most applicable towards the mirror theorems, and we give a few simple applications of the localization theorem.

1.1. Equivariant cohomology. Throughout this section, let X be a smooth, projective variety, and let $T = \{(t_1, \ldots, t_n)\} = (\mathbb{C}^*)^n$ be an algebraic torus acting on X. In the same way that the cohomology ring $H^*(X) = H^*(X, \mathbb{Q})$ captures some of the intersection-theoretic geometry of X, the *equivariant cohomology ring*, denoted $H_T^*(X)$, captures some of the T-equivariant intersection-theoretic geometry of X. The formal definitions are not immediately enlightening, so we choose to forego them; essentially all of the properties necessary for the proofs in these lectures can be gleaned from the following examples. For a more thorough treatment, the reader is directed to [**2**].

The first example concerns the case where the geometry (and thus, the action) is trivial.

EXAMPLE 1.1.1. Let $X = \mathrm{pt}$ be a point. Then $H_T^*(\mathrm{pt})$ is a graded polynomial ring

$$H_T^*(\mathrm{pt}) = \mathbb{Q}[\alpha_1, \ldots, \alpha_n],$$

where $\deg_{\mathbb{R}}(\alpha_i) = 2$. One way to interpret the classes α_i is as Chern classes of equivariant vector bundles. It is a standard consequence of the formal definitions that

$$\alpha_i = c_1(\mathcal{O}_{\alpha_i})$$

where $\mathcal{O}_{\alpha_i}$ is the equivariant line bundle on $X = \mathrm{pt}$ that is geometrically trivial but with T-action on fibers given by $v \mapsto t_i^{-1}v$. In fact, any T-equivariant vector bundle on a point (i.e. any T-representation) uniquely decomposes as a direct sum of tensor products of the line bundles $\mathcal{O}_{\alpha_i}$, and its Chern classes can be determined by the usual properties of Chern classes with respect to tensor products and direct sums.

The second important example concerns the case where the geometry is non-trivial, but the action is trivial.

EXAMPLE 1.1.2. Let T act trivially on X. Then

$$H_T^*(X) = H^*(X) \otimes \mathbb{Q}[\alpha_1, \dots, \alpha_n].$$

If V is a(n equivariantly trivial) rank-r vector bundle on X, then the Euler class of V is the top Chern class:

$$e(V) := c_r(V) \in H^{2r}(X) \subset H_T^{2r}(X).$$

Imposing a T-action on V, such as $v \mapsto t^{-\alpha}v := t_1^{-a_1} \cdots t_n^{-a_n}v$, is the same as tensoring by an equivariant line bundle: $V \otimes \mathcal{O}_\alpha$, where $\alpha = a_1\alpha_1 + \cdots + a_n\alpha_n$. In this case,

$$e(V \otimes \mathcal{O}_\alpha) = c_r(V) + c_{r-1}(V)\alpha + \cdots + c_1(V)\alpha^{r-1} + \alpha^r \in H_T^{2r}(X).$$

In particular, the Euler class is invertible in equivariant cohomology, as long as we formally invert the class α:

$$e^{-1}(V \otimes \mathcal{O}_\alpha) = \frac{1}{\alpha^r} \sum_{0 \le k \le \dim(X)} \left(-\frac{c_1(V)}{\alpha} - \cdots - \frac{c_{r-1}(V)}{\alpha^{r-1}} - \frac{c_r(V)}{\alpha^r} \right)^k.$$

More generally, if T acts trivially on X and V is an equivariant vector bundle such that

$$X = \mathrm{Tot}(V)^T,$$

then $e(V)$ is invertible in $H_T^*(X)$, after formally inverting some of the equivariant parameters.

The final example concerns the standard torus action on projective space, where both the geometry and the action are non-trivial.

EXAMPLE 1.1.3. Let $T = (\mathbb{C}^*)^{n+1}$ act on $\mathbb{P}^n$ by

$$[x_0, \dots, x_n] \mapsto [t_0x_0, \dots, t_nx_n].$$

Then

$$H_T^*(\mathbb{P}^n) = \frac{\mathbb{Q}[H, \alpha_0, \dots, \alpha_n]}{(H - \alpha_0) \cdots (H - \alpha_n)}. \tag{1}$$

In usual cohomology, we interpret H as a first Chern class, $H = c_1(\mathcal{O}_{\mathbb{P}^n}(1))$, and that is essentially the same here. Recall that the total space of $\mathcal{O}_{\mathbb{P}^n}(1)$ can be defined globally by

$$\mathrm{Tot}(\mathcal{O}_{\mathbb{P}^n}(1)) = \frac{(\mathbb{C}^{n+1} \setminus \{0\}) \times \mathbb{C}}{\mathbb{C}^*}, \tag{2}$$

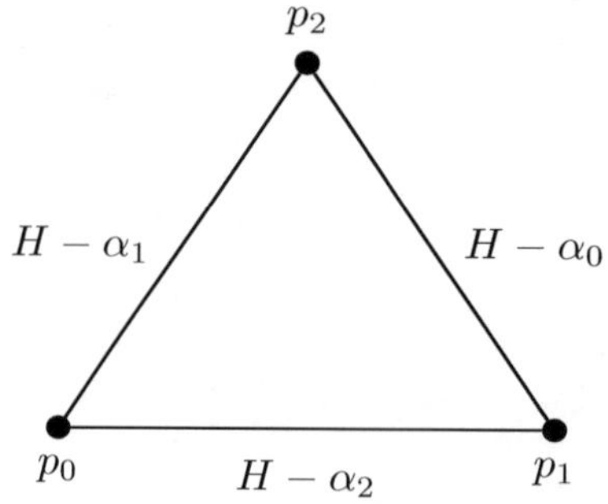

FIGURE 1. The $T = (\mathbb{C}^*)^3$-equivariant structure of $\mathbb{P}^2$.

where $\mathbb{C}^*$ acts on the $n+2$ coordinates in the numerator diagonally; the first $n+1$ coordinates $(x_0, \dots, x_n)$ correspond to the homogeneous coordinates on $\mathbb{P}^n$ and the last coordinate v corresponds to the fiber coordinate. The quotient (2) can be made equivariant in a canonical way:

$$[x_0, \dots, x_n, v] \mapsto [t_0 x_0, \dots, t_n x_n, v],$$

and in the T-equivariant geometry of $\mathbb{P}^n$, we will always take $\mathcal{O}_{\mathbb{P}^n}(1)$ to denote (2) with this T-action, and we define

$$H := c_1(\mathcal{O}_{\mathbb{P}^n}(1)) \in H_T^*(\mathbb{P}^n).$$

Tensoring $\mathcal{O}_{\mathbb{P}^n}(1)$ by $\mathcal{O}_\alpha$ is the same as letting T act on the last coordinate of (2) by $v \mapsto t^{-\alpha} v$. From the description (2), we see that $v = x_i$ descends to a T-equivariant section of $\mathcal{O}_{\mathbb{P}^n}(1) \otimes \mathcal{O}_{-\alpha_i}$. In fact, the set of T-equivariant sections of $\mathcal{O}_{\mathbb{P}^n}(1) \otimes \mathcal{O}_{-\alpha_i}$ are the scalar multiples of $v = x_i$. Since top Chern classes are often interpreted as the (Poincaré dual of) the vanishing of a generic section, we should interpret $H - \alpha_i$ as the vanishing of x_i:

$$H - \alpha_i = [\{x_i = 0\}].$$

Taken a step further, we interpret

$$\prod_{i \in I} (H - \alpha_i) = [\{x_i = 0 : \forall i \in I\}],$$

and the relation in (1) has the natural interpretation that $\{x_i = 0 : \forall i\} = \emptyset$.

If $i_j : p_j \to \mathbb{P}^n$ is the inclusion of the jth coordinate point for $0 \leq j \leq n$, then the description (2) implies that $i_j^* \mathcal{O}_{\mathbb{P}^n}(1) = \mathcal{O}_{\alpha_j}$. Thus, by restricting H to fixed points, we obtain

$$i_j^* H = \alpha_j.$$

Similarly, restricting the tangent bundle to p_j and using the local coordinates $z_k = x_k / x_j$, we obtain

$$i_j^* T\mathbb{P}^n = \bigoplus_{k \neq j} \mathbb{C}\left\{\frac{\partial}{\partial z_k}\right\} = \bigoplus_{k \neq j} \mathcal{O}_{\alpha_j - \alpha_k}.$$

1.2. Localization. Let $i : F \hookrightarrow X$ be a closed subvariety. Via Poincaré duality, there is a well-defined push-forward map $i_* : H^*(F) \to H^{*+2r}(X)$, where r is the complex codimension of F in X. The excess intersection formula tells us that pulling back the image of i by i^* is the same as multiplying by the Euler class of the normal bundle:

$$i^*i_* : H^*(F) \to \mathrm{H}^*(F)$$
$$\phi \mapsto e(N_{F/X})\phi.$$

Clearly, i^*i_* is not an isomorphism, because the Euler class of the normal bundle is not generally invertible. However, if F is the fixed locus of a T-action on X, then we saw in Example 1.1.2 that the Euler class is invertible after inverting (some of) the equivariant parameters. This is the main motivation for the following result.

THEOREM 1.2.1 (Atiyah-Bott [**2**]). *Let X be a smooth projective variety with $T = (\mathbb{C}^*)^n$ action such that $F = X^T$ is smooth. Upon inverting the equivariant parameters $\{\alpha_i\}$, $i_* : H_T^*(F) \to H_T^{*+2r}(X)$ is an isomorphism with inverse*

$$(i_*)^{-1} = \frac{i^*}{e(N_{F/X})}.$$

REMARK 1.2.2. While the fixed locus is required to be smooth, it is almost always reducible, having many connected components. When applying the localization theorem, it is often the case that computations boil down to a clever combinatorial packaging of torus fixed loci.

REMARK 1.2.3. The localization theorem holds in much greater generality than stated here. For the purposes of these notes, it is important to note that it remains true, for example, if *variety* is replaced with *Deligne-Mumford stack.*

Recall that the integration map is defined by $\int_X - := \pi_*(-) : H^*_{(T)}(X) \to H^*_{(T)}(\mathrm{pt})$ where $\pi : X \to \mathrm{pt}$ can be interpreted with or without a T-action on X. As an immediate corollary of the localization theorem, we see that

$$\int_X \phi = \int_X i_*\left(\frac{i^*\phi}{e(N_{F/X})}\right) = \int_F \frac{i^*\phi}{e(N_{F/X})},$$

where the last step uses the fact that $F \to \mathrm{pt}$ factors through X. Thus, we obtain a powerful strategy for computing $\int_X \phi$:

STEP 1. Put a T-action on X.

STEP 2. Choose a "lift" of ϕ to $H_T^*(X)$.

STEP 3. Compute $\int_F \frac{i^*\phi}{e(N_{F/X})}$.

REMARK 1.2.4. In these notes, the classes we integrate will always be given by Chern classes of vector bundles, and the choice of "lift" in Step 2 corresponds to the choice of a T-action on the corresponding vector bundle.

REMARK 1.2.5. Since F often has many components, the integral in Step 3 can be a rather large sum. While each summand is typically computable, it can be highly nontrivial to package the sum in a nice way. However, the choice of lift in Step 2 buys us some flexibility. Often by lifting the class cleverly, you can get it to vanish on some of the fixed loci.

The following examples provide the first applications of this strategy.

EXAMPLE 1.2.6. The Euler characteristic of $\mathbb{P}^n$ is the integral of the Euler class of the tangent bundle:

$$\chi(\mathbb{P}^n) = \int_{\mathbb{P}^n} e(T\mathbb{P}^n).$$

Using the standard action of $(\mathbb{C}^*)^{n+1}$ on $\mathbb{P}^n$ from Example 1.1.3 (or any other torus action with the same fixed points), we compute

$$\int_{\mathbb{P}^n} e(T\mathbb{P}^n) = \sum_{j=0}^{n} \frac{e(i_j^* T\mathbb{P}^n)}{e(N_{p_j/X})} = \sum_{j=0}^{n} \frac{e(i_j^* T\mathbb{P}^n)}{e(i_j^* T\mathbb{P}^n)} = n+1.$$

EXERCISE 1.2.7. Use localization to compute $\chi(\mathrm{Gr}(k,n))$.

EXAMPLE 1.2.8. We know that $\int_{\mathbb{P}^2} H^2 = 1$, so let's try to recover this using localization. If we choose the canonical lift of H, as described in Example 1.1.3, we compute

$$\int_{\mathbb{P}^2} H^2 = \sum_{j=0}^{2} \frac{i_j^* H^2}{e(i_j^* T\mathbb{P}^2)} = \frac{\alpha_0^2}{(\alpha_0-\alpha_1)(\alpha_0-\alpha_2)} + \frac{\alpha_1^2}{(\alpha_1-\alpha_0)(\alpha_1-\alpha_2)}$$

$$+ \frac{\alpha_2^2}{(\alpha_2-\alpha_0)(\alpha_2-\alpha_1)} = \cdots = 1,$$

where the last step required a bit of magical cancellation. If, instead, we lift the class $H \in H^*(\mathbb{P}^2)$ to $H - \alpha_2 \in H_T^*(\mathbb{P}^2)$, then the computation would only involve two summands:

$$\int_{\mathbb{P}^2} H^2 = \cdots = \frac{(\alpha_0-\alpha_2)^2}{(\alpha_0-\alpha_1)(\alpha_0-\alpha_2)} + \frac{(\alpha_1-\alpha_2)^2}{(\alpha_1-\alpha_0)(\alpha_1-\alpha_2)} = \cdots = 1,$$

and the cancellation becomes slightly less magical. Thus, we see how different choices of lifts can change the complexity of the localization computations.

EXERCISE 1.2.9. Find a lift of H^n so that only one fixed point contributes to the computation of $\int_{\mathbb{P}^n} H^n = 1$.

EXERCISE 1.2.10. Let $T = (\mathbb{C}^*)^{n+1}$ act on $\overline{\mathcal{M}}_{0,0}(\mathbb{P}^n, d)$ by post-composing each stable map with the standard action of T on $\mathbb{P}^n$.

(a) Prove that the stable map

$$\mathbb{P}^1 \to \mathbb{P}^n$$

$$[x_0, x_1] \mapsto [x_0^d, y_0^d, 0, \ldots, 0]$$

is a fixed point of the T-action.

(b) What is the rank of the normal bundle to this fixed point?

(c) What is the equivariant Euler class of the normal bundle to this fixed point?

2. Fan-Jarvis-Ruan-Witten theory of the quintic threefold

This lecture introduces FJRW theory and the genus-zero FJRW mirror theorem. We do not aim for completeness in our development of FJRW theory; rather, our intent is to introduce the structures that will be most important for the proof of the mirror theorem. For the reader that would like to learn more, we provide several exercises and specific references to places in the literature where the foundations are developed more thoroughly.

2.1. FJRW theory of the quintic threefold. FJRW invariants are a special class of intersection numbers on *moduli spaces of* 5*-spin curves*. The closed points of the moduli space of 5-spin curves are defined by

$$\overline{\mathcal{M}}^{1/5}_{g,(m_1,\ldots,m_n)} = \{(C;q_1,\ldots,q_n;L;f)\}\,/\sim$$

where

- $(C;q_1,\ldots,q_n)$ is a stable, n-pointed, genus-g orbifold curve, with orbifold structure only at the marks and nodes;
- L is an orbifold line bundle on C such that
  - L has multiplicity m_i at q_i, i.e. near q_i, the total space of L is a quotient by the cyclic isotropy group $\mu_5 = \langle\xi = \mathrm{e}^{2\pi \mathrm{i}/5}\rangle$ acting by $\xi\cdot(x,v) = (\xi x, \xi^{m_i} v)$ with $m_i \in \{1,2,3,4\}$;
  - L satisfies the kissing condition at each node, meaning that the multiplicities on each branch satisfy $m + m' = 0 \mod 5$ (if $m = m' = 0$, then C and L have trivial orbifold structure near that node);
- f is an isomorphism

$$f : L^{\otimes 5} \xrightarrow{\cong} \omega_{C,\log} := \omega_C\left(\sum_i [q_i]\right),$$

where $[q_i]$ is the orbifold divisor of the marked point ($\deg([q_i]) = 1/5$).

An isomorphism of two 5-spin curves $(C;q_1,\ldots,q_n;L;f) \sim (C';q_1',\ldots,q_n';L;f)$ consists of a pair of isomorphisms $g : (C;q_1,\ldots,q_n) \to (C',q_1',\ldots,q_n')$ and $h : g^*L' \to L$ such that $h^*f = g^*f'$. We use $\overline{\mathcal{M}}^{1/5}_{g,n}$ to denote the disjoint union over all possible multiplicity vectors.

REMARK 2.1.1. If this is your first encounter with the moduli spaces of spin curves, then you might naturally be wondering about the necessity of the orbifold structure and the role of the multiplicities. Without orbifold structure at the marked points, a fifth root of $\omega_{C,\log}$ would only exist if $2g-2+n \in 5\mathbb{Z}$, so most of these moduli spaces would be empty. Allowing orbifold structure gives us the flexibility of line bundles with fractional degree and results in a much richer class of moduli spaces with which we can work. More importantly, perhaps, is the observation that, regardless of the orbifold structure at the marked points, the orbifold structure at the nodes is necessary in order to obtain a proper moduli space. For example, if $g = 6$ and $n = 0$, then any smooth curve has a (non-orbifold) fifth root of ω_C, but there does not exist a (non-orbifold) fifth root on any limit point where the curve splits as a $g = 1$ curve meeting a $g = 5$ curve at a node.

The moduli space $\overline{\mathcal{M}}^{1/5}_{g,\vec{m}}$ is a smooth Deligne-Mumford stack of dimension $3g - 3 + n$ [**10**, Theorem 2.2.6], and they each support a *virtual funamental class* of (complex) dimension $2n - \sum m_i$:

$$[\overline{\mathcal{M}}^{1/5}_{g,\vec{m}}]^{\mathrm{vir}} \in H_{4n-2\sum m_i}(\overline{\mathcal{M}}^{1/5}_{g,\vec{m}}).$$

The construction of the virtual class is highly non-trivial, but has been carried out in several equivalent settings [**5**, **6**, **10**, **14**]. FJRW invariants are defined by

$$\langle \phi_{m_1}\psi^{a_1}\cdots\phi_{m_n}\psi^{a_n}\rangle^{\mathrm{FJRW}}_{g,n} := \int_{[\overline{\mathcal{M}}^{1/5}_{g,\vec{m}}]^{\mathrm{vir}}} \psi_1^{a_1}\cdots\psi_n^{a_n} \in \mathbb{Q},$$

where ψ_i is the cotangent line class at the ith marked point on the coarse underlying curve $|C|$.

The general construction of the virtual class is far beyond the scope of these lectures; however, the construction simplifies a great deal in genus zero. In particular, the genus-zero virtual class can be defined by

$$\left[\overline{\mathcal{M}}^{1/5}_{0,\vec{m}}\right]^{\mathrm{vir}} := \left[\overline{\mathcal{M}}^{1/5}_{0,\vec{m}}\right] \cap e\left((R^1\pi_*\mathcal{L}^{\oplus 5})^\vee\right),$$

where $\pi : \mathcal{C} \to \overline{\mathcal{M}}^{1/5}_{0,\vec{m}}$ is the universal curve and $\mathcal{L}$ is the universal line bundle over $\mathcal{C}$.

To gain a slightly better appreciation of the genus-zero virtual class, it is instructive to verify that (a) $R^1\pi_*\mathcal{L}$ is, in fact, a vector bundle, so its Euler class is well-defined, and (b) the virtual dimension is $2n - \sum m_i$. In order to make these verifications, let us first say a few things about orbifold line bundles. Suppose C is a smooth curve. Locally at q_i, sections of L correspond to sections in the orbifold coordinates that are equivariant with respect to the isotropy group action. In other words, local sections are of the form $v(x) = x^{m_i}F(x^5)$ for some polynomial F. The variable x^5 is a local coordinate on the coarse underlying curve. Thus, we see that the orbifold multiplicity m_i imposes a vanishing of degree $m_i/5$ at each orbifold point. We define the round-down of L by

$$|L| := L \otimes \mathcal{O}\left(-\sum m_i[q_i]\right).$$

It follows from the discussion above that $|L|$ is pulled back from the coarse underlying curve $|C|$, and, since rounding up/down by m_i identifies sections, we have $H^i(C,L) = H^i(|C|,|L|)$ for $i = 0,1$. In particular, these observations allow us to import the classical Riemann-Roch theorem to this orbifold setting:

$$H^0(C,L) - H^1(C,L) = \deg(L) + 1 - g - \sum \frac{m_i}{5}.$$

Using these observations, the reader is encouraged to work out the following verifications.

EXERCISE 2.1.2. Prove that $R^1\pi_*\mathcal{L}$ is a vector bundle by showing that $H^0(C,L) = 0$ for any point $(C, q_1, \ldots, q_n, L, f) \in \overline{\mathcal{M}}^{1/5}_{0,\vec{m}}$. (Hint: compute $\deg(|L|)$ on each irreducible component.) In our definition, we do not allow $m_i = 0$. Would your proof still work if one of the m_i is allowed to be zero?

EXERCISE 2.1.3. Use Riemann-Roch to prove that the genus-zero virtual class lies in complex dimension $2n - \sum m_i$.

There are a few additional properties concerning the FJRW invariants that are worth mentioning. First of all, the multiplicity-one insertion ϕ_1 plays a special role in this story. In particular, there is a forgetful map

$$\mathrm{f}: \overline{\mathcal{M}}^{1/5}_{g,(\vec{m},1)} \to \overline{\mathcal{M}}^{1/5}_{g,\vec{m}}.$$

Roughly speaking, this map rounds down L at the last marked point, then forgets that point and stabilizes, if necessary. This only works with ϕ_1 insertions because rounding down a 5-spin bundle at a ϕ_1 insertion produces a 5-spin bundle on the curve without that marked point:

$$L^{\otimes 5} \cong \omega_C\left(\sum_{i=1}^{n+1}[q_i]\right) \Longrightarrow \left(L \otimes \mathcal{O}\left(-[q_{n+1}]\right)\right)^{\otimes 5} \cong \omega_{|C|}\left(\sum_{i=1}^{n}[q_i]\right),$$

where $|C|$ in the right-hand side denotes the coarsening of C at the q_{n+1}.

EXERCISE 2.1.4. Prove that $\mathrm{f}^*\left[\overline{\mathcal{M}}^{1/5}_{0,\vec{m}}\right]^{\mathrm{vir}} = \left[\overline{\mathcal{M}}^{1/5}_{0,(\vec{m},1)}\right]^{\mathrm{vir}}$. Along with the comparison lemma for ψ-classes ([**13**], Lemma 25.2.3), deduce the string and dilaton equations:

$$\langle \phi_{m_1}\psi^{a_1}\cdots\phi_{m_n}\psi^{a_n}\cdot\phi_1\rangle^{\mathrm{FJRW}}_{0,n+1} = \sum_{i=1}^{n}\langle \phi_{m_1}\psi^{a_1}\cdots\phi_{m_i}\psi^{a_i-1}\cdots\phi_{m_n}\psi^{a_n}\rangle^{\mathrm{FJRW}}_{0,n}$$

and

$$\langle \phi_{m_1}\psi^{a_1}\cdots\phi_{m_n}\psi^{a_n}\cdot\phi_1\psi\rangle^{\mathrm{FJRW}}_{0,n+1} = (n-2)\,\langle \phi_{m_1}\psi^{a_1}\cdots\phi_{m_n}\psi^{a_n}\rangle^{\mathrm{FJRW}}_{0,n}.$$

A final property of the genus-zero FJRW invariants that is important for our techniques concerns how the virtual class restricts to the boundary. Since we do not allow multiplicity zero at the marked points, then it would be ideal if the virtual class vanished along nodes where the multiplicities were zero. Essentially, this implies that the genus-zero virtual class restricts to the boundary in a way that is compatible with the gluing map

$$\mathrm{g}: \overline{\mathcal{M}}^{1/5}_{0,(\vec{m}_I)+1} \times \overline{\mathcal{M}}^{1/5}_{0,(\vec{m}_J)+1} \to \overline{\mathcal{M}}^{1/5}_{0,\vec{m}}.$$

This is indeed the case, as the next exercise verifies.

EXERCISE 2.1.5. Let $D \subset \overline{\mathcal{M}}^{1/5}_{0,\vec{m}}$ denote a divisor where the node has multiplicity $m = m' = 0$. Prove that $e\left((R^1\pi_*\mathcal{L}^{\oplus 5})^\vee\right) \cap D = 0$.

REMARK 2.1.6. This exercise is a bit harder than the previous ones. The trick is to use the normalization sequence and show that $R^1\pi_*\mathcal{L}$ contains a sub-line bundle that has a trivial first Chern class (c.f. [**7**], Lemma 4.1.1).

2.2. The genus-zero FJRW mirror theorem. In order to state the mirror theorem, it is useful to interpret FJRW invariants as multi-linear functions on the vector space $V = \mathbb{C}\{\phi_1, \ldots, \phi_4\}$, where we have one basis element for each possible multiplicity. The space V has a non-degenerate pairing defined by

$$(\phi_i, \phi_j) = \langle \phi_1\phi_i\phi_j\rangle^{FJRW}_{0,3} = \frac{1}{5}\delta_{i+j=5},$$

and the dual basis of V is given by $\phi^i = 5\phi_{i-1}$.

The two main players in the mirror theorem are typically called the J-function and the I-function. The J-function is the following formal generating series of genus-zero FJRW invariants:

$$J(\mathbf{t},z) := z\phi_1 + \mathbf{t}(-z) + \sum_{\substack{n\geq 2\\ 1\leq m\leq 4}} \frac{\phi^m}{n!}\left\langle \mathbf{t}(\psi)^n \frac{\phi_m}{z-\psi}\right\rangle^{\mathrm{FJRW}}_{0,n+1},$$

where

- $\mathbf{t}(z) = \sum t_k^m \phi_m z^k \in V[[z]]$ and t_k^m are considered as formal variables,
- $\mathbf{t}(\psi)^n = \mathbf{t}(\psi_1)\cdots\mathbf{t}(\psi_n)$ is expanded linearly, and
- $(z-\psi)^{-1}$ is expanded as a geometric series in ψ.

The I-function is the explicit formal series defined by

$$I(q,z) := z\sum_{d\geq 0}\frac{q^d}{z^d d!}\left(\prod_{\substack{0\leq k<\frac{d+1}{5}\\ \langle k\rangle = \langle\frac{d+1}{5}\rangle}} (kz)^5\right)\phi_{d+1},$$

where $\langle r\rangle$ denotes the fractional part of r. By counting the number of terms in the product, one sees that $I(q,z)$ can be written in the form

$$I(q,z) = I_0(q)z\phi_1 + I_1(q)\phi_2 + I_2(q)\frac{\phi_3}{z} + I_3(q)\frac{\phi_4}{z^2},$$

and we write

$$I_+(q,z) := I_0(q)z\phi_1 + I_1(q)\phi_2,$$

for the part of $I(q,z)$ with non-negative powers of z. The FJRW mirror theorem can be stated as follows.

THEOREM 2.2.1 (Chiodo-Ruan [**7**]). *Set*

$$\tau = I_+(q,-z) + z\phi_1.$$

Then

$$J(\tau,z) = I(q,z).$$

REMARK 2.2.2. If we expand $J(\tau,z)$ as a Laurent polynomial in z and write $J_+(\tau,z)$ for the part with non-negative powers of z, then it is immediate from the definitions that $I_+(q,z) = J_+(\tau,z)$. The content of the theorem is that the coefficients of negative powers of z also agree.

Upon close inspection, the reader might notice that the formulation in Theorem 2.2.1 is not quite the same as that in [**7**]. However, it is not hard to prove that they are equivalent.

EXERCISE 2.2.3. Prove that Theorem 2.2.1 is equivalent to the more standard formulation:

$$J\left(\frac{I_1(q)}{I_0(q)}\phi_2, z\right) = \frac{I(q,z)}{I_0(q)}.$$

(Hint: use the dilaton equation.)

The formulation in the previous exercise immediately determines the genus-zero FJRW invariants with any number of ϕ_2 insertions and one additional insertion, possibly with a ψ-class. However, the next exercise shows that, actually, all genus-zero FJRW invariants are determined by the mirror theorem.

EXERCISE 2.2.4. Prove that the formulation in the previous exercise determines all genus-zero FJRW invariants. (Hint: use the string and dilaton equations along with the formula for the virtual dimension.)

3. Proof of the Fan-Jarvis-Ruan-Witten mirror theorem

In this lecture, we present a proof of the FJRW mirror theorem, as stated in Theorem 2.2.1, by showing that it is a consequence of certain localization relations on auxiliary moduli spaces that have a natural $T = \mathbb{C}^*$ action.

3.1. Auxiliary moduli spaces. Consider moduli spaces

$$\mathcal{G}\overline{\mathcal{M}}^{1/5}_{0,\vec{m},d} = \{(C; q_1, \ldots, q_n; L; D; f; g)\} / \sim$$

where

- $(C; q_1, \ldots, q_n)$ is a pre-stable, n-pointed, genus-zero orbifold curve, with orbifold structure only at the marks and nodes;
- L is an orbifold line bundle on C such that
  - L has multiplicity m_i at q_i, with $m_i \in \{1, 2, 3, 4\}$;
  - L satisfies the kissing condition at each node;
- D is an effective degree-d divisor on C, supported away from the marks and nodes,
- f is an isomorphism

$$f : L^{\otimes 5} \xrightarrow{\cong} \omega_{C,\log} \otimes \mathcal{O}(-D);$$

- $g : C \to \mathbb{P}^1$ is a degree-1 map; and
- all of this data satisfies the stability condition: $\omega_{C,\log} \otimes \mathcal{O}(\epsilon D) \otimes g^*\mathcal{O}_{\mathbb{P}^1}(3)$ is ample for all $\epsilon > 0$.

Isomorphisms are required to preserve the divisor D and commute with the map g. We use $\mathcal{G}\overline{\mathcal{M}}^{1/5}_{g,n}$ to denote the disjoint union over all possible multiplicity vectors.

These conditions are a lot to swallow, so let us parse the main ideas. If $d = 0$, then $\mathcal{G}\overline{\mathcal{M}}^{1/5}_{0,\vec{m},0}$ parametrizes usual 5-spin curves along with an additional degree-one map to $\mathbb{P}^1$. This additional map parametrizes one irreducible component $C_0 \subseteq C$, and the stability condition simply says that C_0 is always stable, regardless of how many special points it has, while every other component is stable if and only if it has at least three special points. If $d \neq 0$, then any non-parametrized component is stable if and only if it satisfies one of the following two conditions (see Figure 2):

(1) it has at least three special points (marked points or nodes), or
(2) it has two special points and intersects $\mathrm{Supp}(D)$ nontrivially.

As in the case of usual 5-spin curves, $\mathcal{G}\overline{\mathcal{M}}^{1/5}_{0,\vec{m},d}$ has a virtual class, defined by

$$\left[\mathcal{G}\overline{\mathcal{M}}^{1/5}_{0,\vec{m},d}\right]^{\mathrm{vir}} := \left[\mathcal{G}\overline{\mathcal{M}}^{1/5}_{0,\vec{m},d}\right] \cap e\left((R^1\pi_*\mathcal{L}^{\oplus 5})^\vee\right),$$

and this virtual class vanishes on divisors where the node has multiplicity zero. Unlike the usual case of 5-spin curves, this moduli space has a torus action. More specifically, define an action of $T = \mathbb{C}^*$ on $\mathcal{G}\overline{\mathcal{M}}^{1/5}_{0,\vec{m},d}$ by post-composing the map $g : C \to \mathbb{P}^1$ with the action $t \cdot [x_0, x_1] = [tx_0, x_1]$ on $\mathbb{P}^1$. This is equivalent to acting directly on $|C_0| \stackrel{g}{=} \mathbb{P}^1$. Let z denote the T-equivariant parameter, so that

$$c_1(T\mathbb{P}^1|_0) = z = -c_1(T\mathbb{P}^1|_\infty),$$

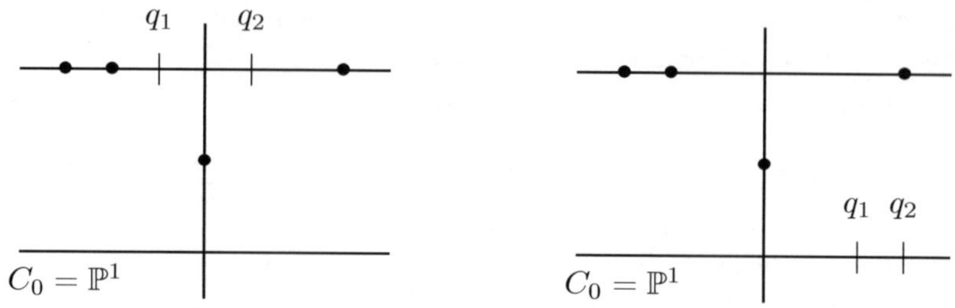

FIGURE 2. The left-hand image depicts a stable element of $\mathcal{G}\overline{\mathcal{M}}^{1/5}_{0,2,4}$, where the dashes represent marked points and the dots represent points in the support of D. The right-hand image depicts an unstable object, because no amount of degree in D will stabilize the top component.

FIGURE 3. The left-hand image depicts a generic T-fixed point of $\mathcal{G}\overline{\mathcal{M}}^{1/5}_{0,n,d}$, where the marked points and support of D are distributed on the two branches attached to C_0 at 0 and ∞. The right-hand image depicts the distinguished fixed locus $\Gamma_0 \subset \mathcal{G}\overline{\mathcal{M}}^{1/5}_{0,n+1,d}$.

where $0 = [1,0]$ and $\infty = [0,1]$. A point of $\mathcal{G}\overline{\mathcal{M}}^{1/5}_{0,\vec{m},d}$ is fixed by the T-action if and only if

$$C_0 \cap \left(\left\{\overline{C \setminus C_0}\right\} \cup \{q_1, \ldots, q_n\} \cup \mathrm{Supp(D)}\right) \subseteq \{0, \infty\}.$$

See Figure 3.

3.2. Unification of the J- and I-functions. Let $i : \Gamma_0 \hookrightarrow \mathcal{G}\overline{\mathcal{M}}^{1/5}_{0,n+1,d}$ be the T-fixed locus where $q_{n+1} = \infty \in C_0$, and define

$$\mathcal{J}(\mathbf{t}, q, z) := -z^2 \sum_{\substack{n \geq 0 \\ d \geq 0 \\ 1 \leq m \leq 4}} \frac{q^d}{n!} \int_{\Gamma_0} \frac{i^* \left[\mathcal{G}\overline{\mathcal{M}}^{1/5}_{0,n+1,d} \cap \mathbf{t}(\psi)^n \cap \phi_m\right]^{\mathrm{vir}}}{e(N_{\Gamma_0})} \phi^m,$$

where, as before, z is viewed as the equivariant parameter of T: $H^*_T(\mathrm{pt}) = \mathbb{Q}[z]$. The notation in the numerator should be interpreted as

$$i^* \left[\mathcal{G}\overline{\mathcal{M}}^{1/5}_{0,n,d} \cap \phi_1 \psi^{a_1} \cap \cdots \cap \phi_{m_n} \psi^{a_n}\right]^{\mathrm{vir}} = i^* \left[\mathcal{G}\overline{\mathcal{M}}^{1/5}_{0,\vec{m},d}\right]^{\mathrm{vir}} \cap \psi_1^{a_1} \cap \cdots \cap \psi_n^{a_n}.$$

The formal series $\mathcal{J}(\mathbf{t}, q, z)$ unifies the J- and I-functions in the following sense.

PROPOSITION 3.2.1. *Setting $q = 0$,*

$$\mathcal{J}(\mathbf{t}, 0, z) = J(\mathbf{t}, z).$$

Setting $\mathbf{t} = 0$,

$$\mathcal{J}(0, q, z) = I(q, z).$$

PROOF. First, consider the case $q = 0$, which amounts to setting $d = 0$. The first three images in Figure 4 depict the three types of fixed loci and how they contribute to $J(\mathbf{t}, z)$. We explain each of these computations in turn.

If $n = 0$, then $\deg(\omega_{C,\log}) = -1$, and the moduli space is non-empty only if $m = 4$, in which case Γ_0 is a point (with an order-5 automorphism). One checks that $H^1(C, L) = 0$, so the virtual class restricts to the usual fundamental class. The normal bundle has a single factor of $-z = c_1(TC_0|_\infty)$ corresponding to moving q_1 away from ∞. Dividing by the automorphism group cancels with the 5 in the dual basis: $\phi^4 = 5\phi_1$. Thus, the coefficient works out to be $z\phi_1$.

If $n = 1$, then Γ_0 is still a point (with an order-5 automorphism). On this locus, $\psi_1 = c_1(TC_0^\vee|_0) = -z$. The virtual class is the usual fundamental class again, and the normal bundle has two contributions z and $-z$ corresponding to moving the two marked points away from 0 and ∞. Putting everything together, the coefficient in this case is $\mathbf{t}(-z)$.

For $n \geq 2$, $\Gamma_0 \cong \overline{\mathcal{M}}_{0,n+1}^{1/5}$. Using the normalization sequence

$$0 \to \mathcal{O}_C \to \mathcal{O}_{C_0} \oplus \mathcal{O}_{\overline{C \setminus C_0}} \to \mathcal{O}_q \to 0,$$

one checks that the virtual class on $\mathcal{G}\overline{\mathcal{M}}_{0,n+1}^{1/5}$ restricts to the virtual class on $\overline{\mathcal{M}}_{0,n+1}^{1/5}$. The normal bundle has two factors z and $-z$ corresponding to moving the node away from 0 and moving q_{n+1} away from ∞, respectively, and one factor of $z - \psi$ corresponding to smoothing the node at 0 (see page 16 of [**12**] for a careful explanation of the node smoothing term). Thus, these coefficients are equal to the $n \geq 2$ terms of $J(\mathbf{t}, z)$.

Next, consider the case $\mathbf{t} = 0$, which amounts to setting $n = 0$. Then Γ_0 is a point (with an order-5 automorphism) where $C = C_0$ and D is entirely supported at $0 \in C_0$ (see Figure 4). Since

$$L^{\otimes 5} = \omega_{C,\log} \otimes \mathcal{O}(-d) \cong \mathcal{O}(-d-1),$$

we see that $L \cong \mathcal{O}\left(-\frac{d+1}{5}\right)$ and $m = -(d+1) \bmod 5$. With respect to the open cover consisting of the complement of 0 and the complement of ∞, a basis of Čech sections is given by

$$H^1(C, L) = \mathbb{C}\left\{x_0^{-k} x_1^{k-\frac{d+1}{5}} : 0 < k < \frac{d+1}{5},\ \langle k \rangle = \left\langle \frac{d+1}{5} \right\rangle\right\}.$$

Since the T-action has weight $-kz$ on each basis element, we see that the virtual class restricts as

$$e\left(H^1(C, L^{\oplus 5})^\vee\right) = \prod_{\substack{0 < k < \frac{d+1}{5} \\ \langle k \rangle = \left\langle \frac{d+1}{5} \right\rangle}} (kz)^5.$$

The normal bundle contributes $-z$ corresponding to moving the marked point away from ∞ and $d!z^d$ corresponding to moving the support of D away from 0. Combining these contributions recovers the I-function. □

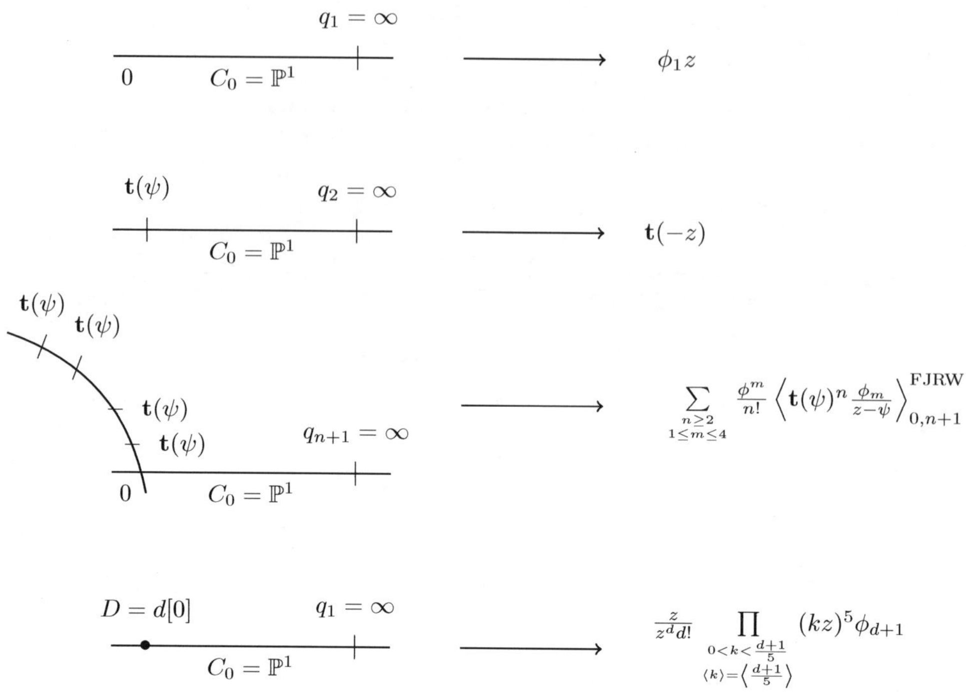

FIGURE 4. Contributions to $\mathcal{J}(\mathbf{t}, q, z)$ when $q = 0$ (first three) and $\mathbf{t} = 0$ (last one).

3.3. Proof of the mirror theorem. We now prove the FJRW mirror theorem.

PROOF OF THEOREM 2.2.1. It suffices to prove

$$\mathcal{J}(t, q, z) = \mathcal{J}(t + \tau, 0, z) \tag{3}$$

where

$$t = \sum_{1\leq m\leq 4} t^m \phi_m \qquad \text{and} \qquad \tau = I_+(q, -z) + \phi_1 z.$$

If (3) holds, then setting $t = 0$ recovers Theorem 2.2.1 via Proposition 3.2.1. Notice that (3) is true after setting $q = 0$ and it is true upon restricting to non-negative powers of z. Our strategy is to show that both sides of (3) satisfy the same recursion, which determines the coefficients recursively in (d, n) (lexicographic ordering).

Begin with the left-hand side, and consider the equivariant integrals

$$\int_{\left[\mathcal{G}\overline{\mathcal{M}}^{1/5}_{0,(\vec{m},m),d}\right]^{\mathrm{vir}}} \mathrm{ev}_D^*(0) \cdot \mathrm{ev}_{n+1}^*(\infty) \in H_T^*(\mathrm{pt}) = \mathbb{Q}[z] \tag{4}$$

where $\mathrm{ev}_D^*(0)$ denotes the equivariant class where $g^{-1}(0) \cap \mathrm{Supp}(D) \neq \emptyset$ and $\mathrm{ev}_{n+1}^*(\infty)$ denotes the equivariant class where $q_{n+1} \in g^{-1}(\infty)$. By definition, (4) is polynomial in z. Therefore, if we invert z and compute the integral by localization, we obtain relations among the coefficients of negative powers of z.

To exhibit how these relations can be useful, let us suppose that we are interested in computing the $q^d t^{m_1} \cdots t^{m_n} \phi^m$-coefficient of $\mathcal{J}(t, q, z)$. Computing the integral (4) by localization, the fixed locus Γ_0 contributes d times the coefficient we are interested in (there is a dz from the restriction of $\mathrm{ev}_D^*(0)$ and a $-z$ from the restriction of $\mathrm{ev}_{n+1}^*(\infty)$). Moreover, the contribution of every other fixed locus (see Figure 3) is determined recursively because the components over 0 and ∞ either have degree $< d$ (and any number of special points), or degree $= d$ but with fewer special points. Thus, with the lexicographic ordering on (d, n), this gives an effective recursion determining $\mathcal{J}(t, q, z)$ from $q = 0$ and the coefficients of the non-negative powers of z.

Inserting these relations as coefficients of the appropriate generating series, we can package them efficiently as the following statement:

$$\left(\frac{\partial}{\partial q}\mathcal{J}(t, q, z), \frac{\partial}{\partial t^m}\mathcal{J}(t, q, -z)\right)[z^{-k}] = 0 \quad \forall k > 0 \text{ and } 1 \le m \le 4, \tag{5}$$

where the notation $(-)[z^{-k}]$ denotes the coefficient of z^{-k} in the expression $(-)$. The discussion above implies that the relations (5) determine the series $\mathcal{J}(t, q, z)$ from $\mathcal{J}(t, 0, z)$ and $\mathcal{J}_+(t, q, z)$.

Now consider the right-hand side, where we investigate the equivariant integrals

$$\int_{\left[\mathcal{G}\overline{\mathcal{M}}^{1/5}_{0,(\vec{m},m)+1,0}\right]^{\mathrm{vir}}} \left(\mathrm{ev}_{n+2}^*(0)\frac{\partial \tau}{\partial q}\right) \cdot \mathrm{ev}_{n+1}^*(\infty). \tag{6}$$

Computing (6) by localization and packaging in the appropriate generating series as before, we find that

$$\left(\frac{\partial}{\partial q}\mathcal{J}(t+\tau, 0, z), \frac{\partial}{\partial t^m}\mathcal{J}(t+\tau, 0, -z)\right)[z^{-k}] = 0 \quad \forall k > 0 \text{ and } 1 \le m \le 4, \tag{7}$$

and the relations (7) determine the z^{-k} coefficients of $\mathcal{J}(t+\tau, 0, z)$ from $q = 0$ and the coefficients of the non-negative powers of z.

Since the recursions (5) and (7) are identical, and they determine the series from $q = 0$ and the coefficients of the non-negative powers of z, then the two sides of (3) are equal. □

4. Gromov-Witten theory of the quintic threefold

In this lecture, we introduce GW invariants of the quintic threefold and state the genus-zero GW mirror theorem. We also apply localization to provide an explicit algorithm for computing any genus-zero GW invariant of the quintic threefold.

4.1. GW theory of the quintic threefold. GW invariants are a special class of intersection numbers on *moduli spaces of stable maps*. The closed points of the moduli space of stable maps are defined by

$$\overline{\mathcal{M}}_{g,n}(Q, d) = \{(C; q_1, \ldots, q_n; f)\} / \sim$$

where

- $(C; q_1, \ldots, q_n)$ is a pre-stable, n-pointed, genus-g curve;
- $f : C \to Q$ is a map of degree $d \in H_2(\mathbb{P}^4) = \mathbb{Z}$; and
- this data satisfies the stability condition: $\omega_{C,\log} \otimes f^*\mathcal{O}_{\mathbb{P}^4}(3)$ is ample.

An isomorphism of two points is an isomorphism of pointed curves that commutes with the maps to Q. The stability condition is equivalent to only allowing points with finitely-many automorphisms.

The moduli spaces $\overline{\mathcal{M}}_{g,n}(Q,d)$ are not smooth, even as Deligne-Mumford stacks, and they can have many components of higher-than-expected dimension. It is a highly non-trivial but fundamental fact that $\overline{\mathcal{M}}_{g,n}(Q,d)$ supports a virtual fundamental class

$$\left[\overline{\mathcal{M}}_{g,n}(Q,d)\right]^{\mathrm{vir}} \in H_{2n}\left(\overline{\mathcal{M}}_{g,n}(Q,d)\right),$$

which was constructed in [**3**]. For $\varphi_1,\ldots,\varphi_n \in H^*(Q)$, GW invariants are defined by

$$\langle \varphi_1\psi^{a_1}\cdots\varphi_n\psi^{a_n}\rangle^{\mathrm{GW}}_{g,n,d} := \int_{[\overline{\mathcal{M}}_{g,n}(Q,d)]^{\mathrm{vir}}} \mathrm{ev}_1^*(\varphi_1)\psi_1^{a_1}\cdots\mathrm{ev}_n^*(\varphi_n)\psi_n^{a_n} \in \mathbb{Q},$$

where $\mathrm{ev}_i : \overline{\mathcal{M}}_{g,n}(Q,d) \to Q$ evaluates the map f at the ith marked point, and ψ_i is the cotangent line class at the ith marked point on the pre-stable curve C.

In order to state the genus-zero mirror theorem, we require the GW J- and I-functions. Let $V = H^*(Q) = \mathbb{Q}\{\phi_i\}$, where $\{\phi_i\}$ is a basis of cohomology and let $\{\phi^i\}$ denote the dual basis under the Poincaré pairing. Then for $\mathbf{t}(z) \in V[[z]]$, the J-function is defined by

$$J(\mathbf{t},z) := z + \mathbf{t}(-z) + \sum_{n,d,i} \frac{q^d}{n!}\left\langle \mathbf{t}(\psi)^n \frac{\phi_i}{z-\psi}\right\rangle^{\mathrm{GW}}_{0,n+1,d} \phi^i,$$

and the I-function is defined by

$$I(q,z) := z\sum_{d\geq 0} q^d \frac{\prod_{k=1}^{5d}(5H+kz)}{\prod_{k=1}^{d}(H+kz)^5},$$

where H is the restriction of the hyperplane class from $H^*(\mathbb{P}^4)$. The I-function should be expanded as a polynomial in $H \in H^*(Q)$. Similarly to the FJRW case, we can write

$$I(q,z) = I_0(q)z + I_1(q)H + I_2(q)\frac{H^2}{z} + I_3(q)\frac{H^3}{z^3},$$

and we define

$$I_+(q,z) := I_0(q)z + I_1(q)H.$$

The mirror theorem can be stated in the following way.

THEOREM 4.1.1 (Givental [**11**]). *Set*

$$\tau = I_+(q,-z) + z\phi_1.$$

Then

$$J(\tau,z) = I(q,z).$$

Exactly as in the FJRW case, the dilaton equation can be used to prove that Theorem 4.1.1 is equivalent to the more standard formulation of the mirror theorem:

$$J\left(\frac{I_1(q)}{I_0(q)}\phi_2, z\right) = \frac{I(q,z)}{I_0(q)},$$

and the string and dilaton equation, along with the formula for the virtual dimension, show that the mirror theorem determines all genus-zero FJRW invariants of the quintic.

4.2. Computing genus-zero GW invariants. Similarly to the FJRW setting, the general construction of the virtual class is far beyond the scope of these lectures. However, the situation greatly simplifies in genus zero. This is due to the fact that genus-zero stable maps to the quintic embed into $\overline{\mathcal{M}}_{0,n}(\mathbb{P}^4,d)$, which is a smooth Deligne-Mumford stack, and the push-forward of the virtual class can be described concretely as an Euler class:

$$i_*[\overline{\mathcal{M}}_{0,n}(Q,d)]^{\text{vir}} = [\overline{\mathcal{M}}_{0,n}(\mathbb{P}^4,d)] \cap e(R^0\pi_*f^*\mathcal{O}_{\mathbb{P}^4}(5)),$$

where $\pi : \mathcal{C} \to \overline{\mathcal{M}}_{0,n}(\mathbb{P}^4,d)$ is the universal curve and $f : \mathcal{C} \to \mathbb{P}^4$ is the universal map. Therefore, if $\varphi_1,\ldots,\varphi_n \in H^*(\mathbb{P}^4)$, the projection formula implies that

(8)
$$\langle \varphi_1\psi^{a_1}\cdots\varphi_n\psi^{a_n}\rangle^{\text{GW}}_{0,n,d} = \int_{[\overline{\mathcal{M}}_{0,n}(\mathbb{P}^4,d)]} (\text{ev}_1^*(\varphi_1)\psi_1^{a_1}\cdots\text{ev}_n^*(\varphi_n)\psi^{a_n})\, e(R^0\pi_*f^*\mathcal{O}_{\mathbb{P}^4}(5)),$$

where we implicitly restrict φ_i to $H^*(Q)$ in the left-hand side. Since $\overline{\mathcal{M}}_{0,n}(\mathbb{P}^4,d)$ has an action of the torus $T = (\mathbb{C}^*)^5$, the invariants (8) can be computed by localization, and we now provide a brief overview of how this works out. These techniques are developed more generally by Graber and Pandharipande [**12**].

Each component of the T-fixed locus in $\overline{\mathcal{M}}_{0,n}(\mathbb{P}^4,d)$ can be indexed by a decorated tree. Such a tree Γ has vertices $V(\Gamma)$ and edges $E(\Gamma)$. Each vertex v is labeled with an index $j_v \in \{0,\ldots,4\}$ and has legs indexed by $I_v \subset \{1,\ldots,n\}$, and corresponds to a connected component of $f^{-1}(p_{j_v})$ containing the fixed points $\{q_i : i \in I_v\}$. Each edge e is labeled with a positive integer d_e and corresponds to a connected multiple cover of the corresponding T-invariant line in $\mathbb{P}^4$. For example, the fixed locus in Exercise 1.2.10 is indexed by the graph with two vertices labeled 0 and 1 connected by an edge labeled d. Figure 5 gives a more general example of a localization graph and its corresponding fixed locus.

Let $i_\Gamma : F_\Gamma \hookrightarrow \overline{\mathcal{M}}_{0,n}(\mathbb{P}^4,d)$ denote the component of the fixed locus indexed by Γ, and let $G_{n,d}$ denote the set of all decorated graphs that index fixed loci in $\overline{\mathcal{M}}_{0,n}(\mathbb{P}^4,d)$. Then the localization theorem implies that

$$(8) = \sum_{\Gamma\in G_{n,d}} \frac{1}{|\text{Aut}(\Gamma)|}\int_{F_\Gamma} \frac{i_\Gamma^*\left(\text{ev}_1^*(\varphi_1)\psi_1^{a_1}\cdots\text{ev}_n^*(\varphi_n)\psi^{a_n}e(R^0\pi_*f^*\mathcal{O}_{\mathbb{P}^4}(5))\right)}{e(N_{F_\Gamma})}.$$

By normalizing the curves in the fixed locus F_Γ, the numerator can be computed as a product of terms coming from vertices, edges, and flags (i.e. nodes). The normal bundle has a factor of $R^0\pi_*f^*T\mathbb{P}^4$ from deforming the map f which can also be computed using the normalization sequence, along with a factor from smoothing each node.

For each edge $e \in E$, let j_e and j_e' denote the labels on the two adjacent vertices. Let E_v denote the set of edges adjacent to v, and let $F(\Gamma)$ denote the set of flags $\{(v,e) : e \in E_v\}$. Carefully computing the localization contributions to each fixed locus, one derives the following formula. We refer the reader to [**13**, Chapter 27] for a careful derivation.

THEOREM 4.2.1. *The GW invariant* (8) *is equal to*

$$\sum_{\Gamma\in G_{n,d}} \frac{1}{|\text{Aut}(\Gamma)|} \prod_{v\in V(\Gamma)} V(j_v,I_v,E_v) \prod_{e\in E(\Gamma)} E(d_e,j_e,j_e') \prod_{(v,e)\in F(\Gamma)} F(j_v)$$

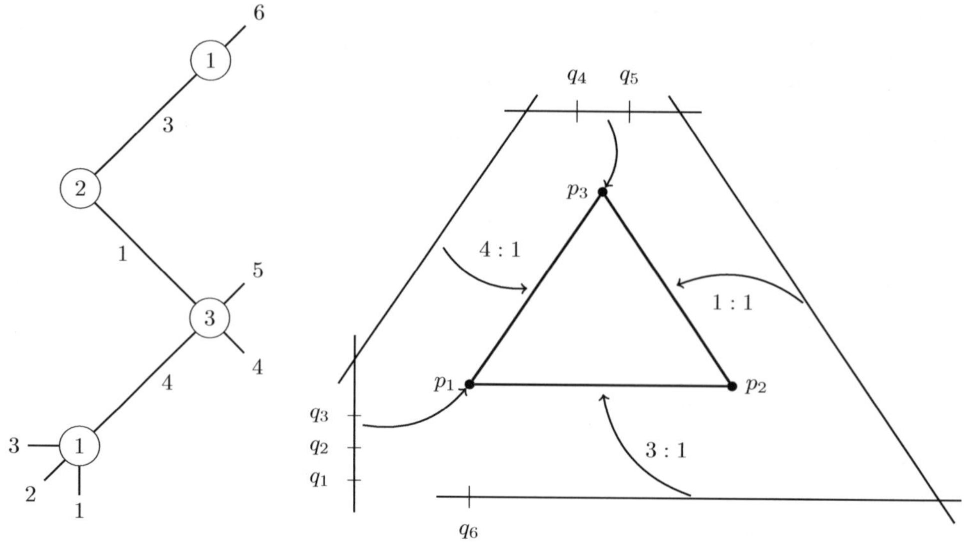

FIGURE 5. The left-hand image is an example of a localization graph, and the right-hand image is a schematic description of the maps in the corresponding fixed locus. The inner triangle in the right-hand image represents the $\mathbb{P}^2$ inside $\mathbb{P}^4$ spanned by the points p_1, p_2, and p_3. The maps in this fixed locus contract two components of the curve (the one supporting marked points q_1, q_2, and q_3, and the one supporting marked points q_4 and q_5). The non-contracted components are all mapped onto the corresponding T-invariant lines in $\mathbb{P}^4$ as multiple covers of the indicated degree, where any special points are totally ramified over the two corresponding T-fixed points in $\mathbb{P}^4$.

where

$$V(j_v, I_v, E_v) = \frac{5\alpha_j}{\prod_{j'\neq j}(\alpha_j - \alpha_{j'})} \int_{\overline{\mathcal{M}}_{0,\mathrm{val}(v)}} \frac{\prod_{k\in I_v} i_j^*\varphi_k \psi_k^{a_k}}{\prod_{e\in E_v}\left(\frac{\alpha_j - \alpha_{j'_e}}{d_e} - \psi_e\right)},$$

$$E(d,j,j') = \frac{1}{d}\frac{\prod_{k=0}^{5d}\left(5\alpha_j + k\frac{\alpha_{j'}-\alpha_j}{d}\right)}{\prod_{l=0}^{4}\prod_{\substack{k=0\\(l,k)\neq(j,0)}}^{d}\left(\alpha_j - \alpha_l + k\frac{\alpha_{j'}-\alpha_j}{d}\right)},$$

and

$$F(j) = \left(\frac{5\alpha_j}{\prod_{j'\neq j}(\alpha_j - \alpha_{j'})}\right)^{-1},$$

where, in the vertex terms, we make special conventions for the unstable moduli spaces:

$$\int_{\overline{\mathcal{M}}_{0,2}} \frac{\psi_1^k}{z-\psi_2} = (-z)^k \quad \textit{and} \quad \int_{\overline{\mathcal{M}}_{0,1}} \frac{1}{z-\psi_1} = z.$$

If you have never computed ψ-class integrals on $\overline{\mathcal{M}}_{0,n}$, the following exercise is a fun application of the string equation.

EXERCISE 4.2.2. Prove that

$$\int_{\overline{\mathcal{M}}_{0,n}} \psi_1^{a_1} \cdots \psi_n^{a_n} = \binom{n-3}{a_1, \ldots, a_n}.$$

Along with Theorem 4.2.1, Exercise 4.2.2 provides an algorithm to compute any genus-zero GW invariant of the quintic threefold. As a first example, the interested reader can recover the following classical result.

EXERCISE 4.2.3. A smooth quintic threefold contains 2875 lines.

All of this seems rather promising; however, the complexity of the graph sums grows at an enormous rate and computations very quickly become inconceivable, even with the most powerful computers at our disposal. If you are not convinced, take a stab at the following exercise.

EXERCISE 4.2.4. Enumerate the T-fixed loci in $\overline{\mathcal{M}}_{0,0}(\mathbb{P}^4, d)$ for $d = 2, 3, 4$.

Therefore, in order to prove an explicit result such as Theorem 4.1.1, we need to find and exploit recursive combinatorial structures in the localization graphs.

5. Proof of the Gromov-Witten mirror theorem

In this lecture, we prove the genus-zero GW mirror theorem for the quintic threefold by studying recursions that arise from the combinatorial structure of the localization graphs that were described in the previous lecture.

PROOF OF THEOREM 4.1.1. We begin with the left-hand side of Theorem 4.1.1. Pushing forward to $H^*(\mathbb{P}^4)$, we obtain

$$i_*J(\tau, z) = 5Hz + 5H\tau(-z) + \sum_{\substack{n,d \\ 0 \le i \le 4}} \frac{q^d}{n!} \left\langle \tau(\psi)^n \frac{H^i}{z - \psi} \right\rangle_{0,n+1,d}^{\mathrm{GW}} H^{4-i},$$

where we can safely include the summand $i = 4$ because the corresponding invariants vanish. Lifting to equivariant cohomology, then restricting to the fixed point basis, we can write

$$J_j(\tau, z) := \frac{i_j^* i_* J(\tau, z)}{e(N_{p_j/\mathbb{P}^4})} = \frac{5\alpha_j}{\prod_{j' \neq j}(\alpha_j - \alpha_{j'})}(z + \tau_j(-z)) + \sum_{n,d} \frac{q^d}{n!} \left\langle \tau(\psi)^n \frac{\rho_j}{z - \psi} \right\rangle_{0,n+1,d}^{\mathrm{GW}},$$

where $\tau_j(z)$ is obtained from $\tau(z)$ by replacing H with α_j and $\rho_j = \frac{\prod_{j' \neq j}(H - \alpha_{j'})}{\prod_{j' \neq j}(\alpha_j - \alpha_{j'})}$ denotes the equivariant cohomology class that restricts to the unit at the jth fixed point. The correlators in J_j can be computed by the localization formulas of the previous lecture. Each contributing graph has a distinguished vertex v_0 that supports the $(n + 1)$th marked point. We break the graphs up into two types (see Figure 6):

(1) Graphs where $\mathrm{val}(v_0) = 2$; and
(2) Graphs where $\mathrm{val}(v_0) > 2$.

Notice that every graph of type (2) can be obtained in a unique way by taking some number of graphs of type (1), identifying their distinguished vertices as a single vertex, and possibly adding some additional points at the new distinguished vertex. Therefore, away from the vertex v_0, contributions from graphs of type (2) are equal to sums of contributions of graphs of type (1). We now investigate what happens at the vertex v_0.

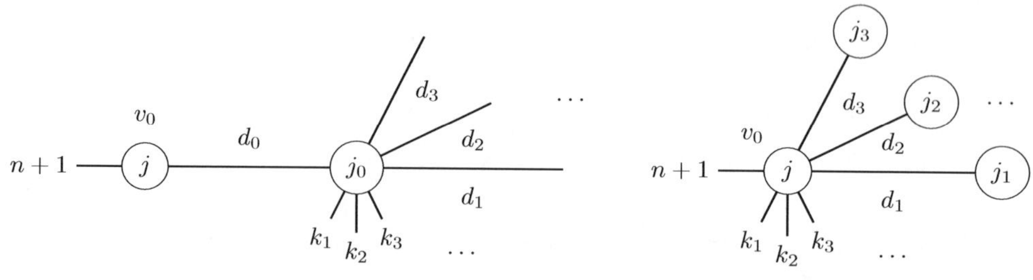

FIGURE 6. Graphs of type (1) (left) and type (2) (right).

For graphs of type (1), the contribution of the vertex v_0 is

$$\frac{5\alpha_j}{\prod_{j'\neq j}(\alpha_j - \alpha_{j'})} \frac{1}{z + \frac{\alpha_j - \alpha_{j_0}}{d_0}}. \tag{9}$$

This expression can be expanded as a power series in z. For graphs of type (2), the contribution of the vertex v_0 is

$$\frac{5\alpha_j}{\prod_{j'\neq j}(\alpha_j - \alpha_{j'})} \int_{\overline{\mathcal{M}}_{0,\mathrm{val}(v_0)}} \frac{\prod_{k\in I_v} \tau_j(\psi_k)}{\prod_{e\in E_v}\left(\frac{\alpha_j - \alpha_{j'_e}}{d_e} - \psi_e\right)} \frac{1}{z - \psi_{n+1}}, \tag{10}$$

where the denominator is expanded as a power series in the ψ-classes. Therefore, in order to compute the contribution from a graph of type (2), one can compute the contributions from the corresponding graphs of type (1), multiply them each by $F(j)$ to cancel the pre-factor in (9), replace each z with the appropriate $-\psi_e$, and then compute the integral over $\overline{\mathcal{M}}_{0,\mathrm{val}(v_0)}$ as in (10).

To summarize the above discussion more succinctly, define the power series $T_j(z)$ by the formula

$$T_j(-z) := F(j) \cdot (\text{contributions of all graphs of type (1)}).$$

Then

$$J_j(\tau, z) = \frac{5\alpha_j}{\prod_{j'\neq j}(\alpha_j - \alpha_{j'})}\left(z + \tau_j(-z) + T_j(-z) + \sum_{n\geq 2}\frac{1}{n!}\int_{\overline{\mathcal{M}}_{0,n+1}} \frac{(\tau_j(\psi) + T_j(\psi))^n}{z - \psi_{n+1}}\right), \tag{11}$$

where (by (9)) we can write

$$T_j(-z) = \sum_{\substack{j'\neq j\\ d>0}} \frac{T_j^{d,j'}}{z - \frac{\alpha_{j'} - \alpha_j}{d}}. \tag{12}$$

In addition, by removing the vertex v_0 and the edge e_0 from graphs of type (1), we see that

$$T_j^{d,j'} = q^d F(j) E(d,j,j') F(j') J_{j'}\left(\tau, z = \frac{\alpha_j - \alpha_{j'}}{d}\right). \tag{13}$$

Equations (11), (12), and (13) determine $J(\tau, z)$ from $\tau(z)$. To see why, notice that the $q^{\leq d}$ terms in (11) are determined by the $q^{\leq d}$ terms in $T_j(z)$ (and $\tau_j(z)$), and the edge-removal recursion (13) computes $q^{\leq d}$ terms of $T_j(z)$ as $q^{<d}$ terms of $J_{j'}(\tau, z)$.

Thus, in order to prove Theorem 4.1.1, we must show that

(14)
$$I_j(q,z) := \frac{i_j^* i_* J(\tau,z)}{e(N_{p_j/\mathbb{P}^4})} = z\sum_{d\geq 0} q^d \frac{\prod_{k=0}^{5d}(5\alpha_j + kz)}{\prod_{l=0}^{4}\prod^{d}_{\substack{k=0\\(l,k)\neq(j,0)}}(\alpha_j - \alpha_l + kz)}$$

can also be written in the form (11) for some $\widetilde{T}_j^{d,j'}$ that satisfy the edge-removal recursion (13). Since (14) has the same simple poles as (12), along with higher-order poles at $z=0$, then the partial fractions decomposition implies that

$$I_j(q,z) = \frac{5\alpha_j}{\prod_{j'\neq j}(\alpha_j - \alpha_{j'})}\left(z + \tau_j(-z) + \widetilde{T}_j(-z) + \mathcal{O}(z^{-1})\right)$$

where

$$\widetilde{T}_j(-z) = \sum_{\substack{j'\neq j\\ d>0}} \frac{\widetilde{T}_j^{d,j'}}{z - \frac{\alpha_{j'}-\alpha_j}{d}}.$$

Moreover, the next exercise, which is a direct computation, verifies that the q-series $\widetilde{T}_j^{d,j'}$ satisfy (13).

EXERCISE 5.0.1. Prove that

$$\mathrm{Res}_{z=\frac{\alpha_{j'}-\alpha_j}{d}} I_j(q,z) = q^d E(d,j,j')F(j')I_{j'}\left(q, z=\frac{\alpha_{j'}-\alpha_j}{d}\right).$$

Therefore, the only thing that is left to verify is that $I_j(q,z)$ has the form (11); in other words, we need to check that

(15)
$$I_j(q,z) = \frac{5\alpha_j}{\prod_{j'\neq j}(\alpha_j-\alpha_{j'})}\left(z + \tau_j(-z) + \widetilde{T}_j(-z) + \sum_{n\geq 2}\frac{1}{n!}\int_{\overline{\mathcal{M}}_{0,n+1}} \frac{(\tau_j(\psi)+\widetilde{T}_j(\psi))^n}{z-\psi_{n+1}}\right).$$

This would verify that both J and I satisfy the same recursion on the degree of q, so they are determined from the $\tau_j(z)$ in exactly the same way. This last check follows the same arguments as the proof of Theorem 2.2.1. For the reader's convenience, we outline the main steps and leave the details as exercises.

Define moduli spaces

$$\mathcal{G}\overline{\mathcal{M}}_{0,n,d} = \{(C; q_1,\ldots,q_n; D; g)\} / \sim$$

where

- $(C;q_1,\ldots,q_n)$ is a pre-stable, n-pointed, genus-zero curve;
- D is an effective degree-d divisor on C, supported away from the marks and nodes,
- $g: C\to\mathbb{P}^1$ is a degree-1 map; and
- stability: $\omega_{C,\log}\otimes\mathcal{O}(\epsilon D)\otimes g^*\mathcal{O}_{\mathbb{P}^1}(3)$ is ample for all $\epsilon>0$.

Define a T-equivariant virtual class:

$$[\mathcal{G}\overline{\mathcal{M}}_{0,n,d}]_j^{\mathrm{vir}} := \frac{e(R^0\pi_*(\mathcal{O}(5\mathcal{D})\otimes\mathcal{O}_{5\alpha_j}))}{e(R^0\pi_*(\oplus_{j'\neq j}\mathcal{O}(\mathcal{D})\otimes\mathcal{O}_{\alpha_j-\alpha_{j'}}))}\cap[\mathcal{G}\overline{\mathcal{M}}_{0,n,d}].$$

The moduli spaces $\mathcal{G}\overline{\mathcal{M}}_{0,n,d}$ have a $\mathbb{C}^*$-action, and a distinguished fixed locus $i_{\Gamma_0}: \Gamma_0\hookrightarrow\mathcal{G}\overline{\mathcal{M}}_{0,n,d}$. Define

$$\mathcal{J}_j(t,q,z) := -z^2\sum_{n,d\geq 0}\frac{q^d}{n!}\frac{i_{\Gamma_0}^*[\mathcal{G}\overline{\mathcal{M}}_{0,n+1,d}]_j^{\mathrm{vir}}\cap t(\psi)^n}{e(N_{\Gamma_0})}.$$

where $t(z) = \sum_{k\geq 0} t_k z^k$. The following two exercises are straightforward modifications of the arguments of Lecture 3, and they finish the proof of Theorem 4.1.1.

EXERCISE 5.0.2. Define

$$\widetilde{\tau}_j(z) := I_j(q, -z) + z = \tau_j(z) + \widetilde{T}_j(z).$$

Prove that

$$\mathcal{J}_j(\widetilde{\tau}_j, 0, z) = \frac{5\alpha_j}{\prod_{j'\neq j}(\alpha_j - \alpha_{j'})}\left(z + \tau_j(-z) + \widetilde{T}_j(-z) + \sum_{n\geq 2}\frac{1}{n!}\int_{\overline{\mathcal{M}}_{0,n+1}}\frac{(\tau_j(\psi) + \widetilde{T}_j(\psi))^n}{z - \psi_{n+1}}\right).$$

and

$$\mathcal{J}_j(0, q, z) = I_j(q, z).$$

EXERCISE 5.0.3. Prove that

$$\mathcal{J}_j(t_0 + \widetilde{\tau}_j, 0, z) = \mathcal{J}_j(t_0, q, z).$$

□

Acknowledgments

Special thanks to the organizers of the Snowbird workshop *Crossing the Walls in Enumerative Geometry* for the invitation to speak; to the basketball and hot tub crew for providing a good deal of evening entertainment while we were at Snowbird; and to Emily Clader for many conversations related to these lectures and for feedback on a preliminary version of these notes.

References

[1] Alejandro Adem, Johann Leida, and Yongbin Ruan, *Orbifolds and stringy topology*, Cambridge Tracts in Mathematics, vol. 171, Cambridge University Press, Cambridge, 2007. MR2359514

[2] M. F. Atiyah and R. Bott, *The moment map and equivariant cohomology*, Topology **23** (1984), no. 1, 1–28, DOI 10.1016/0040-9383(84)90021-1. MR721448

[3] K. Behrend and B. Fantechi, *The intrinsic normal cone*, Invent. Math. **128** (1997), no. 1, 45–88, DOI 10.1007/s002220050136. MR1437495

[4] Philip Candelas, Xenia C. de la Ossa, Paul S. Green, and Linda Parkes, *A pair of Calabi-Yau manifolds as an exactly soluble superconformal theory*, Nuclear Phys. B **359** (1991), no. 1, 21–74, DOI 10.1016/0550-3213(91)90292-6. MR1115626

[5] Huai-Liang Chang, Jun Li, and Wei-Ping Li, *Witten's top Chern class via cosection localization*, Invent. Math. **200** (2015), no. 3, 1015–1063, DOI 10.1007/s00222-014-0549-5. MR3348143

[6] Alessandro Chiodo, *The Witten top Chern class via K-theory*, J. Algebraic Geom. **15** (2006), no. 4, 681–707, DOI 10.1090/S1056-3911-06-00444-9. MR2237266

[7] Alessandro Chiodo and Yongbin Ruan, *Landau-Ginzburg/Calabi-Yau correspondence for quintic three-folds via symplectic transformations*, Invent. Math. **182** (2010), no. 1, 117–165, DOI 10.1007/s00222-010-0260-0. MR2672282

[8] Tom Coates, Alessio Corti, Hiroshi Iritani, and Hsian-Hua Tseng, *A mirror theorem for toric stacks*, Compos. Math. **151** (2015), no. 10, 1878–1912, DOI 10.1112/S0010437X15007356. MR3414388

[9] David A. Cox and Sheldon Katz, *Mirror symmetry and algebraic geometry*, Mathematical Surveys and Monographs, vol. 68, American Mathematical Society, Providence, RI, 1999. MR1677117

[10] H. Fan, T. Jarvis, and Y. Ruan, *The Witten equation and its virtual fundamental cycle*, arXiv:0712.4025, 2007.

[11] Alexander Givental, *A mirror theorem for toric complete intersections*, Topological field theory, primitive forms and related topics (Kyoto, 1996), Progr. Math., vol. 160, Birkhäuser Boston, Boston, MA, 1998, pp. 141–175. MR1653024

[12] T. Graber and R. Pandharipande, *Localization of virtual classes*, Invent. Math. **135** (1999), no. 2, 487–518, DOI 10.1007/s002220050293. MR1666787

[13] Kentaro Hori, Sheldon Katz, Albrecht Klemm, Rahul Pandharipande, Richard Thomas, Cumrun Vafa, Ravi Vakil, and Eric Zaslow, *Mirror symmetry*, Clay Mathematics Monographs, vol. 1, American Mathematical Society, Providence, RI; Clay Mathematics Institute, Cambridge, MA, 2003. With a preface by Vafa. MR2003030

[14] Alexander Polishchuk and Arkady Vaintrob, *Algebraic construction of Witten's top Chern class*, Advances in algebraic geometry motivated by physics (Lowell, MA, 2000), Contemp. Math., vol. 276, Amer. Math. Soc., Providence, RI, 2001, pp. 229–249, DOI 10.1090/conm/276/04523. MR1837120

[15] Dustin Ross and Yongbin Ruan, *Wall-crossing in genus zero Landau-Ginzburg theory*, J. Reine Angew. Math. **733** (2017), 183–201, DOI 10.1515/crelle-2015-0005. MR3731328

Department of Mathematics, San Francisco State University, 1600 Holloway Avenue, San Francisco, California 94132

Email address: rossd@sfsu.edu

Contemporary Mathematics
Volume **763**, 2021
https://doi.org/10.1090/conm/763/15339

A brief introduction to cosection localization and P-fields

Wei-Ping Li

1. Introduction

The seminal paper [**COGP**] by Candelas, dela Ossa, Green, and Parkes plays a fundamental role in the development of mirror symmetry and Gromov-Witten (GW) theory. understanding and computing the Gromov-Witten invariants of the quintic Calabi-Yau (CY) threefold in $\mathbb{P}^4$ has been one of the key research topics since the appearance of the paper [**COGP**]. On the mathematical side, the perfect obstruction theory and the theory of virtual cycles were set up by J. Li and G. Tian in [**LT1**] and by K. Behrend and B. Fantechi in [**BF**]. These theories were used to define the Gromov-Witten invariants in algebraic geometry in [**B**] and [**LT1**].

Computations of the GW invariants of the CY quintic threefold started with Kontsevich's paper [**K**], where he proposed considering the torus action on $\mathbb{P}^4$, which induces a torus action on the moduli space of stable maps to the quintic using torus localization (see [**GP99**]) and the fact that the obstruction sheaf is a bundle. This proposal was successfully carried out by Givental in [**G**] and Lian-Liu-Yau in [**LLY**] to calculate the genus zero GW invariants of the quintic among many more general CY manifolds. This method meets some serious difficulties when one pushes to genus one computations. One of the difficulties is that the obstruction sheaf fails to be a vector bundle. Zinger overcame this difficulty by defining the reduced genus one Gromov-Witten invariants in [**Z1**]. In [**VZ**], Vakil and Zinger used blowups to resolve the singularity of the ghost component and the cone to get a vector bundle. The relation between reduced genus one GW invariants and the original GW invariants was established by Zinger in [**Z1**], J. Li and Zinger in [**LZ**]. The genus one Gromov-Witten invariants of Calabi-Yau hypersurfaces were finally calculated by Zinger in [**Z2**].

On the physical side, there has been much more advancement in the structure of higher genus GW invariants. Bershadsky, Cecotti, Ooguri, and Vafa in [**BCOV**] established the Feynman graph sum rule; Yamaguchi and Yau in [**YY**] proposed a finite generation conjecture for the GW potential function F_g, which implies the convergence of F_g.

Recently, several new techniques have been developed for studying moduli problems. One is the cosection localization method by Kiem and J. Li in [**KL13**], and the other is the p-fields interpretation of GW invariants by Chang and J. Li in [**CL**].

Partially supported by Hong Kong GRF16301515 and GRF 16303518.

The combination of these two methods gives a new interpretation of GW invariants and a new algebraic geometric definition of Fan-Jarvis-Ruan-Witten (FJRW) invariants developed by Fan, Jarvis, and Ruan in [**FJR1**]. This new interpretation enables us to set up a geometric platform, called mixed-spin-p (MSP) fields, linking the GW invariants of the quintic CY threefold with the FJRW invariants of the quintic polynomial. This new moduli stack can help us to better understand the higher genus GW invariants, thus establishing mathematically the BCOV Feynman graph sum rule and solving the Yamaguchi-Yau finite generation conjecture and the Yamaguchi-Yau functional equation.

In this paper, we will go through some basics of obstruction theory in §1. We will touch a bit of defining virtual cycles by the intrinsic normal cone in §2. We will discuss the method of cosection localization in §3 and the method of p-fields in §4.

We will spend more time explaining key concepts and theorems using a simple example where we have a vector bundle E over a smooth variety W, with a section s such that the space we are interested in is the zero locus $s^{-1}(0)$ of s. Even though the moduli problems we are studying are generally stacks , we will discuss our basic examples and key lemmas for schemes only for the purpose of focusing on essential ingredients. For example, we will go into detail about the normal cone, since we will use it in our calculations several times. We will discuss the obstruction theory in details for the basic example, and we will prove the key lemmas for the cosection localization carefully so that readers can understand why and how the cosection localization works.

For the deformation-obstruction theory and virtual cycles, one can read the papers [**LT1**], [**BF**], and the survey paper [**LT2**]. One can also read the book by Hartshorne [**H**] for the part on deformation-obstruction theory. For the cotangent complexes and stacks, one can read the book [**Ill71**] and the Stacks Project.

2. Virtual cycles, obstruction theory (local), and normal cones

We start with a simple example. We will use this example to illustrate key ideas in various sections.

2.1. A basic example.

EXAMPLE 2.1. Let E be a vector bundle of rank r over a smooth projective m-dimensional variety W with the projection map $\pi\colon E \to W$; let s be a section of the bundle and let $X = s^{-1}(0)$ be the zero locus of the section s. We want to have a "correct" definition of the "fundamental class" of X, or rather the "virtual cycle" of X, denoted by $[X]^{\mathrm{vir}}$. The word "correct" means that the fundamental class is X itself in the generic case and invariant under the variation of the section s.

When the section s is transversal to the zero section of E, the fundamental class of X is represented by X itself in the Chow group $A_{m-r}(W)$, that is,

$$[X]^{\mathrm{vir}} = [X] \in A_{m-r}(W).$$

Clearly, from the property of characteristic classes, the Euler class $e(E)$ of E is the fundamental class $[X]$, where $e(E) \in A^r(W) = A_{m-r}(W)$.

For the general case where s is not transversal to the zero section of E, we see that X may have a dimension bigger than $m - r$, or even if X has dimension $m - r$, X may have a non-reduced scheme structure. How do we define $[X]^{\mathrm{vir}}$?

The definition should at least include the transversal case as a special case, and we require that $[X]^{\mathrm{vir}}$ remains the same as a class in $A_{m-r}(W)$ when we vary the section s. This requirement forces $[X]^{\mathrm{vir}}$ to be the Euler class $e(E)$ of E, in other words,

$$[X]^{\mathrm{vir}} = e(E) \in A_{m-r}(W).$$

□

In many situations, such as defining invariants via moduli spaces, it is better to have $[X]^{\mathrm{vir}} \in A_{m-r}(X)$. Based on previous discussions, we should look at a class $[X]^{\mathrm{vir}} \in A_{m-r}(X)$ such that

$$\iota_*[X]^{\mathrm{vir}} = e(E) \in A_{m-r}(W),$$

where $\iota\colon X \to W$ denotes the natural inclusion map. The way to do it is to use the localized Euler class (see [**F**]). Let's discuss briefly the definition, and one can consult Fulton's book [**F**] for a more precise and in-depth explanation.

Because X is a closed subscheme of W, it has an ideal sheaf I_X. The normal cone of X in W is defined by

$$C_{X\subset W} = \operatorname{Spec}\left(\bigoplus_{n=0}^{\infty} \frac{I_X^n}{I_X^{n+1}}\right).$$

When X is smooth, the normal cone is the normal bundle

$$N_{X\subset W} = \operatorname{Spec}\left(\bigoplus_{n=0}^{\infty} \operatorname{Sym}^n \frac{I_X}{I_X^2}\right).$$

The normal cone $C_{X\subset W}$ is a subcone of $E|_X$. Let

$$s_E\colon X \to E|_X$$

be the zero section of E. Consider the Gysin map

$$s_E^!\colon A_*(E|_X) \cong A_{*-r}(X).$$

We define the localized Euler class

$$e_{\mathrm{loc}}(E,s) = s_E^!(C_{X\subset W}) \in A_{m-r}(X).$$

Now we define the virtual cycle of X to be

$$[X]^{\mathrm{vir}} \colon = e_{\mathrm{loc}}(E,s).$$

Now, $\iota_*\big(e_{\mathrm{loc}}(E,s)\big)$ is a class in $A_{m-r}(W)$. By our previous requirement of virtual cycles, we should have

$$\iota_*\big(e_{\mathrm{loc}}(E,s)\big) = e(E).$$

This is indeed the case from Prop.14.1 (a) in [**F**].

2.2. Normal cones. We discuss in detail the concept of normal cones. The explicit expression of normal cones can help us to understand the cosection localization in a later section.

Let's look at MacPherson's graph construction of the normal cone. With the same setup as the previous subsection, consider the following injective map

$$W \times \mathbb{C}^* \to \mathbb{P}(E \oplus \mathbb{C}_W) \times \mathbb{C}, \tag{2.1}$$
$$(x,t) \to \big([s(x),t],t\big),$$

where $\mathbb{C}_W = W \times \mathbb{C}$, and recall that s is a section of E. Under the identification

$$\mathbb{P}(E \oplus \mathbb{C}_W)|_{\{t \neq 0\}} \times \mathbb{C} \cong E \times \mathbb{C},$$

the map (2.1) can be rewritten as

$$W \times \mathbb{C}^* \to E \times \mathbb{C}, \quad (x,t) \to \big(t^{-1}s(x), t\big).$$

We blow up $W \times \mathbb{C}$ along $X \times 0$ to get $\widetilde{W \times \mathbb{C}}$ and obtain a morphism

$$\widetilde{W \times \mathbb{C}} \to \mathbb{P}(E \oplus \mathbb{C}_W) \times \mathbb{C}. \tag{2.2}$$

The closure of the image of $W \times \mathbb{C}^*$ in $\mathbb{P}(E \oplus \mathbb{C}_W) \times \mathbb{C}$ is the image of the morphism (2.2). Let $\widetilde{W}$ be the strict transform of $W \times 0$ in $\widetilde{W \times \mathbb{C}}$, which is also the blowup of W along X. Define

$$M^0_{X/W} \colon = \widetilde{W \times \mathbb{C}} - \widetilde{W} \to \mathbb{P}(E \oplus \mathbb{C}_W) \times \mathbb{C}.$$

Let $\rho \colon M^0_{X/W} \to \mathbb{C}$ be the natural projection map. Then the fiber $\rho^{-1}(t)$ is W for $t \neq 0$, and we have the trivial imbedding

$$X \times \mathbb{C}^* \to W \times \mathbb{C}^*,$$

since the blowup only changes the central fiber.

On the other hand, $\rho^{-1}(0) = C_{X \subset W}$ is exactly the normal cone of X in W with X imbedded in the cone $C_{X \subset W}$ as the zero section. Therefore, the map

$$X \times \mathbb{C} \to M^0_{X/W}$$

gives the deformation of the imbedding of X in W to the imbedding of X in the normal cone $C_{X \subset W}$ as the zero section. We call $M^0_{X/W}$ the space of the deformation of W to the normal cone $C_{X/W}$.

To understand the construction better and see why $\rho^{-1}(0)$ is exactly the normal cone $C_{X \subset W}$ of X in W, we can look at the local model following [**F**].

Let $W = \mathrm{Spec}B$, let $E = \mathbb{C}_W^{\oplus r}$, let $s = (f_1, \ldots, f_r)$, and $I = (f_1, \ldots, f_r) \subset B$ be the ideal of X. The blowup and the map (2.2) can be described by

$$\mathrm{Proj} \bigoplus_{n=0}^{\infty} (I,t)^n \to \mathrm{Proj}\big(B[x_1, \ldots, x_r, T]\big) \times \mathbb{C}.$$

Algebraically, it can be described by

$$B[x_1, \ldots, x_r, T] \otimes \mathbb{C}[t] \to S^\bullet \colon = \bigoplus_{n=0}^{\infty} (I,t)^n,$$

$$x_i \to f_i, \quad T \to t, \quad t \to t.$$

Let S^n be the group of degree n homogeneous elements in $S^\bullet$:

$$S^n = (I,t)^n = I^n + I^{n-1}t + \ldots + It^{n-1} + Bt^n + Bt^{n+1} + \ldots.$$

Away from the infinity, we have

$$\big(S^\bullet\big)_{(t)} = \ldots \oplus I^n t^{-n} \oplus \ldots \oplus It^{-1} \oplus B \oplus Bt \oplus \ldots.$$

Geometrically, $\big(S^\bullet\big)_{(t)}$ is the coordinate ring of $M^0_{X/W}$; equivalently, we have

$$\mathrm{Spec}\big((S^\bullet)_{(t)}\big) = M^0_{X/W}.$$

Restricting to the fiber $t = 0$, we obtain a surjective map

$$B[x_1, \dots, x_r, T]_{(T)} \otimes \frac{\mathbb{C}[t]}{(t)} \longrightarrow \frac{(S^\bullet)_{(t)}}{(t)(S^\bullet)_{(t)}} = \bigoplus_n \frac{I^n}{I^{n+1}} t^{-n},$$

which induces the homomorphisms

$$B\left[\frac{x_1}{T}, \dots, \frac{x_r}{T}\right] \longrightarrow \frac{B}{I}\left[\frac{x_1}{T}, \dots, \frac{x_r}{T}\right] \longrightarrow \bigoplus_n \frac{I^n}{I^{n+1}} t^{-n} = \bigoplus_n \frac{I^n}{I^{n+1}},$$
$$\frac{x_i}{T} \to \frac{f_i}{t}.$$

Back to spaces, we get the embeddings

$$C_{X\subset W} = \operatorname{Spec}\left(\bigoplus_n \frac{I^n}{I^{n+1}}\right) \to E|_X \to E.$$

Now we see that $\rho^{-1}(0)$ is the normal cone $C_{X\subset W}$, which is a subcone of $E|_X$.

2.3. Obstruction theory. Consider the Example in §2.1: the set $X \subset W$ is given by the zero locus $s^{-1}(0)$ of a nonzero section s of a rank-r vector bundle $\pi\colon E \to W$. In this case, X can also be written as the zero locus of $(s, 0)\colon W \to E \oplus \mathbb{C}_W$. Clearly,

$$e(E \oplus \mathbb{C}_W) \neq e(E).$$

Thus, the virtual cycle $[X]^{\mathrm{vir}}$ depends on how X is defined. The obstruction theory provides a venue for a correct definition of the virtual cycle. We can look at the basic example again to get ideas of how the obstruction theory works.

EXAMPLE 2.2. Consider the local case where $W = \operatorname{Spec}B$, $B = \mathbb{C}[x_1, \dots, x_m]$, $E = \mathbb{C}_W^{\oplus r}$, $s(x) = \big(f_1(x), \dots, f_r(x)\big)$, $I = (f_1, \dots, f_r)$, and $X = \operatorname{Spec}(B/I)$, which is the zero locus of the section s. Assume, without loss of generality, $s(\vec{0}) = 0$. Consider an exact sequence

$$0 \to J \to A' \to A \to 0,$$

where A, A' are artinian rings and $J \cdot m_{A'} = 0$ for the maximal ideal $m_{A'}$ of A'. Here, we have $A'/m_{A'} \cong \mathbb{C} \cong A/m_A$. Let p be the point of $\operatorname{Spec}A$ corresponding to the maximal ideal m_A and let p' be the point of $\operatorname{Spec}A'$ corresponding to the maximal ideal $m_{A'}$. Consider the following problem of extensions. Given the maps below:

$$\begin{array}{ccc} (\operatorname{Spec}A, p) & \xrightarrow{\Phi} & (X, \vec{0}) \\ {\scriptstyle\iota}\downarrow & & \\ (\operatorname{Spec}A', p'), & & \end{array}$$

where ι is the closed immersion, is there a map Φ' from $\operatorname{Spec}A'$ to X extending the map Φ? Writing in the algebraic setup, we obtain the extension problem of, given a map φ, finding a map $\varphi_{A'}$ from B/I to A' to make the diagram commutative:

$$\begin{array}{ccc} A & \xleftarrow{\ \varphi\ } & B/I \\ {\scriptstyle\pi}\uparrow & \swarrow{\scriptstyle\varphi_{A'}} & \\ A'. & & \end{array}$$

Since W is smooth, or since B is a free polynomial ring, the map

$$B \to B/I \to A$$

can be lifted to the map $\tilde{\varphi}\colon B \to A'$. Specifically, we choose $x_i' \in A'$ such that $\pi(x_i') = \varphi([x_i]) \in A$, where $[x_i] \in B/I$ is the image of $x_i \in B$. Then $x_i' \in m_{A'}$. Define the map $\tilde{\varphi}$ by letting $\tilde{\varphi}(x_i) = x_i'$. We remark that for a different choice $\tilde{x}_i'$, we have $\tilde{x}_i' - x_i' \in J$. We also have

$$\pi\big(f_i(\vec{x}')\big) = f_i\big(\pi(x_1'), \ldots, \pi(x_i')\big) = \varphi\big(f_i(x_1, \ldots, x_m) \mod I\big) = 0.$$

Thus, we have $f_i(x_1', \ldots, x_m') \in J$.

For simplicity, we let $\vec{x}' = (x_1', \ldots, x_m')$. We want to see if we can find a map $\varphi_{A'}$ making the following diagram commutative:

$$\begin{array}{ccccc} B & \longrightarrow & B/I & & \\ & \searrow^{\tilde{\varphi}} & \downarrow{\scriptstyle\varphi_{A'}} & \searrow^{\varphi} & \\ & & A' & \longrightarrow & A. \end{array}$$

It is easy to see that, if

$$f_i(x_1', \ldots, x_m') = 0 \quad \text{for all } i, \tag{2.3}$$

then $\tilde{\varphi}$ will map the ideal I to zero. Thus, we get the induced map $\varphi_{A'}$. However, (2.3) may not hold. Since there is some freedom to choose x_i', the remedy is to change the map $\tilde{\varphi}$ to $\tilde{\varphi}_\delta$ by changing the choice of x_i'; that is, we change x_i' to $x_i' + \delta_i$ for $\delta_i \in J$. Thus, we hope to define the map

$$\tilde{\varphi}_\delta(x_i)\colon = x_i' + \delta_i, \quad \delta_i \in J$$

such that for all i, we have

$$f_i(x_1' + \delta_1, \ldots, x_m' + \delta_m) = 0. \tag{2.4}$$

Notice that, by the Taylor expansion, we have

$$f_i(x_1' + \delta_1, \ldots, x_m' + \delta_m) = f_i(x_1', \ldots, x_m') + \sum_{j=1}^{m} \frac{\partial f_i}{\partial x_j}(x_1', \ldots, x_m')\delta_j,$$

since $\delta_\ell \cdot \delta_k \in J^2 \subset J \cdot m_{A'} = 0$. Therefore, a map $\tilde{\varphi}_\delta$ satisfying the property (2.4) exists if we can find $\delta_1, \ldots, \delta_m \in J$ such that

$$\sum_{j=1}^{m} \frac{\partial f_i}{\partial x_j}(x_1', \ldots, x_m')\delta_j = -f_i(x_1', \ldots, x_m') \in J \quad \text{for all } i. \tag{2.5}$$

Note that for any $g \in \mathbb{C}[x_1, \ldots, x_m]$, by the Taylor expansion, we have

$$\begin{aligned} g(x_1', \ldots, x_m') =& g(\vec{0}) + \sum_{j=1}^{m} \frac{\partial g}{\partial x_j}(\vec{0}) x_j' + \text{higher order terms}, \\ g(x_1', \ldots, x_m') \cdot \delta_i =& g(\vec{0})\delta_i + \sum_{j=1}^{m} \frac{\partial g}{\partial x_j}(\vec{0}) x_j' \delta_i + \delta_i \cdot (\text{higher order terms}) \\ =& g(\vec{0})\delta_i, \text{ since } x_j'\delta_i \in m_{A'} \cdot J = 0. \end{aligned}$$

Therefore, (2.5) is equivalent to the following equation:

$$\sum_{j=1}^{m} \frac{\partial f_i}{\partial x_j}(\vec{0})\delta_j = -f_i(x_1', \ldots, x_m') \in J \quad \text{for all } i. \tag{2.6}$$

Let $ds|_{\vec{0}}$ denote the linear transformation

$$\begin{pmatrix} \frac{\partial f_1}{\partial x_1}(\vec{0}) & \cdots & \frac{\partial f_1}{\partial x_m}(\vec{0}) \\ \vdots & \vdots & \vdots \\ \frac{\partial f_r}{\partial x_1}(\vec{0}) & \cdots & \frac{\partial f_r}{\partial x_m}(\vec{0}) \end{pmatrix} \cdot \begin{pmatrix} \delta_1 \\ \vdots \\ \delta_m \end{pmatrix} = \begin{pmatrix} \sum_{i=1}^m \frac{\partial f_1}{\partial x_i}(\vec{0})\delta_i \\ \vdots \\ \sum_{i=1}^m \frac{\partial f_r}{\partial x_i}(\vec{0})\delta_i \end{pmatrix}.$$

The linear system (2.6) with unknowns δ_i may not always have solutions. We can phrase this question in the following way: the linear system (2.6) has solutions if and only if

$$\left[\begin{pmatrix} -f_1(\vec{x}') \\ \vdots \\ -f_r(\vec{x}') \end{pmatrix}\right] = 0 \in \operatorname{Coker}\big(ds|_{\vec{0}} \colon T_W|_{\vec{0}} \to E|_{\vec{0}}\big) \otimes J.$$

Naturally, we can define the so called obstruction class to be

$$\operatorname{ob}(A', J, \varphi) = \left[\begin{pmatrix} -f_1(\vec{x}') \\ \vdots \\ -f_r(\vec{x}') \end{pmatrix}\right] \in \operatorname{Coker}\big(ds|_{\vec{0}} \colon T_W|_{\vec{0}} \to E|_{\vec{0}}\big) \otimes J. \tag{2.7}$$

By the same argument as above, $\operatorname{ob}(A', J, \varphi)$ doesn't depend upon the choice of x_i'.

The computations above tell us that the concept of the obstruction characterizes the lifting problem

$$\operatorname{ob}(A', J, \varphi) = 0$$

if and only if (2.6) has a solution if and only if (2.4) has a solution if and only if $\varphi_{A'}$ exists.

We also know more. If $\operatorname{ob}(A', J, \varphi) = 0$, the set of possible liftings $\varphi_{A'}$ is identified non naturally with $\operatorname{Ker}\big(ds|_{\vec{0}}\big)$, and, more precisely, is a $\operatorname{Ker}\big(ds|_{\vec{0}}\big)$-torsor. □

Now we can give a definition of the obstruction theory.

DEFINITION 2.3. Let (X, p) be a pointed space. Let A be an artinian ring, let $I \subset A$ be an ideal with $I \cdot m_A = 0$, and let ϕ be a morphism $\phi\colon \operatorname{Spec} A/I \to X$ such that $\phi(0) = p$, where 0 is the closed point of $\operatorname{Spec} A/I$ corresponding to the maximal ideal m_A. An obstruction theory for (X, p) with coefficients in a finite dimensional $\mathbb{C}$-vector space V is an association of a class $\mathrm{ob}(A, I, \phi) \in V \otimes I$ to each triple (A, I, ϕ) as above, such that

(1) $\mathrm{ob}(A, I, \phi) = 0$ if and only if there is an extension $\tilde{\phi}\colon \mathrm{Spec}A \to X$ of ϕ;
(2) $\mathrm{ob}(A, I, \phi)$ satisfies a base change property.

In Example 2.2, we saw an obstruction theory of $(X, \vec{0})$ with coefficients in the vector space $\mathrm{Coker}(ds|_{\vec{0}})$.

Why does the obstruction theory lead to the correct definition of virtual cycles?

Let (X, p) be a pointed space with an obstruction theory in a vector space V. Let $\hat{X}$ be the formal completion of X at p and let $n = \dim T_pX$. Take a basis $\{z_1, \dots, z_n\}$ of the cotangent space T_p^*X. Kuranishi's construction gives a pair (ι, Φ)

$$\iota\colon \hat{X} \longrightarrow \hat{W} = \mathrm{Spec}\mathbb{C}[[z_1, \dots, z_n]], \quad \Phi\colon \hat{W} \longrightarrow V$$

such that $\hat{X}$ is the zero locus of Φ. If we define a section

$$s(z_1, \dots, z_n) = \big((z_1, \dots, z_n), \Phi(z_1, \dots, z_n)\big) \in \hat{W} \times V$$

of the vector bundle $\hat{W} \times V$ over $\hat{W}$, then the obstruction theory, defined via the section s as discussed in Example 2.2, agrees with the given obstruction theory. Therefore, we can use the localized Euler class to define the virtual cycle of (X, p). Since we only use Kuranishi's method to illustrate a geometric realization of a space X with an obstruction theory as the zero locus of a section of a vector bundle, and we won't use it to define the virtual cycle, we won't discuss it in detail.

However, both Definition 2.3 and Kuranishi's method are local in nature, i.e., we only consider the germ $\hat{X}$ of X at p. To get a global treatment, we need the perfect obstruction theory defined as follows:

DEFINITION 2.4. A perfect obstruction theory for a scheme Y is given by a perfect complex $\mathcal{T}^\bullet\colon [T^0 \to T^1]$ in $D^b(X)$ such that for any triple (A, I, ϕ), where A is a $\mathbb{C}$-algebra, where $I \subset A$ is an ideal such that $I^2 = 0$, and where $\phi\colon \mathrm{Spec}A/I \to X$ is a morphism, there exists $\mathrm{ob}(A, I, \phi) \in h^1(\mathcal{T}^\bullet \otimes_{\mathcal{O}_X} I)$ such that

(1) $\mathrm{ob}(A, I, \phi) = 0$ if and only if there exists an extension $\tilde{\phi}\colon \mathrm{Spec}A \to X$ of ϕ;
(2) If $\mathrm{ob}(A, I, \phi) = 0$, the space of all extensions $\tilde{\phi}$ are parameterized by $h^0(\mathcal{T}^\bullet \otimes_{\mathcal{O}_X} I)$;
(3) $\mathrm{ob}(A, I, \phi)$ satisfies some base change property.

A perfect complex is a complex which is locally quasi-isomorphic to a bounded complex of locally free sheaves of finite ranks. For the perfect complex $\mathcal{T}^\bullet\colon [T^0 \to T^1]$ in the definition, essentially we can locally replace $[T^0 \to T^1]$ by a complex of locally free sheaves of finite ranks $[\mathcal{E}^0 \to \mathcal{E}^1]$ and replace $h^i(\mathcal{T}^\bullet \otimes_{\mathcal{O}_X} I)$ by the kernel and cokernel of the morphism $\mathcal{E}^0 \otimes_{\mathcal{O}_X} I \to \mathcal{E}^1 \otimes_{\mathcal{O}_X} I$ for $i = 0, 1$ respectively.

In Example 2.2, we see that $X = s^{-1}(0)$ has a perfect obstruction theory. The perfect complex can be chosen as

$$\mathcal{T}^\bullet\colon [\Omega_W^\vee|_X \xrightarrow{ds} E|_X].$$

It gives an obstruction theory by the computations in Example 2.2.

In the next section, we will discuss how to define the virtual cycle of X that is endowed with a perfect obstruction theory.

3. Virtual cycles via intrinsic normal cone (global)

From the discussion in the previous section, we see that a pointed space (X, p), endowed with an obstruction theory with coefficients in a vector space V, can be written as a zero locus of a section of a vector bundle over a bigger smooth space, at least formally. Such X has a fundamental class as defined in the previous section. However, since the Kuranishi method is local, and hence the normal cone is local, to glue these normal cones is a hard problem, which was solved by J. Li and G. Tian in [**LT1**]. On the other hand, given a scheme X, there is a complex $\mathbb{L}_X^\bullet$ in the derived category $D(\mathcal{O}_{X_{et}})$, which is a generalization of the sheaf of Kähler differentials of X. The deformation-obstruction theory of X can be understood and stated via the cotangent complex (see §10 in [**H**]). In [**BF**], K. Behrend and B. Fantechi used cotangent complex to define an intrinsic normal cone for X equipped with a perfect obstruction theory. Furthermore under a mild condition of the perfect obstruction theory, one can define the virtual cycle $[X]^{\mathrm{vir}}$, in other words, the fundamental class of X.

In this section, we follow the approach in [**BF**] to define virtual cycles. Therefore, we discuss cotangent complex $\mathbb{L}^\bullet$ and its truncate $\mathbb{L}^{\bullet \geq -2}$ in detail. We state briefly the setup and the results on obstruction theory via cotangent complex. Finally, we introduce intrinsic normal cones and the construction of virtual cycles via intrinsic normal cones.

3.1. Cotangent complex. Given a smooth nonsingular variety X, the tangent bundle T_X of X, or equivalently, its dual T_X^*, contains a lot of important geometric information about X. When X is not smooth, the cotangent bundle T_X^* is replaced by the sheaf of Kähler differentials Ω_X. For a morphism of schemes $f\colon X \to Y$, we have the sheaf of relative Kähler differentials $\Omega_{X/Y}$ on X. Sheaves of relative Kähler differentials satisfy the following property: for morphisms of schemes,

$$X \xrightarrow{f} Y \xrightarrow{g} Z,$$

we have an exact sequence

$$f^*\Omega_{Y/Z} \to \Omega_{X/Z} \to \Omega_{X/Y} \to 0. \tag{3.1}$$

Just as Ω_X is not locally free when X is not nonsingular, the leftmost map in the exact sequence (3.1) is not injective in general, and, hence, the exact sequence is not a short exact sequence. One way to resolve the failure of (3.1) being a short exact sequence is to expand our horizon to derived categories. The sheaf of relative Kähler differentials $\Omega_{X/Y}$ is replaced by the cotangent complex $\mathbb{L}_{X/Y}^\bullet \in D(\mathcal{O}_{X_{et}})$, and the exact sequence (3.1) is replaced by the distinguished triangle

$$Lf^*\mathbb{L}_{Y/Z}^\bullet \to \mathbb{L}_{X/Z}^\bullet \to \mathbb{L}_{X/Y}^\bullet \to Lf^*\mathbb{L}_{Y/Z}^\bullet[1]. \tag{3.2}$$

Under some mild conditions, $\mathbb{L}_{X/Y}^\bullet$ can be represented by a complex

$$\dots L^{-3} \to L^{-2} \to L^{-1} \to L^0,$$

such that $H^0(\mathbb{L}_{X/Y}^\bullet) = \Omega_{X/Y}$. When the morphism $f\colon X \to Y$ is smooth,

$$\mathbb{L}_{X/Y}^\bullet = \Omega_{X/Y}.$$

Cotangent complexes are stable under base change. For the cartesian diagram

$$\begin{array}{ccc} X' & \xrightarrow{u} & X \\ \downarrow & & \downarrow \\ Y' & \longrightarrow & Y, \end{array}$$

we have $Lu^*\mathbb{L}^\bullet_{X/Y} = \mathbb{L}^\bullet_{X'/Y'}$.

How do we see the cotangent complex? In fact, for the deformation-obstruction theory in the next subsection, we don't need all the information for the cotangent complex; rather, we just need a truncated part: $\mathbb{L}^{\bullet \geq -1}$. The Lichtenbaum-Schlessinger complex provides an explicit description for $\mathbb{L}^{\bullet \geq -1}$.

Let's describe it in the local setup (see [**H**]). Let $A \to B$ be a homomorphism of rings. Choose a polynomial ring $R = A[x]$, where $x = \{x_i\}$ is a set of free variables, such that B is a quotient of R as an A-algebra. Let I be the kernel of $R \to B$, let F be a free R-module with a surjection $j \colon F \to I$, and let Q be the kernel of the map j:

$$0 \to I \to R \to B \to 0, \quad 0 \to Q \to F \xrightarrow{j} I \to 0.$$

Let F_0 be the submodule of F generated by $j(a)b - j(b)a$ for $a, b \in F$. Since $j(F_0) = 0$, we have $F_0 \subset Q$. Define $L_0 = \Omega_{R/A} \otimes_R B$, let $L_1 = F \otimes_R B = F/IF$, and let $L_2 = Q/F_0$. Let $d^{-2} \colon L_2 \to L_1$ be the map induced by the map $Q \to F$, and let $d^{-1} \colon L_1 \to L_0$ be the composition of the map $L_1 = F/IF \to I/I^2$ induced by j and the differential $\delta \colon I/I^2 \to \Omega_{R/A} \otimes_R B = L_0$. Then $d^{-1} \circ d^{-2} = 0$, so we obtain a complex

$$L^\bullet \colon L_2 \to L_1 \to L_0.$$

One can use the cohomological index by letting $L^{-i} = L_i$. We used L_i to follow the notations in [**H**].

We can verify that all modules L_i are B-modules, and that all maps are homomorphisms of B-modules.

In the construction of $L^\bullet$, the different choices of R, F, and the map j give different $L^\bullet$. However, for any B-module M, for all i, the groups $h^i\big(\mathrm{Hom}_B(L^\bullet, M)\big)$, which control deformation and obstruction, are independent of the choices of R, F, and j (see §3 in [**H**]).

If we let $X = \mathrm{Spec}B$, $Y = \mathrm{Spec}A$, and the morphism $X \to Y$ be induced from the homomorphism $A \to B$, then $\mathbb{L}^{\bullet \geq -2}_{X/Y}$ is isomorphic to $L^\bullet$.

EXAMPLE 3.1. Let $A \to B$ be a surjection. We can take $R = A$ and hence $L_0 = 0$. Tensoring the exact sequence

$$0 \to Q \to F \to I \to 0$$

with B, we obtain an exact sequence

$$Q \otimes_A B \to F \otimes_A B \to I/I^2 \to 0.$$

Since the map $Q \otimes_A B \to F \otimes_A B$ can be factored through a surjection $Q \otimes_A B \to Q/F_0$, we obtain an exact sequence

$$L_2 \xrightarrow{d^{-2}} L_1 \to I/I^2 \to 0.$$

Thus $\mathrm{Coker}(d^{-2}) = I/I^2$.

For a closed immersion of schemes $Z \hookrightarrow X$, the above discussion is its local version. Let $\mathcal{I}$ be the sheaf of ideals of Z in X. From the discussion above, we have

$$\mathbb{L}_{Z/X}^{\bullet \geq -1} = [\mathrm{Coker}(d^{-2}) \to L^0] = \mathcal{I}/\mathcal{I}^2[1].$$

□

EXAMPLE 3.2. Let k be a field. Consider a finitely generated k-algebra B and the ring homomorphism $A = k \to B$. Let $R = k[x_1, \ldots, x_n]$ such that $B = R/I$. For the homomorphism $A \to B$, we have the corresponding complex

$$L_2 \xrightarrow{d^{-2}} L_1 \to L_0,$$

where $L_0 = \Omega_R \otimes_R B$ and an exact sequence

$$Q \otimes_A B \to F \otimes_A B \to I/I^2 \to 0.$$

Thus, $\mathrm{Coker}(d^{-2}) = I/I^2$, and

$$\mathbb{L}_{B/A}^{\bullet \geq -1} = [I/I^2 \to \Omega_R \otimes_R B].$$

Thus, for an affine scheme X of finite type over k, let X be a closed subscheme of a smooth affine scheme W. Let $\mathcal{I}$ be the sheaf of ideals of X in W. Then the truncated cotangent complex is

$$\mathbb{L}_X^{\bullet \geq -1} = [\mathcal{I}/\mathcal{I}^2 \to \Omega_W|_X].$$

□

3.2. Cotangent complexes and deformation-obstruction theory. Let X be a compact smooth manifold. We know that $H^2(X, T_X)$ is the obstruction space for the deformation of complex structures on X. When $H^2(X, T_X) = 0$, the tangent space of the moduli space of complex structures is $H^1(X, T_X)$. Thus the tangent bundle T_X, or equivalently, the sheaf of Kähler differentials Ω_X, controls the deformation-obstruction of X.

When X is singular, the cotangent complex $\mathbb{L}_X^\bullet$ controls the deformation-obstruction of X (see [**Ill71**]).

Let X be a scheme of finite type over a field k. Let T and $\overline{T}$ be schemes, and let T be a closed subscheme of $\overline{T}$ with the ideal sheaf $\mathcal{J}$ such that $\mathcal{J}^2 = 0$. Consider the lifting a morphism $g\colon T \to X$ to a morphism $\overline{g}\colon \overline{T} \to X$:

$$\begin{array}{ccc} T & \xrightarrow{g} & X \\ \downarrow & \nearrow_{\overline{g}} & \\ \overline{T}. & & \end{array}$$

From the morphisms in the diagram and the computation in Example 3.1, we obtain

(1) The morphism g gives $g^*\mathbb{L}_X^\bullet \to \mathbb{L}_T^\bullet$.
(2) The morphism $T \to \overline{T}$ gives $\mathbb{L}_T^\bullet \to \mathbb{L}_{T/\overline{T}}^\bullet$.
(3) Since $\mathcal{J}^2 = 0$, we obtain $\mathbb{L}_{T/\overline{T}} \to \mathbb{L}_{T/\overline{T}}^{\bullet \geq -1} = \mathcal{J}[1]$.

Combining the three morphisms together, we get a morphism

$$g^*\mathbb{L}_X^\bullet \to \mathcal{J}[1],$$

or equivalently, we obtain

$$\omega(g) \in \operatorname{Hom}\big(g^*\mathbb{L}_X^\bullet, \mathcal{J}[1]\big) = \operatorname{Ext}^1\big(g^*\mathbb{L}_X^\bullet, \mathcal{J}\big).$$

We have the following facts, which give a treatment of the obstruction theory of X via the cotangent complex of X:

- g extends to $\overline{g}\colon \overline{T} \to X$ if and only if $\omega(g) = 0$.
- If $\omega(g) = 0$, the set of extensions forms a torsor under $\operatorname{Hom}\big(g^*\mathbb{L}_X^\bullet, \mathcal{J}\big)$.

The above facts can be understood in the local situation by Example 2.2.

3.3. Intrinsic normal cones. The intrinsic normal cone of a scheme (or a DM stack in general) was introduced by Behrend and Fantechi in [**BF**]. Let X be a scheme of finite type over a field k. The intrinsic normal cone $\mathfrak{C}_X$ has a local presentation as follows: Let U be an affine scheme of finite type over k with an étale morphism to X, let M be a smooth affine scheme of finite type over k, and let $f\colon U \to M$ be a local immersion. Let $\mathcal{I}$ be the sheaf of ideals of U in M. Consider the morphism

$$\mathbb{L}_X^\bullet|_U \to \mathbb{L}_X^{\bullet \geq -1}|_U = [\mathcal{I}/\mathcal{I}^2 \to f^*\Omega_M],$$

whose dual is

$$[f^*T_M \to (\mathcal{I}/\mathcal{I}^2)^* = N_{U\subset M}] \to \mathbb{L}_X^{\bullet\vee}|_U.$$

The intrinsic normal cone $\mathfrak{C}_X|_U$ is the quotient stack $[C_{U\subset M}/f^*T_M]$. Note that the cone $C_{U\subset M}$ is a subcone of the vector bundle $N_{U\subset M}$ and that f^*T_M acts on $N_{U\subset M}$ via addition preserving the cone $C_{U\subset M}$.

DEFINITION 3.3. The morphism $\phi\colon \mathbb{E}^\bullet \to \mathbb{L}_X^\bullet$, where $\mathbb{E}^\bullet \in D\big(\mathcal{O}_{X_{\text{ét}}}\big)$, is called a perfect obstruction theory if $h^0(\varphi)$ is an isomorphism and $h^{-1}(\phi)$ is surjective, and if $\mathbb{E}^\bullet$ satisfies the conditions, where $h^i(\mathbb{E}^\bullet) = 0$ for all $i > 0$ and $i \leq -2$, where $h^i(\mathbb{E}^\bullet)$ is coherent, and where $\mathbb{E}^i$ is étale locally quasi-isomorphic to locally free sheaves.

It has been proven (Theorem 4.5 in [**BF**]) that Definition 3.3 is equivalent to Definition 2.4.

Suppose there exists $\mathbb{F}^\bullet = [F^{-1} \to F^0]$, isomorphic to $\mathbb{E}^\bullet$ in $D\big(\mathcal{O}_{\text{ét}}\big)$, where F^{-1} and F^0 are locally free sheaves on X. Such $\mathbb{F}^\bullet$ is called a global resolution of $\mathbb{E}^\bullet$. Not every $\mathbb{E}^\bullet$ admits a global resolution. Therefore, the existence of a global resolution of $\mathbb{E}^\bullet$ is a necessary condition for the construction of virtual cycles.

ϕ induces a morphism

$$[f^*T_M \to N_{C\subset M}] \to \mathbb{L}_X^{\bullet\vee} \to \mathbb{E}^\bullet,$$

which induces a morphism

$$\mathfrak{C}_X \to h^1/h^0\big(\mathbb{E}^{\bullet\vee}\big) = [F^{-1\vee}/F^{0\vee}].$$

Consider the fiber product

$$\begin{array}{ccc} C(\mathbb{F}^\bullet) & \longrightarrow & F^{-1\vee} \\ \downarrow & & \downarrow \\ \mathfrak{C}_X & \longrightarrow & h^1/h^0(\mathbb{E}^{\bullet\vee}) = [F^{-1\vee}/F^{0\vee}]. \end{array}$$

$C(\mathbb{F}^\bullet)$ is a subcone of $F^{-1\vee}$. The virtual cycle of X with respect to the obstruction theory $\mathbb{E}^\bullet$ is defined as the cycle $s^!_{F^{-1\vee}} C(\mathbb{F}^\bullet) \in A_*(X)$.

4. Cosection localization

The properties of an obstruction theory for a scheme X lead to the properties of virtual cycles defined by the obstruction theory. Cosection localization is a method of studying virtual cycles when the obstruction sheaf (see Example below) admits a morphism to $\mathcal{O}_X$. Cosection localization theorem states that if the morphism is surjective, then the virtual cycle is zero; if the morphism is not surjective, then the virtual cycle is supported on the subset of points over which the morphism fails to be surjective. This property of virtual cycles helps to prove vanishing results for the Gromov-Witten invariants and plays a fundamental role in the p-field treatment of the Gromov-Witten invariants and Fan-Jarvis-Ruan-Witten invariants.

4.1. The basic example and key lemmas. Back to our basic example. Consider a vector bundle E over a manifold W. Let s be a section of E and let $X = s^{-1}(0)$ be the zero locus of s. From Example 2.2, we know that X has a perfect obstruction theory

$$\big[(T_W \xrightarrow{ds} E)|_X\big],$$

whose obstruction sheaf is

$$Ob_X: = \operatorname{coker}\big(ds|_X\big).$$

Suppose that the obstruction sheaf has a surjective cosection:

$$Ob_X \xrightarrow{\sigma} \mathcal{O}_X \to 0.$$

Thus, we get a surjection

$$\overline{\sigma} = \sigma \circ \mathrm{pr}\colon E|_X \to \mathcal{O}_X,$$

where $\mathrm{pr}\colon E|_X \to Ob_X$ is the natural map.

The following three key lemmas illustrate why and how cosection localization works. Thus, we will give detailed proofs of the first two lemmas.

LEMMA 4.1. *Let the setup be as above. For any germ $\varphi\colon \mathrm{Spec}(k[[\xi]]) \to W$ with $\varphi(0) \in X$ and $\mathrm{im}(\varphi) \not\subset X$, the section $\overline{\sigma} \circ s \circ \varphi \in k[[\xi]]$ satisfies $\overline{\sigma} \circ s \circ \varphi \in \xi \cdot \varphi^* I_X$.*

PROOF. We know that $\varphi^* I_X$ is the ideal generated by $s_1 \circ \varphi, \ldots, s_r \circ \varphi$, where $s = (s_1, \ldots, s_r)$ is a local description of s under a local trivialization of E near $\varphi(0)$. Let $\varphi^* I_X = (\xi^n)$. Thus, there exists a vector $v \in E|_{\varphi(0)}$ such that

$$s \circ \varphi = \xi^n \cdot v + O(\xi^{n+1}).$$

Then we obtain

$$\varphi^* ds = d(s \circ \varphi) = d(\xi^n \cdot v) + O(\xi^n) = n\xi^{n-1} \cdot v d\xi + O(\xi^n).$$

Define $\overline{\varphi}\colon \varphi^{-1}(X) = \{\xi^n = 0\} \to X$. From

$$T_W|_X \xrightarrow{ds} E|_X \xrightarrow{\text{pr}} Ob_X \to 0,$$

if we apply $\overline{\varphi}^*$, we obtain

$$\overline{\varphi}^* T_W|_X \xrightarrow{\overline{\varphi}^* ds} \overline{\varphi}^* E|_X \xrightarrow{\overline{\varphi}^* \text{pr}} \overline{\varphi}^* Ob_X \to 0.$$

Since $\text{pr} \circ ds = 0$, we have

$$\overline{\varphi}^*(\overline{\sigma}) \circ \overline{\varphi}^* ds = \overline{\varphi}^* \sigma \circ \overline{\varphi}^* \text{pr} \circ \overline{\varphi}^* ds = 0.$$

Since $\overline{\varphi}^* ds = n\xi^{n-1} \cdot v d\xi$, we have $\overline{\sigma}(v) = 0$. Hence

$$\overline{\sigma} \circ s \circ \varphi = O(\xi^{n+1}) \in \xi \cdot \varphi^* I_X.$$

□

LEMMA 4.2. *With the same setup, we have $C_{X\subset W} \subset Ker(\overline{\sigma}) \subset E$.*

PROOF. We use the graph construction of the normal cone studied in §2.2. We see that $C_{X\subset W} \subset E$ is a specialization of the section $t^{-1} \cdot s \subset E$ as $t \to 0$. Recall that

$$\Gamma\colon = \{(t^{-1}s, t) \in E \times \mathbb{C}^*\}, \text{ and } \Gamma_t = \{t^{-1}s\}.$$

Let $\overline{\Gamma}$ be the closure of Γ in $E \times \mathbb{C}$. Then $\overline{\Gamma}_0 \subset E$ is the normal cone $C_{X\subset W}$, which is of pure dimension. Let Z be an irreducible component, and let $v \in Z$ be a general element. Choose a smooth affine curve $(S, 0)$ with a point 0 on it and a map

$$\rho\colon (S, 0) \to (\overline{\Gamma}, v) \text{ such that } \rho(S/0) \subset \Gamma.$$

Define several induced maps

$$\rho_W\colon S \xrightarrow{\rho} E \times \mathbb{C} \to E \to W, \text{ with } \rho_{\mathbb{C}}\colon S \xrightarrow{\rho} E \times \mathbb{C} \to \mathbb{C}.$$

The map $\rho_{\mathbb{C}}$ dominates $\mathbb{C}$. Let ξ be a uniforminzing parameter of S at the point 0. Then

$$\rho_{\mathbb{C}}^* t = \xi^n$$

for some $n > 0$, and we have

$$t^{-1} s \circ \rho_W = \xi^{-n} s \circ \rho_W\colon S - 0 \to E.$$

Since the map approaches v as $\xi \to 0$, as we defined, we have

$$s \circ \rho_W = v\xi^n + O(\xi^{n+1}).$$

Thus, we obtain

$$\overline{\sigma} \circ s \circ \rho_W = \overline{\sigma}(v)\xi^n + O(\xi^{n+1}).$$

Suppose $Z \not\subset \text{Ker}(\overline{\sigma})$, then $v \neq 0$ and $\rho_W^* I_X = (\xi^n)$. Applying Lemma 4.1 to ρ_W, we obtain

$$\overline{\sigma} \circ s \circ \rho_W \in \xi \cdot \rho_W^* I_X = (\xi^{n+1}).$$

Thus, $\overline{\sigma}(v) = 0$. □

We also have a lemma below which we won' t prove here.

LEMMA 4.3. *Given an exact sequence*

$$0 \to F_1 \xrightarrow{\iota} F_2 \to F_3 \to 0$$

of vector bundles, let $r = rankF_3$ and $\alpha \in A_* F_1$. Then

$$s^!_{F_2}(\iota_* \alpha) = c_r(F_3) \cap s^!_{F_1}(\alpha).$$

Now we can see that if we have a cosection σ of the obstruction sheaf Ob_X, the normal cone $C_{X\subset W} \subset E|_X$ is in fact contained in a smaller set $\mathrm{Ker}(\overline{\sigma})$ by Lemma 4.2. We also have an exact sequence

$$0 \to \mathrm{Ker}(\overline{\sigma}) \to E|_X \xrightarrow{\overline{\sigma}} \mathcal{O}_X \to 0.$$

Applying Lemma 4.3, we obtain that the virtual cycle $[X]^{\mathrm{vir}}$ of X, defined as $s_E^! C_{X\subset W} \in A_*X$, is in fact equal to zero.

4.2. Localized gysin map. Now we consider the general situation where the morphism from the obstruction to $\mathcal{O}_X$ is not surjective. In fact, we study an even more general case where the obstruction sheaf has a meromorphic cosection. We explain that we can construct a localized virtual cycle contained in the degenerate locus of the cosection, where the cosection is undefined or not surjective and where its image in A_*X is the virtual cycle of X.

Let $V \to X$ be a vector bundle. Consider the meromorphic cosection

$$\sigma\colon V|_U \twoheadrightarrow \mathcal{O}_X|_U,$$

for an open subset $U \subset X$ over which σ is well-defined and surjective, i.e., for a point $p \notin U$, the map $\sigma|_p$ is either undefined or not surjective. Let

$$\begin{aligned} X(\sigma)\colon &= X - U, \\ G\colon &= \mathrm{Ker}\{\sigma|_U\colon V|_U \to \mathcal{O}_U\} \subset V|_U, \\ V(\sigma)\colon &= V|_{X(\sigma)} \cup G. \end{aligned}$$

Back to our basic example. We have a rank-r vector bundle E over a smooth manifold W with a section s. Let $X = s^{-1}(0)$ be the zero locus of s, and $V = E|_X$. We know that the normal cone $C_{X\subset W}$ is contained in V. When the obstruction sheaf Ob_X has a meromorphic cosection σ, by the previous subsection, $C_{X\subset W}$ is contained in $V(\sigma)$. We define a localized Gysin map

$$s^!_{E,\sigma}\colon A_*V(\sigma) \to A_*X(\sigma)$$

as follows.

For a class $B \in Z_d\big(V(\sigma)\big)$, if $B \subset V|_{X(\sigma)}$, we define

$$s^!_{E,\sigma}\colon = s^!_{V|_{X(\sigma)}}(B);$$

if $B \cap U \neq \emptyset$, take a proper morphism ρ from a variety M to X such that $\rho^*\sigma$ extends to a surjective map

$$\tilde{\sigma}\colon \tilde{V}\colon = \rho^*V \twoheadrightarrow \mathcal{O}_M(D),$$

where D is a Cartier divisor on M. Let

$$\rho(\sigma)\colon = \rho|_{\mathrm{Supp}D}\colon \mathrm{Supp}D \to X(\sigma),$$

$$\tilde{G}\colon = \mathrm{Ker}\tilde{\sigma} \subset \tilde{V}.$$

Let $\tilde{\rho}\colon \tilde{V} \to V$ be the natural map. There exists $\tilde{B} \subset \tilde{G}$ such that

$$\tilde{\rho}_*\big([\tilde{B}]\big) = k[B]$$

for some integer k. Motivated by Lemma 4.3, we define

$$s^!_{E,\sigma}\big([B]\big) = k^{-1}\rho(\sigma)_*\big([D]\cdot s^!_{\tilde{G}}[\tilde{B}]\big) \in A_{d-*}X(\sigma).$$

In the definition above, there are various choices, such as the variety M and the existence of $\tilde{B}$, and thus, one needs to verify that the definition of the localized Gysin map is independent of these choices. If we let

$$\iota\colon X(\sigma) \to X,$$

be the inclusion, we have

$$\iota_*(s^!_{E,\sigma}) = [X]^{\mathrm{vir}}.$$

In the most general case, we have the main theorem of Kiem-J. Li.

THEOREM 4.4. *Let X/S be a Deligne-Mumford stack endowed with a relative perfect obstruction theory. Suppose there is a surjective homomorphism $\sigma\colon Ob_X \to \mathcal{O}_U$ where U is an open subset of X. Let $X(\sigma) = X - U$ be the degeneracy locus of σ. Then X has a localized virtual cycle*

$$[X]^{\mathrm{vir}}_{\mathrm{loc}} \in A_* X(\sigma)$$

such that $\iota_*\big([X]^{\mathrm{vir}}_{\mathrm{loc}}\big) = [X]^{\mathrm{vir}}$ where ι is the inclusion map.

Here and in the next section, relative perfect obstruction theory is used. It is just the relative version of perfect obstruction theory. We won't give a precise definition. However, we will see it in a bit more detail in the next section.

5. p-fields and applications

In this section, we discuss briefly the key ideas of p-fields in the work of H. L. Chang and J. Li in [**CL**] and skip all proofs. We explain an application of p-fields theory, which is the p-fields definition of FJRW invariants. The importance of the p-fields formulation of GW invariants and FJRW invariants is discussed via the introduction to the mixed-spin-p fields theory.

5.1. p-fields for stable maps. Let Q be the quintic Calabi-Yau threefold in $\mathbb{P}^4$:

$$Q = \{w = x_1^5 + \ldots + x_5^5 = 0\}.$$

Consider the moduli stack of genus g degree d stable maps to $\mathbb{P}^4$ with p-fields:

$$\overline{\mathfrak{M}}_g(\mathbb{P}^4, d)^p\colon = \{[C, f, \rho] \mid [C, f] \in \overline{\mathfrak{M}}_g(\mathbb{P}^4, d),\ \rho \in H^0\big(\omega_C \otimes f^*\mathcal{O}_{\mathbb{P}^4}(-5)\big)\}.$$

Here $f\colon C \to \mathbb{P}^4$ is a degree-d stable map from a genus g curve C to $\mathbb{P}^4$, and ρ is called a p-field. We can also rephrase the moduli stack as

$$\overline{\mathfrak{M}}_g(\mathbb{P}^4, d)^p = \{C, \mathcal{L}, 0 \neq \varphi = (\varphi_1, \ldots, \varphi_5) \in H^0(\mathcal{L}^{\oplus 5}), \rho \in H^0(\mathcal{L}^{-5} \otimes \omega_C)\}.$$

Let $\mathcal{D}_g$ be the stack of genus g curves C with line bundles $\mathcal{L}$. Let

$$\mathcal{P}\colon = \overline{\mathfrak{M}}_g(\mathbb{P}^4, d)^p$$

for simplicity. We have a natural map $\mathcal{P} \to \mathcal{D}_g$ by forgetting sections φ and ρ. Let $\mathcal{C}$ be the universal curve over $\mathcal{P}$ with a projection $\pi_{\mathcal{P}}$ to $\mathcal{P}$, and let $\mathcal{L}_{\mathcal{P}}$ be the universal line bundle over $\mathcal{C}$. We can write a relative perfect obstruction theory for $\mathcal{P}/\mathcal{D}_g$:

$$\mathbb{L}^*_{\mathcal{P}/\mathcal{D}_g} \to R^*\pi_{\mathcal{P}*}\big(\mathcal{L}_{\mathcal{P}}^{\oplus 5} \oplus \mathcal{L}_{\mathcal{P}}^{-5} \otimes \omega_{\mathcal{C}/\mathcal{P}}\big). \tag{5.1}$$

The relative obstruction sheaf is

$$Ob_{\mathcal{P}/\mathcal{D}_g} = R^1\pi_{\mathcal{P}*}\big(\mathcal{L}_{\mathcal{P}}^{\oplus 5} \oplus \mathcal{L}_{\mathcal{P}}^{-5} \otimes \omega_{\mathcal{C}/\mathcal{P}}\big). \tag{5.2}$$

The most important observation is that $Ob_{\mathcal{P}/\mathcal{D}_g}$ has a cosection σ. It is defined fiberwise as follows: Given a point $\xi = (C, \mathcal{L}, \varphi, \rho) \in \mathcal{P}$, the fiber of $Ob_{\mathcal{P}/\mathcal{D}_g}$ over ξ is

$$Ob_{\mathcal{P}/\mathcal{D}_g}|_\xi = H^1\big(C, \mathcal{L}^{\oplus 5}\big) \oplus H^1\big(C, \mathcal{L}^{-5} \otimes \omega_C\big).$$

Taking a vector v in the fiber,

$$v = \big((\dot\varphi_1, \ldots, \dot\varphi_5), \dot\rho\big) \in Ob_{\mathcal{P}/\mathcal{D}_g}|_\xi.$$

Define

$$\sigma|_\xi\big((\dot\varphi_1, \ldots, \dot\varphi_5), \dot\rho\big) := \dot\rho \sum_{i=1}^{5} \varphi_i^5 + 5\rho \sum_{i=1}^{5} \varphi_i^4 \dot\varphi_i. \tag{5.3}$$

Why does the cosection take the form above?

Take a function

$$pw = p(x_1^5 + \ldots + x_5^5),$$

with each variable considered as a function of the time variable t. If we take the derivative of pw with respect to t and use the physics notation, we get the form (5.3) of the cosection.

The degeneracy locus of σ, which is the set of points ξ over which σ is not surjective, is determined by

$$\dot\rho \sum_{i=1}^{5} \varphi_i^5 + 5\rho \sum_{i=1}^{5} \varphi_i^4 \dot\varphi_i = 0 \text{ for all } (\dot\varphi_1, \ldots, \dot\varphi_5).$$

We obtain the description of the degeneracy locus as

$$\{\xi \in \mathcal{P} \,|\, \sum_{i=1}^{5} \varphi_i^5 = 0, \rho = 0\} = \overline{\mathfrak{M}}_g(Q, d).$$

Thus, the degeneracy locus is just the moduli stack $\overline{\mathfrak{M}}_g(Q, d)$ of stable maps to the quintic Q, which is proper. Note that the original moduli stack of p-fields $\mathcal{P}$ is not proper due to the presence of sections φ and p.

The relative obstruction sheaf $Ob_{\mathcal{P}/\mathcal{D}_g}$ admits a map to the absolute obstruction sheaf $Ob_{\mathcal{P}}$, and the cosection σ can be lifted to a cosection of $Ob_{\mathcal{P}}$. Using the technique in §4, we can define a localized virtual cycle of $\mathcal{P}$ as

$$[\overline{\mathfrak{M}}_g(\mathbb{P}^4, d)^p]^{\mathrm{vir}}_{\mathrm{loc}} \in A_* \overline{\mathfrak{M}}_g(Q, d).$$

If we use this virtual cycle to define an invariant of the quintic Q, denoted by $N_g(d)^p$, and let $N_g(d)$ be the Gromov-Witten invariants of the quintic, then we have the following theorem:

THEOREM 5.1 (Chang-J. Li). *The invariants defined by the virtual cycle $[\overline{\mathfrak{M}}_g(\mathbb{P}^4, d)^p]^{\mathrm{vir}}_{\mathrm{loc}}$ equal the Gromov-Witten invariants of the quintic Q up to a sign, that is,*

$$N_g(d)^p = (-1)^{5d+1-g} N_g(d).$$

Let's outline the road map of the proof of the theorem.

(1) Let $\mathfrak{M}_g$ be the stack of pre-stable curves. Recall that the Gromov-Witten invariants are defined using the virtual cycle from the perfect obstruction theory

$$\big(R\pi_* f^* T_Q\big)^* \to \mathbb{L}^{\bullet \geq -1}_{\overline{\mathfrak{M}}_g(Q,d)/\mathfrak{M}_g},$$

where π is the map from the universal curve $\mathcal{C}$ to $\overline{\mathfrak{M}}_g(Q,d)$ and $f\colon \mathcal{C} \to Q$ is the universal stable map. Firstly, we describe stable maps in $\overline{\mathfrak{M}}_g(Q,d)$ via sections of line bundles. There exists a relative perfect obstruction theory for $\overline{\mathfrak{M}}_g(Q,d) \to \mathcal{D}_g$. Then we compare the relative obstruction theory of $\overline{\mathfrak{M}}_g(Q,d) \to \mathcal{D}_g$ with that of $\overline{\mathfrak{M}}_g(Q,d) \to \mathfrak{M}_g$ using the new description. Then we prove that the virtual cycles defined by two relative obstruction theories are equal in $A_*\overline{\mathfrak{M}}_g(Q,d)$.

(2) Construct a degeneration. Let M^0 be the space of the deformation of $\mathbb{P}^4$ to the normal bundle N of $Q \subset \mathbb{P}^4$. Then M^0 admits a map τ to $\mathbb{C}$. The fiber of τ over a point $t \neq 0$ is isomorphic to $\mathbb{P}^4$, and the fiber of τ over the point 0 is isomorphic to the normal bundle $N = N_{Q\subset\mathbb{P}^4}$. Consider the family version of stable maps to M^0 with p-fields $\overline{\mathfrak{M}}_g(M^0,d)^p$. Study the relative obstruction theory of $\overline{\mathfrak{M}}_g(M^0,d)^p/\mathcal{D}_g \times \mathbb{C}$.

(3) Consider a family version of cosection similar to the case we studied above. The cosection localization of Keim-Li in §4 constructs a cosection localized virtual cycle $[\overline{\mathfrak{M}}_g(M^0,d)^p]^{\mathrm{vir}}_{\mathrm{loc}}$. We can check that, for $c \in \mathbb{C}$, we have

$$[\overline{\mathfrak{M}}_g(M^0,d)^p]^{\mathrm{vir}}_{\mathrm{loc}}|_c = [\overline{\mathfrak{M}}_g(\mathbb{P}^4,d)^p]^{\mathrm{vir}}_{\mathrm{loc}} \text{ when } c \neq 0,$$

$$[\overline{\mathfrak{M}}_g(M^0,d)^p]^{\mathrm{vir}}_{\mathrm{loc}}|_0 = [\overline{\mathfrak{M}}_g(N,d)^p]^{\mathrm{vir}}_{\mathrm{loc}}.$$

(4) Compare the obstruction theory of $\overline{\mathfrak{M}}_g(N,d)^p$ with that of $\overline{\mathfrak{M}}_g(Q,d)$ and show that the invariants defined by virtual cycles from the respective obstruction theories are equal up to a sign.

The p-fields theory developed by H.L. Chang and J. Li in [C-L] is a mathematical treatment and a generalization of Guffin and Sharpe's work in [**GS**]. Guffin and Sharpe constructed the genus zero Landau-Ginzburg (LG) model for $(K_{\mathbb{P}^1}, w)$ and showed that the invariants defined in this model equal the genus zero Gromov-Witten invariants of the quintic CY threefold define by $w = 0$. The extra p-fields introduced in the new setup are sections in $H^0(C, \mathcal{L}^{\otimes -5} \otimes \omega_C)$, which is dual to $H^1(C, \mathcal{L}^{\otimes 5})$. Thus, the obstruction space $H^1(C, \mathcal{L}^{\otimes 5})$ in the original setup of stable maps is a part of the moduli data. As we know, genus zero Gromov-Witten invariants of the quintic CY threefold were calculated successfully due to the "hyperplane property," i.e., the calculation is reduced to calculating the Euler class of certain bundles. This can be done in genus zero due to the fact that the obstruction $H^1(C, \mathcal{L}^{\otimes 5}) = 0$. For genus one and higher, it is easy to see that the obstruction $H^1(C, \mathcal{L}^{\otimes 5})$ may not be zero over some components of the moduli stack of stable maps. The p-fields method provides a way to overcome the failure of the hyperplane property. We will explain this issue when we introduce mixed-spin-p fields in the last subsection. However, the new moduli stack is not proper and thus cannot be used to define invariants directly. Fortunately, Kiem-Li's cosection localization method provides a perfect tool for handling the situation. The potential function pw happens to provide the cosection of the obstruction sheaf of the moduli stack; the degeneracy locus of the cosection is proper, and hence, we can define invariants from the localized virtual cycle of the obstruction theory.

5.2. p-fields formulation of FJRW invariants. The p-fields treatment of Gromov-Witten invariants can be considered as stable maps to $K_{\mathbb{P}^4}$. The canonical line bundle $K_{\mathbb{P}^4}$ is a geometric invariant quotient in the quotient stack

$$[\mathbb{C}^5 \times \mathbb{C}/\mathbb{C}^*], \quad t\cdot(x_1,\ldots,x_5,p) = (tx_1,\ldots,tx_5,t^{-5}z) \text{ for } t \in \mathbb{C}^*.$$

A map from a curve C to $[\mathbb{C}^5 \times \mathbb{C}/\mathbb{C}^*]$ can be described by a line bundle $\mathcal{L}$ over C and sections $(\varphi_1, \dots, \varphi_5) \in H^0(C, \mathcal{L})$ and $\rho \in H^0(C, \mathcal{L}^{-5})$. Since the stability condition of geometric invariant theory (GIT) for $K_{\mathbb{P}^4}$ is given by $(x_1, \dots, x_5) \neq 0$, or

$$K_{\mathbb{P}^4} = \big(\mathbb{C}^5 \times \mathbb{C} - (0, \dots, 0) \times \mathbb{C}\big)/\mathbb{C}^*,$$

the map from C to $K_{\mathbb{P}^4}$ satisfies $(\varphi_1, \dots, \varphi_5) \neq 0$. We modify the last section ρ by requiring $\rho \in H^0(C, \mathcal{L}^{-5} \otimes \omega_C)$. One of the reasons to make such a modification is to have the obstruction sheaf with a cosection. Thus, this explains the setup of the p-field treatment for Gromov-Witten theory.

The quotient stack $[\mathbb{C}^5 \times \mathbb{C}/\mathbb{C}^*]$ has another GIT quotient

$$\big(\mathbb{C}^5 \times \mathbb{C} - \mathbb{C}^5 \times \{0\}\big)/\mathbb{C}^*,$$

which is the orbifold $[\mathbb{C}^5/\mathbb{Z}_5]$.

Thus, a map from a curve C to the GIT quotient can be described by

$$\big(C, \mathcal{L}, (\varphi_1, \dots, \varphi_5) \in H^0(C, \mathcal{L}^{\oplus 5}), \rho \in H^0(C, \mathcal{L}^{-5})\big).$$

Again, we modify the ρ section by requiring $\rho \in H^0(C, \mathcal{L}^{-5} \otimes \omega_C)$. Therefore we obtain

$$\big(C, \mathcal{L}, (\varphi_1, \dots, \varphi_5) \in H^0(C, \mathcal{L}^{\oplus 5}), \rho \in H^0(C, \mathcal{L}^{-5} \otimes \omega_C)\big).$$

The stability condition of the GIT implies that ρ is nowhere zero, in other words, $\mathcal{L}^{\otimes 5} \cong \omega_C$. Hence we get a 5-spin curve $(C, \mathcal{L})$, and, as a consequence, we require C to be an orbifold curve (see [**AJ**]). We obtain a moduli stack $\overline{\mathfrak{M}}_g^{1/5,5p}$ of 5-spin curves of genus g with five p-fields $\varphi_1, \dots, \varphi_5$. This stack is not proper. Let $\overline{\mathfrak{M}}_g^{1/5}$ be the moduli stack of 5-spin curves of genus g. This moduli stack is proper. There is a natural forgetful morphism from $\overline{\mathfrak{M}}_g^{1/5,5p}$ to $\overline{\mathfrak{M}}_g^{1/5}$, which is just to forget the sections. There is a perfect relative obstruction theory for $\overline{\mathfrak{M}}_g^{1/5,5p}/\overline{\mathfrak{M}}_g^{1/5}$. In fact, let $\mathcal{C}$ be the universal curve with $\pi_{\mathcal{C}}$ as the projection from $\mathcal{C}$ to $\overline{\mathfrak{M}}_g^{1/5,5p}$, and let $\mathcal{L}_{\mathcal{C}}$ be the universal line bundle. Then, the relative obstruction theory is given by

$$\mathbb{L}^\bullet_{\overline{\mathfrak{M}}_g^{1/5,5p}/\overline{\mathfrak{M}}_g^{1/5}} \to R^*\pi_{\mathcal{C}*}\big(\mathcal{L}_{\mathcal{C}}^{\oplus 5} \oplus \mathcal{L}_{\mathcal{C}}^{-5} \otimes \omega_{\mathcal{C}/\overline{\mathfrak{M}}_g^{1/5,5p}}\big).$$

The obstruction sheaf is

$$Ob_{\overline{\mathfrak{M}}_g^{1/5,5p}/\overline{\mathfrak{M}}} = R^1\pi_{\mathcal{C}*}\big(\mathcal{L}_{\mathcal{C}}^{\oplus 5} \oplus \mathcal{L}_{\mathcal{C}}^{-5} \otimes \omega_{\mathcal{C}/\overline{\mathfrak{M}}_g^{1/5,5p}}\big).$$

Exactly the same as the setup in §5.1, we obtain a cosection of the obstruction whose degeneracy locus, in this case, is the moduli stack $\overline{\mathfrak{M}}_g^{1/5}$ of 5-spin curves, which is proper. Hence we can define invariants. It is proved in [**CLL15**] that these invariants are exactly the FJRW invariants defined by Fan-Jarvis-Ruan in [**FJR1**] and by Witten in [**W**].

5.3. Mixed-spin-p fields. Since $K_{\mathbb{P}^4}$ and $\mathbb{C}^5/\mathbb{Z}_5$ are two GIT quotients of the stack $[\mathbb{C}^5 \times \mathbb{C}/\mathbb{C}^*]$, one naturally should consider the variations of the GIT quotients, which gives rise to wall crossings. However, the Artin stacky point $[0/\mathbb{C}^*]$ poses a great difficulty. One can read [**FJR2**] and subsequent papers related to GLSM theory. Another way to work around this difficulty is to use the master space technique. In our case, we introduce a GIT quotient

$$(\mathbb{C}^5 \times \mathbb{C} \times \mathbb{P}^1 - S)/\mathbb{C}^*, \quad S\colon = \{(x_i = 0 = u_1) \cup (p = 0 = u_2)\},$$

where $\mathbb{C}^*$ action is given by

$$t \cdot (x_1, \ldots, x_5, p, [u_1, u_2]) \colon = (tx_1, \ldots, tx_5, t^{-5}p, [tu_1, u_2]).$$

We consider the maps from a curve C to the GIT quotient, again with the section ρ modified by ω_C. Specifically, we study tuples

$$\xi = (C, \mathcal{L}, \mathcal{N}, \varphi_1, \ldots, \varphi_5, \rho, [\nu_1, \nu_2]),$$

where C is an orbifold curve, where $\mathcal{L}$ and $\mathcal{N}$ are line bundles on C, where $\varphi_1, \ldots, \varphi_5$ are sections of $\mathcal{L}$, where ρ is a section of $\mathcal{L}^{-5} \otimes \omega_C$, where ν_1 is a section of $\mathcal{L} \otimes \mathcal{N}$, and where ν_2 is a section of $\mathcal{N}$. ξ is called a mixed-spin-p field (MSP for short).

The stability condition of the GIT quotient gives rise to the following conditions:

(1) $(\varphi_1, \ldots, \varphi_5, \nu_1)$ is nowhere zero;
(2) (ρ, ν_2) is nowhere zero;
(3) (ν_1, ν_2) is nowhere zero.

These stability conditions lead us to the following observation:

If $\nu_1 = 0$, then ν_2 is nowhere zero and hence $\mathcal{N} \cong \mathcal{O}_C$ and ν_2 is a constant, and $(\varphi_1, \ldots, \varphi_5)$ is nowhere zero. We obtain exactly the moduli stack of stable maps to $\mathbb{P}^4$ with p-fields, which is discussed in §5.1 and is responsible for the GW invariants. If $\nu_2 = 0$, then ν_1 is nowhere zero and hence $\mathcal{N} \cong \mathcal{L}^*$; the section ρ is nowhere zero and hence we obtain a 5-spin curve $\mathcal{L}^{\otimes 5} \cong \omega_C$. This is exactly the moduli stack discussed in §5.2 and is responsible for the FJRW invariants. Therefore, the moduli stack of MSP fields provides a geometric platform to contain the GW sector and LG sector.

This new moduli stack has a perfect obstruction theory whose obstruction sheaf admits a cosection. The setup is the same as in §5.1 with extra line bundles $\mathcal{L} \otimes \mathcal{N}$ and $\mathcal{N}$ added due to the appearance of sections ν_1 and ν_2. In fact, the cosection takes the same form, unaffected by the appearance of the line bundles $\mathcal{L} \otimes \mathcal{N}$ and $\mathcal{N}$. The degeneracy locus of the cosection is proper (see [**CLLL**]), even though this is not immediately seen to be the case. The degeneracy locus provides a platform to understand the GW and FJRW invariants. Guo and Ross in [**GR**] used the MSP setup to calculate genus one FJRW invariants of the quintic polynomial. Chang, Guo, Li, and Zhou in [**CGLZ**] used MSP to calculate the genus one GW invariants of the quintic CY threefold. A slight generalization of the MSP was constructed in [**CGLZ**] and used to prove Yamaguchi-Yau finite generation conjecture, holomorphic anomaly equation conjecture, and BCOV Feynman rules for any genus GW invariants of the quintic CY threefold in [**CGL1**, **CGL2**].

Now we can explain how the p-field formulation of GW invariants can overcome the failure of hyperplane property for higher genus computations of GW invariants. The failure of hyperplane property prevents us from using the torus localization method to calculate GW invariants. Due to the fact that the p-field formulations of GW invariants and FJRW invariants are almost identical, we are able to construct the MSP field theory, which is a bigger platform to include the GW invariants and FJRW invariants. The moduli stack of MSP fields admits torus actions, and thus, we can do torus localization calculations (see [**CLLL4**]). As shown in [**CGLZ**, **CGL1**, **CGL2**], it is effective in calculating higher genus GW invariants.

It is fair to say that the p-fields method combined with the cosection localization technique provides a new way to study moduli problems from the Gromov-Witten theory and the Landau-Ginzburg model. It has been shown to be an effective way to understand the intrinsic structures of the GW invariants of the quintic CY threefold. For a survey of MSP method, one can see the survey paper [**CLLL17a**]. To understand the physical background of the MSP theory, one can see the survey paper [**CLLL17b**].

Acknowledgments

These are the lecture notes for the May 2018 Snowbird Workshop, "Crossing the Walls in Enumerative Geometry," held at Snowbird, Utah. I would like to thank Tyler Jarvis and Nathan Priddis for the opportunity to participate in the two-week workshop. I would like to thank J. Li and H. L. Chang, from whom I learned a lot about deformation-obstruction theory and virtual cycles. I would also like to thank Y. B. Ruan, who shared his ideas on CY-LG correspondence. Finally, I would like to thank the referee and an English editor who provided valuable comments, which improved the exposition of the notes.

References

[AJ] Dan Abramovich and Tyler J. Jarvis, *Moduli of twisted spin curves*, Proc. Amer. Math. Soc. **131** (2003), no. 3, 685–699, DOI 10.1090/S0002-9939-02-06562-0. MR1937405

[B] K. Behrend, *Gromov-Witten invariants in algebraic geometry*, Invent. Math. **127** (1997), no. 3, 601–617, DOI 10.1007/s002220050132. MR1431140

[BF] K. Behrend and B. Fantechi, *The intrinsic normal cone*, Invent. Math. **128** (1997), no. 1, 45–88, DOI 10.1007/s002220050136. MR1437495

[BCOV] M. Bershadsky, S. Cecotti, H. Ooguri, and C. Vafa, *Kodaira-Spencer theory of gravity and exact results for quantum string amplitudes*, Comm. Math. Phys. **165** (1994), no. 2, 311–427. MR1301851

[COGP] Philip Candelas, Xenia C. de la Ossa, Paul S. Green, and Linda Parkes, *A pair of Calabi-Yau manifolds as an exactly soluble superconformal theory*, Nuclear Phys. B **359** (1991), no. 1, 21–74, DOI 10.1016/0550-3213(91)90292-6. MR1115626

[CGL1] H.-L. Chang, S. Guo, and J. Li, *Polynomial structure of Gromov-Witten potential of quintic 3-folds via NMSP*, `arXiv:1809.11058`.

[CGL2] H.-L. Chang, S. Guo, and J. Li, *BCOV's Feynman rule of quintic 3-folds via NMSP*, `arXiv:1810.00394`.

[CGLZ] H.-L. Chang, S. Guo, W.-P. Li, and J. Zhou, *Genus one GW invariants of quintic threefolds via MSP localization*, IMRN.

[CL] Huai-Liang Chang and Jun Li, *Gromov-Witten invariants of stable maps with fields*, Int. Math. Res. Not. IMRN **18** (2012), 4163–4217, DOI 10.1093/imrn/rnr186. MR2975379

[CLL15] Huai-Liang Chang, Jun Li, and Wei-Ping Li, *Witten's top Chern class via cosection localization*, Invent. Math. **200** (2015), no. 3, 1015–1063, DOI 10.1007/s00222-014-0549-5. MR3348143

[CLLL] Huai-Liang Chang, Jun Li, Wei-Ping Li, and Chiu-Chu Melissa Liu, *Mixed-spin-P fields of Fermat polynomials*, Camb. J. Math. **7** (2019), no. 3, 319–364, DOI 10.4310/CJM.2019.v7.n3.a3. MR4010064

[CLLL17a] Huai-Liang Chang, Jun Li, Wei-Ping Li, and Chiu-Chu Melissa Liu, *A survey on mixed spin P-fields*, Chin. Ann. Math. Ser. B **38** (2017), no. 4, 869–882, DOI 10.1007/s11401-017-1101-x. MR3673172

[CLLL17b] Huai-Liang Chang, Jun Li, Wei-Ping Li, and Chiu-Chu Melissa Liu, *On the mathematics and physics of mixed spin P-fields*, String-Math 2015, Proc. Sympos. Pure Math., vol. 96, Amer. Math. Soc., Providence, RI, 2017, pp. 47–73. MR3751813

[CLLL4] H-L. Chang, J. Li, W.-P. Li, and C.-C. Liu, *An effective theory of GW and FJRW invariants of quintics Calabi-Yau manifolds*, preprint, `arXiv:1603.06184`.

[dJ] A. J. de Jong et. al., *The stacks project.*

[FJR1] Huijun Fan, Tyler Jarvis, and Yongbin Ruan, *The Witten equation, mirror symmetry, and quantum singularity theory*, Ann. of Math. (2) **178** (2013), no. 1, 1–106, DOI 10.4007/annals.2013.178.1.1. MR3043578

[FJR2] Huijun Fan, Tyler J. Jarvis, and Yongbin Ruan, *A mathematical theory of the gauged linear sigma model*, Geom. Topol. **22** (2018), no. 1, 235–303, DOI 10.2140/gt.2018.22.235. MR3720344

[F] William Fulton, *Intersection theory*, Ergebnisse der Mathematik und ihrer Grenzgebiete (3) [Results in Mathematics and Related Areas (3)], vol. 2, Springer-Verlag, Berlin, 1984. MR732620

[G] Alexander B. Givental, *Equivariant Gromov-Witten invariants*, Internat. Math. Res. Notices **13** (1996), 613–663, DOI 10.1155/S1073792896000414. MR1408320

[GP99] T. Graber and R. Pandharipande, *Localization of virtual classes*, Invent. Math. **135** (1999), no. 2, 487–518, DOI 10.1007/s002220050293. MR1666787

[GR] S. Guo and D. Ross, *The Genus-One Global Mirror Theorem for the Quintic Threefold.*

[GS] Josh Guffin and Eric Sharpe, *A-twisted Landau-Ginzburg models*, J. Geom. Phys. **59** (2009), no. 12, 1547–1580, DOI 10.1016/j.geomphys.2009.07.014. MR2583793

[H] Robin Hartshorne, *Deformation theory*, Graduate Texts in Mathematics, vol. 257, Springer, New York, 2010. MR2583634

[HKQ09] M.-x. Huang, A. Klemm, and S. Quackenbush, *Topological string theory on compact Calabi-Yau: modularity and boundary conditions*, Homological mirror symmetry, Lecture Notes in Phys., vol. 757, Springer, Berlin, 2009, pp. 45–102. MR2596635

[Ill71] Luc Illusie, *Complexe cotangent et déformations. I* (French), Lecture Notes in Mathematics, Vol. 239, Springer-Verlag, Berlin-New York, 1971. MR0491680

[KL13] Young-Hoon Kiem and Jun Li, *Localizing virtual cycles by cosections*, J. Amer. Math. Soc. **26** (2013), no. 4, 1025–1050, DOI 10.1090/S0894-0347-2013-00768-7. MR3073883

[K] Maxim Kontsevich, *Enumeration of rational curves via torus actions*, The moduli space of curves (Texel Island, 1994), Progr. Math., vol. 129, Birkhäuser Boston, Boston, MA, 1995, pp. 335–368, DOI 10.1007/978-1-4612-4264-2_12. MR1363062

[LT1] Jun Li and Gang Tian, *Virtual moduli cycles and Gromov-Witten invariants of algebraic varieties*, J. Amer. Math. Soc. **11** (1998), no. 1, 119–174, DOI 10.1090/S0894-0347-98-00250-1. MR1467172

[LT2] Jun Li and Gang Tian, *A brief tour of GW invariants*, Surveys in differential geometry: differential geometry inspired by string theory, Surv. Differ. Geom., vol. 5, Int. Press, Boston, MA, 1999, pp. 543–569, DOI 10.4310/SDG.1999.v5.n1.a9. MR1772279

[LZ] Jun Li and Aleksey Zinger, *On the genus-one Gromov-Witten invariants of complete intersections*, J. Differential Geom. **82** (2009), no. 3, 641–690. MR2534991

[LLY] Bong H. Lian, Kefeng Liu, and Shing-Tung Yau, *Mirror principle. I*, Asian J. Math. **1** (1997), no. 4, 729–763, DOI 10.4310/AJM.1997.v1.n4.a5. MR1621573

[VZ] Ravi Vakil and Aleksey Zinger, *A desingularization of the main component of the moduli space of genus-one stable maps into* $\mathbb{P}^n$, Geom. Topol. **12** (2008), no. 1, 1–95, DOI 10.2140/gt.2008.12.1. MR2377245

[W] Edward Witten, *Phases of $N = 2$ theories in two dimensions*, Nuclear Phys. B **403** (1993), no. 1-2, 159–222, DOI 10.1016/0550-3213(93)90033-L. MR1232617

[YY] Satoshi Yamaguchi and Shing-Tung Yau, *Topological string partition functions as polynomials*, J. High Energy Phys. **7** (2004), 047, 20, DOI 10.1088/1126-6708/2004/07/047. MR2095047

[Z1] Aleksey Zinger, *Reduced genus-one Gromov-Witten invariants*, J. Differential Geom. **83** (2009), no. 2, 407–460. MR2577474

[Z2] Aleksey Zinger, *The reduced genus 1 Gromov-Witten invariants of Calabi-Yau hypersurfaces*, J. Amer. Math. Soc. **22** (2009), no. 3, 691–737, DOI 10.1090/S0894-0347-08-00625-5. MR2505298

Department of Mathematics, Hong Kong University of Science and Technology, Hong Kong

Email address: `mawpli@ust.hk`

Contemporary Mathematics
Volume **763**, 2021
https://doi.org/10.1090/conm/763/15326

Virtual classes for hypersurfaces via two-periodic complexes

Mark Shoemaker

ABSTRACT. These expository notes are based on a series of lectures given at the May 2018 Snowbird workshop, Crossing the Walls in Enumerative Geometry. We give an introductory treatment of the notion of a virtual fundamental class in algebraic geometry, and describe a new construction of the virtual fundamental class for Gromov–Witten theory of a hypersurface. The results presented here are based on joint work with I. Ciocan-Fontanine, D. Favero, J. Guéré, and B. Kim.

CONTENTS

Introduction

0.1. Setup. Fix X to be a smooth projective variety. The moduli space $\overline{\mathscr{M}}_{g,n}(X,d)$ parametrizes maps

$$f\colon C \to X,$$

where C is a smooth or nodal curve of genus g with n marked points, f is a regular map of degree d, and the automorphisms of C which preserve f are required to be finite. One considers integrals over these moduli spaces, with the hopes of obtaining useful invariants of the variety X.

A special class of such integrals yield what are known as *Gromov–Witten invariants* of X. In particularly simple cases these invariants give counts of curves in X, although in general such a direct enumerative interpretation is not possible. As predicted by physicists, these integrals should satisfy a number of remarkable properties. For instance, they should be invariant under deformations of X, and

2010 *Mathematics Subject Classification.* Primary 14N35; Secondary 53D45, 14E16.
This work was partially supported by NSF grant DMS-1708104.

should be compatible with the following natural maps between the moduli spaces curves

$$\begin{aligned}
\text{for}\colon \overline{\mathscr{M}}_{g,n+1} &\to \overline{\mathscr{M}}_{g,n},\\
\text{gl}_1\colon \overline{\mathscr{M}}_{g,n+2} &\to \overline{\mathscr{M}}_{g+1,n},\\
\text{gl}_2\colon \overline{\mathscr{M}}_{g_1,n_1+1} \times \overline{\mathscr{M}}_{g_2,n_2+1} &\to \overline{\mathscr{M}}_{g_1+g_2,n_1+n_2}.
\end{aligned} \tag{0.1}$$

in a precise sense.

In attempting to rigorously define Gromov–Witten invariants, a serious technical challenge arises. In contrast to $\overline{\mathscr{M}}_{g,n}$, the moduli space of stable maps $\overline{\mathscr{M}}_{g,n}(X,d)$ is generally not smooth and often contains many irreducible components of different dimension. The fundamental class of $\overline{\mathscr{M}}_{g,n}(X,d)$ is not well-defined; it is therefore unclear how one should "integrate" over $\overline{\mathscr{M}}_{g,n}(X,d)$.

0.2. The virtual fundamental class. The solution is the notion of a *virtual fundamental class*, a choice of element in the Chow group of $\overline{\mathscr{M}}_{g,n}(X,d)$ which behaves in many ways like a fundamental class. To be more precise, one hopes to construct an element of $A_*\left(\overline{\mathscr{M}}_{g,n}(X,d)\right)$ which satisfies a number of properties analogous to those of a fundamental class, together with additional properties reflecting the goals of the previous section. In particular one requires:

(1) the virtual fundamental class is supported entirely in a single degree of $A_*\left(\overline{\mathscr{M}}_{g,n}(X,d)\right)$, the so-called *expected dimension* (see Section 2.3);
(2) the virtual fundamental class is compatible with the various gluing and forgetful maps of (0.1) (see [**2**, Section 2] for precise statements);
(3) Gromov–Witten invariants of X, defined by integrating certain classes against the virtual fundamental class, are invariant under deformations of X.

It is not clear a-priori that such a class should exist, and in general there is no guarantee that a choice of such a class is unique. The construction of the virtual fundamental class has been a fundamental challenge in Gromov–Witten theory (as well as other curve counting theories). Unfortunately any construction of the virtual fundamental class is necessarily technical, especially in its full generality. As such, the virtual fundamental class is one of the more formidable hurdles for graduate students learning the theory.

Constructing the virtual fundamental class has now been successfully carried out in a number of different ways. In Gromov–Witten theory this has been done by Behrend–Fantechi in [**3**], Fukaya–Ono in [**17**], Li–Tian in [**25**], Ruan in [**31**], and others.

0.3. Summary. These expository notes have two complementary goals. The first is to give an introduction to the concept of a virtual fundamental class. This is done in Section 1, where we explain the connection to more classical intersection theory. In Section 2 we detail the case of $\overline{\mathscr{M}}_{g,n}(\mathbb{P}^r,d)$ where the construction can be simplified.

The second goal of these notes is to present a new method, developed by Ciocan-Fontanine, Favero, Guéré, Kim, and the author in [**11**], of constructing a virtual class for the Gromov–Witten theory of a hypersurface using two-periodic complexes of vector bundles. This is done in Sections 3 and 4.

0.4. Audience. These notes are intended for an audience who has some familiarity with Gromov–Witten theory, but has perhaps not yet learned the technical details of the virtual fundamental class. While this may seem to be a rather specific collection of readers, in our experience it includes a large proportion of graduate students learning Gromov–Witten theory. We assume readers have a familiarity with algebraic geometry at the level of at least the first three chapters of Hartshorne [**21**]. A sufficient pre-requisite for the Gromov–Witten theory material is, e.g., the excellent introductory text by Kock and Vainsencher [**23**].

On the other hand, while Sections 1 and 2 are certainly well-known to many, we hope that the later sections will be of interest to experts in the field, as they provide a new construction of the virtual fundamental class on the space of maps to a hypersurface.

0.5. A note on generality. The method described in these notes is a particular case of a more general construction developed in [**11**]. The setting in which these techniques apply is that of a *gauged linear sigma model* (GLSM), which simultaneously generalizes Gromov–Witten theory of complete intersections as well as FJRW theory [**15, 30**] of homogeneous singularities. GLSMs were first described in physics by Witten in [**32**], while the first mathematical definition was given by Fan–Jarvis–Ruan in [**16**].

In the particular case of FJRW theory, Polishchuk–Vaintrob used two-periodic complexes (and more generally the derived category of factorizations) to give an algebraic construction of a virtual class in [**29, 30**]. The paper [**11**] was inspired by the work in [**16**] and [**30**]. In these notes we use two-periodic complexes in a manner analogous to that of [**29**] to define Gromov–Witten-like invariants for hypersurfaces. For a more detailed discussion of connections to other results, see Section 5.

Although we do not make further mention of the GLSM, it is in the background of everything that follows. We hope that by focusing on a special case, these notes will provide a simplified roadmap for readers interested in [**11**].

1. The zero locus of a section

In this section we describe a relatively simple setting in which a virtual fundamental class arises, the zero locus of a section of a vector bundle. This is referred to as the *basic model* in [**28**, Appendix B]. In this setting, a virtual fundamental class can be defined using the techniques of intersection theory and is given by a more familiar object, the refined Euler class. We will see in the next section that the virtual class for $\overline{\mathscr{M}}_{g,n}(\mathbb{P}^r, d)$ is in fact a particular case of this example. The material of this section is described in detail in [**18**, Chapter 6].

Let X be a smooth variety of dimension n and let $E \to X$ be a vector bundle of rank r. Let $s \in \Gamma(X, E)$ be a section, and define $Z := Z(s)$ to be the scheme-theoretic zero locus of the section s. In this setting, Z has an *expected dimension*, given by $n - r$. If s is a regular section of E, then this is the dimension of Z.

We hope to construct a class $[Z]^{\text{vir}}$ in $A_*(Z)$, called the virtual fundamental class, satisfying the following properties:

(1) the class $[Z]^{\text{vir}}$ is supported in the expected dimension: $[Z]^{\text{vir}} \in A_{n-r}(Z)$;

(2) the class $[Z]^{\text{vir}}$ is invariant under deformations in the following sense: if s' is another section of E such that $Z' = Z(s') \subset Z$, then $[Z]^{\text{vir}}$ and $[Z']^{\text{vir}}$ are equal in $A_*(Z)$;

(3) if $s(X)$ intersects the zero section of E transversally, then $[Z]^{\text{vir}} = [Z]$.

Such a class gives an invariant of Z which remembers the fact that Z arose from a section of E. If s is not transverse to the zero section, one may attempt to construct such a class by deforming s to some s' which is transverse, and such that $Z' = Z(s') \subset Z$. If such a deformation is possible, one could define the virtual fundamental class of Z to be $[Z'] \in A_*(Z)$. In many settings, however, such a section does not exist. One motivation for the deformation property (2) above is computation. Sometimes one can deform the section s to a section s' for which the virtual class is easier to compute.

If $s \in \Gamma(X, E)$ is not a regular section, then $\dim(Z)$ will be strictly greater than the expected dimension $n - r$. We explain below how to construct a class in the Chow group of Z which satisfies the properties given above.

EXAMPLE 1.1. Let us consider what is in some sense the worst possible case, that is, $s \in \Gamma(X, E)$ is the zero section $0 \in \Gamma(X, E)$. In this case we define

$$[Z]^{\text{vir}} = e(E) \in A_{n-r}(X),$$

where $e(E)$ denotes the *Euler class* (the top Chern class) of E [**18**, Chapter 3]. If X is compact, then under the cycle map $A_*(X) \to H_*(X) \cong H^*(X)$, $e(E)$ maps to the topological Euler class of, e.g. [**4**].

If there exists a section $s' \in \Gamma(X, E)$ transverse to the zero section, and $Z' = Z(s')$ is the zero locus, then the class $[Z'] \in A_{n-r}(X)$ is equal to $e(E)$. In this sense $e(E)$ is the correct replacement for $[X]$, as it coincides with the class obtained by deforming s into s' and taking the fundamental class of the deformed zero locus.

For a more general section $s \in \Gamma(X, E)$, not necessarily regular or zero, there is a notion of a *refined Euler class* $e(E, s)$ in $A_{n-r}(Z)$. It agrees with the usual Euler class in the sense that under the pushforward map

$$i_*\colon A_*(Z) \to A_*(X),$$

$e(E, s)$ maps to $e(E)$.

DEFINITION 1.2. Given $Z = Z(s)$ as above, the *normal cone* $C_Z X$ is defined to be the limit of the graph of the section $\lambda \cdot s$ as $\lambda \mapsto \infty$. More precisely, consider the map

$$\begin{aligned} X \times \mathbb{C} &\to E \times \mathbb{C}^* \\ (x, \lambda) &\mapsto \lambda s(x). \end{aligned}$$

Let Γ denote the image of this map and let $\overline{\Gamma}$ denote its closure in $E \times \mathbb{P}^1$. Define $C_Z X$ to be the intersection $\text{tot}(E) \times \{\infty\}$ with $\overline{\Gamma}$.

The scheme $\overline{\Gamma} \to \mathbb{C}$ deforms the subvariety $s(X) \subset E$ to $C_Z X$. If $s(x) \neq 0$ for a given point x, $\lambda s(x)$ will go to infinity as λ goes to ∞. As a consequence, $C_Z X$ is contained entirely in $E|_Z$. In [**18**, Appendix B.6.6] it is shown that $C_Z X$ is of pure dimension n.

The projection $\pi|_Z \colon E|_Z \to Z$ is flat, and so defines a map

$$\begin{aligned}\pi|_Z^* \colon A_k(Z) \to & A_{k+r}(E|_Z)\\ [V] \mapsto & [\pi|_Z^{-1}(V)].\end{aligned}$$

It is a (nontrivial) fact that this map is an isomorphism. We denote its inverse by $0|_Z^* \colon A_{k+r}(E|_Z) \to A_k(Z)$, as we may view it as pullback by the zero section $0|_Z \colon Z \to E|_Z$. With this setup, one can define the following:

DEFINITION 1.3. The refined Euler class of $E \to X$ with respect to $s \in \Gamma(X, E)$ is the pullback of the normal cone of Z in X via the zero section:

$$e(E, s) := 0|_Z^* \left([C_Z X]\right).$$

The pullback preserves codimension, so this class lies in $A_{n-r}(Z)$.

DEFINITION 1.4. Define the virtual class

$$[Z]^{vir} := e(E, s).$$

It is an exercise to check that this class satisfies the properties desired of a virtual fundamental class, and agrees with the previous definition in the special case where s is transverse to the zero section or $s = 0$.

We will see below that the virtual class for the moduli space of stable maps to projective space is in fact a special case of this construction.

2. Stable maps to projective space

In this section we construct a virtual fundamental class on $\overline{\mathscr{M}}_{g,n}(\mathbb{P}^r, d)$ using the refined Euler class.

2.1. Spaces of sections. We begin by defining a virtual fundamental class for a space of sections of a vector bundle over a family of curves. We define the moduli space of sections below and realize it as the zero locus of a section of a vector bundle. In Section 2.2 we will see that $\overline{\mathscr{M}}_{g,n}(\mathbb{P}^r, d)$ itself is a space of sections.

Let $\pi \colon \mathcal{C} \to S$ be a flat (and proper) family of pre-stable curves lying over a smooth variety (or stack) S. Let $\mathcal{V} \to \mathcal{C}$ be a vector bundle on $\mathcal{C}$. For each geometric point $s \in S$, the fiber of π is a (at worst) nodal curve C_s, equipped with a vector bundle $V_s \to C_s$.

$$\begin{array}{ccc} V_s & \hookrightarrow & \mathcal{V} \\ \downarrow & & \downarrow \\ C_s & \hookrightarrow & \mathcal{C} \\ \downarrow{\scriptstyle \pi_s} & & \downarrow{\scriptstyle \pi} \\ s & \hookrightarrow & S. \end{array}$$

DEFINITION 2.1. Let $\pi \colon \mathcal{C} \to S$ and $\mathcal{V} \to \mathcal{C}$ be defined as in the introduction to this section. Define the *space of sections* of $\mathcal{V}$ to be

$$\operatorname{tot}(\pi_* \mathcal{V}) := \operatorname{Spec}\left(\operatorname{Sym}\left(\mathbb{R}^1 \pi_*(\omega_\pi \otimes \mathcal{V}^\vee)\right)\right),$$

where ω_π is the relative dualizing sheaf for π.

Given a sheaf $\mathcal{A}$ over S, recall that the s-points of $\operatorname{Spec}(\operatorname{Sym}(\mathcal{A}))$ are the elements of $\hom(\mathcal{A}, \mathcal{O}_s)$. With this we observe that

$$
\begin{aligned}
\operatorname{tot}(\pi_*\mathcal{V})|_s &= \hom\left(\mathbb{R}^1\pi_*(\omega_\pi \otimes \mathcal{V}^\vee)_s, \mathcal{O}_s\right) \\
&= \hom\left(\mathbb{R}^1\pi_{s*}(\omega_{\pi_s} \otimes V_s^\vee), \mathcal{O}_s\right) \\
&= \Gamma(C_s, V_s),
\end{aligned}
\tag{2.1}
$$

justifying the terminology space of sections. Here the second equality is by base change (see [**20**], or [**27**, Theorem 1.2] for a summary of the case at hand). The third equality is Serre duality. In fact a stronger statement is true.

DEFINITION 2.2. [**5**, Definition 2.1] Define the contravariant functor

$$\sec_{\mathcal{V}} \colon \text{Schemes} \to \text{Set}$$

which sends a scheme T to the set of pairs (ρ, p), where $\rho\colon T \to S$ is a morphism and $p \in H^0(\mathcal{C}_T, \rho^*\mathcal{L})$ where $\mathcal{C}_T = T \times_S \mathcal{C}$ and $\rho^*\mathcal{L} = \mathcal{C}_T \times_{\mathcal{C}} \mathcal{L}$. A morphism between (p, ρ) and (p', ρ') is given by a morphism $\tau\colon \mathcal{C}_T \to \mathcal{C}_T$ such that $\rho' \circ \tau = \rho$ and under the induced isomorphism $\rho^*\mathcal{L} \to \tau^*\rho'^*\mathcal{L}$, $p = \tau^*(p')$.

By a similar argument to (2.1) above, one obtains the following.

PROPOSITION 2.3. [**5**, Proposition 2.2] *The stack* $\operatorname{tot}(\pi_*\mathcal{V})$ *represents the functor* $\sec_{\mathcal{V}}$.

REMARK 2.4. The notation $\operatorname{tot}(\pi_*\mathcal{V})$ is an abuse of notation as the space of sections is not in general a vector bundle and $\pi_*\mathcal{V}$ is not locally free. We hope this will not cause confusion.

The space $\operatorname{tot}(\pi_*\mathcal{V})$ is not usually smooth and in fact is not even of pure dimension. Nevertheless we can exploit the ideas of Section 1 to construct a Chow class on $\operatorname{tot}(\pi_*\mathcal{V})$ of relative dimension $\chi(C_s, V_s)$ over S. We will use this in the next section to define a virtual class for $\overline{\mathscr{M}}_{g,n}(\mathbb{P}^r, d)$.

First, we embed $\mathcal{V}$ into a π-acyclic vector bundle $\mathcal{A} \to \mathcal{C}$.

LEMMA 2.5. *If* $\pi\colon \mathcal{C} \to S$ *is projective, there exists an embedding of* $\mathcal{V}$ *into a vector bundle* $\mathcal{A}$ *satisfying* $\mathbb{R}^1\pi_*(\mathcal{A}) = 0$.

PROOF. Let $\mathcal{O}(1) \to \mathcal{C}$ be a π-relatively ample line bundle. Then for sufficiently large n,

$$\mathbb{R}^1\pi_*(\mathcal{V}^\vee(n)) = \mathbb{R}^1\pi_*(\mathcal{O}(n)) = 0$$

and the map

$$\pi^*(\pi_*(\mathcal{V}^\vee(n))) \to \mathcal{V}^\vee(n)$$

is surjective. Twisting by $\mathcal{O}(-n)$, we obtain the surjective map

$$\pi^*(\pi_*(\mathcal{V}^\vee(n)))(-n) \to \mathcal{V}^\vee.$$

Dualizing this map gives an embedding $\mathcal{V}$ into

$$\mathcal{A} := \pi^*(\pi_*(\mathcal{V}^\vee(n))^\vee)(n).$$

□

Given such an embedding $\mathcal{V} \hookrightarrow \mathcal{A}$, let $\mathcal{B}$ denote the cokernel. Define

$$A := \pi_*(\mathcal{A}),$$
$$B := \pi_*(\mathcal{B}).$$

The short exact sequence

$$0 \to \mathcal{V} \to \mathcal{A} \to \mathcal{B} \to 0$$

induces a long exact sequence

$$\begin{aligned} 0 \longrightarrow \pi_*\mathcal{V} \longrightarrow A \longrightarrow B \\ \longrightarrow \mathbb{R}^1\pi_*\mathcal{V} \longrightarrow \mathbb{R}^1\pi_*\mathcal{A} \longrightarrow \mathbb{R}^1\pi_*\mathcal{B} \longrightarrow 0. \end{aligned} \tag{2.2}$$

Note, however, that $\mathbb{R}^1\pi_*\mathcal{A} = 0$ by construction and so $\mathbb{R}^1\pi_*\mathcal{B} = 0$ as well. Because the Euler characteristic is constant in flat families, the fibers of A and B have constant rank. By Grauert's criterion [**21**, 12.9], they are vector bundles. We conclude that the two-term complex $[A \to B]$ is a resolution of $\mathbb{R}\pi_*\mathcal{V}$ by vector bundles.

Consider the total space of A,

$$X := \mathrm{tot}(A),$$

and the map $\tau\colon X \to S$ which forgets the section of $\mathcal{A}$. Define $E := \tau^*(B)$ over X. The map $A \to B$ induces a section $s \in \Gamma(X, E)$. By (2.2) the zero section $Z(s) \hookrightarrow X$ is exactly $\mathrm{tot}(\pi_*\mathcal{V})$. We are now in the situation of Section 1. We proceed as before.

DEFINITION 2.6. Define the virtual fundamental class of the space of sections $\mathrm{tot}(\pi_*\mathcal{V})$ to be

$$[\mathrm{tot}(\pi_*\mathcal{V})]^{\mathrm{vir}} := e(E, s) \in A_*(\mathrm{tot}(\pi_*\mathcal{V})).$$

Note in the above that $[\mathrm{tot}(\pi_*\mathcal{V})]^{\mathrm{vir}} \in A_k(\mathrm{tot}(\pi_*\mathcal{V}))$, where

$$\begin{aligned} k &= \dim(\mathrm{tot}(A)) - \mathrm{rank}(E) \\ &= \dim(S) + \mathrm{rank}(A) - \mathrm{rank}(B) \\ &= \dim(S) + \chi(C_s, V_s), \end{aligned}$$

for s a geometric point of S.

One can check that the virtual class is well-defined in the following sense.

PROPOSITION 2.7. *The class* $[\mathrm{tot}(\pi_*\mathcal{V})]^{\mathrm{vir}}$ *is independent of the choice of embedding* $\mathcal{V} \to \mathcal{A}$.

PROOF. This follows from more general constructions such as in [**3**]. However in this simple case one can prove it more easily. The sketch of the proof is the following:

(1) Reduce to the case of two embeddings $\mathcal{V} \to \mathcal{A}$ and $\mathcal{V} \to \mathcal{A}'$, where the second map factors through an embedding $\mathcal{A} \to \mathcal{A}'$. This can be accomplished by embedding $\mathcal{A}$ and $\mathcal{A}'$ in a common larger π-acyclic vector bundle $\widetilde{\mathcal{A}}$.
(2) By a deformation argument, reduce to the case that $\mathcal{A}' = \mathcal{A} \oplus \mathcal{A}''$ and the embedding $\mathcal{A} \to \mathcal{A}'$ is given by $(\mathrm{id}_{\mathcal{A}}, 0)$ (see [**18**, Proposition 18.1] for such an argument).
(3) In this simple case, one observes from the definition that $e(E', s')$ and $e(E, s)$ coincide. □

2.2. Stable maps as a space of sections. In this section we realize $\overline{\mathscr{M}}_{g,n}(\mathbb{P}^r, d)$ as (an open subset of) a space of sections and use the results of Section 2.1 to construct the virtual fundamental class. We conclude by showing that this construction agrees with the definition of Behrend and Fantechi in terms of the intrinsic normal cone [**1**, **3**].

Recall that the moduli space $\overline{\mathscr{M}}_{g,n}(\mathbb{P}^r, d)$ is the stack representing families of genus g curves with n marked points, together with a degree d map from the curve to $\mathcal{C}$. More precisely, a family of (genus g, n-marked, degree d) stable maps over S is given by a diagram:

$$\begin{array}{ccc} & \mathcal{C} & \xrightarrow{f} \mathbb{P}^r \\ \{\sigma_i\}_{i=1}^n & \uparrow\downarrow \pi & \\ & S & \end{array}$$

where:

(1) π is a flat family of prestable curves of genus g. Included in this is the condition that the sections $\{\sigma_i\}_{1\leq i\leq n}$ are disjoint from each other and from the nodes of C_s;
(2) f restricts on each fiber to a degree d map $f_s \colon C_s \to \mathbb{P}^r$;
(3) the line bundle $\omega_{\pi,\log} \otimes f^*(\mathcal{O}_{\mathbb{P}^r}(3))$ is ample, where $\omega_{\pi,\log}$ is defined as

$$\omega_{\pi,\log} := \omega_\pi \otimes \mathcal{O}_\mathcal{C}\left(\sum_{i=1}^n \sigma_i(S)\right).$$

There is a natural notion of equivalence of such families. See [**19**] for a construction of $\overline{\mathscr{M}}_{g,n}(\mathbb{P}^r, d)$, and a proof that the coarse moduli space is projective. We argue below that $\overline{\mathscr{M}}_{g,n}(\mathbb{P}^r, d)$ can be realized as (a substack of) a space of sections, this time over a smooth Artin stack.

DEFINITION 2.8. Define the contravariant functor $\sec^\circ_{\mathcal{L}^{\oplus r+1}} \colon$ Schemes $\to$ Set which sends a scheme S to the set of equivalence classes of families over S of the form:

$$\begin{array}{ccc} & & \mathcal{L}^{\oplus r+1} \\ & s \nearrow \swarrow & \\ & \mathcal{C} & \\ \sigma_i & \uparrow\downarrow \pi & \\ & S & \end{array}$$

where:

(1) π is a flat family of prestable curves of genus g. The sections $\{\sigma_i\}_{1\leq i\leq n}$ are disjoint from each other and from the nodes of C_s;
(2) $\mathcal{L}$ is a line bundle on $\mathcal{C}$, of degree d on each fiber C_s;
(3) The section $s \in \Gamma(\mathcal{C}, \mathcal{L}^{\oplus r+1})$ is nowhere vanishing;
(4) The line bundle $\omega_{\pi,\log} \otimes \mathcal{L}^{\otimes 3}$ is ample.

A morphism between two such families $(\mathcal{C}, \mathcal{L}, s)$ and $(\mathcal{C}', \mathcal{L}', s')$ over S is given by a morphism of pre-stable curves $\phi\colon \mathcal{C} \to \mathcal{C}'$ together with an isomorphism $\mathcal{L} \to \phi^*(\mathcal{L}')$ which identifies s with $\phi^*(s')$.

LEMMA 2.9. *$\overline{\mathscr{M}}_{g,n}(\mathbb{P}^r, d)$ represents the functor* $\sec^\circ_{\mathcal{L}^{\oplus r+1}}$.

This is well-known and follows from the correspondence between maps $T \to \mathbb{P}^r$ and $r+1$-tuples of sections of a line bundle $L \to T$. See [**10**, Section 2.2] for further details and for a generalization to toric varieties.

Proof of Lemma 2.9. Recall that projective space $\mathbb{P}^r$ is a fine moduli space for the functor which sends a scheme S to the set of equivalence classes of the following data $(L \to S, s \in \Gamma(S, L^{\oplus r+1}))$, where $L \to S$ is a line bundle and s is nowhere vanishing. Here the collections $(L \to S, s \in \Gamma(S, L^{\oplus r+1}))$ and $(L' \to S, s' \in \Gamma(S, L'^{\oplus r+1}))$ are equivalent if there is an isomorphism $L \to L'$ identifying s with s'.

Given a family of curves $\mathcal{C} \to S$, a family of degree d maps $f \colon \mathcal{C} \to \mathbb{P}^r$ corresponds to (an equivalence class of) a line bundle $\mathcal{L} \to \mathcal{C}$ and a section $s \in \Gamma(\mathcal{C}, \mathcal{L}^{\oplus r+1})$. Under this equivalence, $\mathcal{L}$ is identified with $f^*(\mathcal{O}_{\mathbb{P}^r}(-1))$. Thus condition (4) for the functor $\sec^{\circ}_{\mathcal{L}^{\oplus r+1}}$ corresponds to condition (3) for the functor of stable maps. □

Forgetting the section s, we obtain a morphism from $\overline{\mathscr{M}}_{g,n}(\mathbb{P}^r, d)$ to another stack, defined below.

Definition 2.10. Define a family of (genus g, n-marked) prestable curves with a degree d line bundle over S to be given by the following data:

$$\begin{array}{c} \mathcal{L} \\ \downarrow \\ \mathcal{C} \\ {\scriptstyle \sigma_i}\uparrow\downarrow{\scriptstyle \pi} \\ S \end{array}$$

where $\mathcal{L}$ is degree d on each fiber and condition (1) is satisfied. Define $\mathfrak{Bun}_{g,n,d}$, to be the stack representing such families.

Define $\mathfrak{Bun}^{\circ}_{g,n,d} \subset \mathfrak{Bun}_{g,n,d}$ to be the open substack consisting of families such that Condition 4 above is also satisfied.

Note that all families in $\mathfrak{Bun}^{\circ}_{g,n,d}$ have, in addition to automorphisms of the underlying curve $\mathcal{C}$, a $\mathbb{C}^*$-family of automorphisms obtained by the scaling automorphisms of $\mathcal{L}$.

Proposition 2.11. [**12**, Proposition 2.1.1] *The stack $\mathfrak{Bun}_{g,n,d}$ is a smooth Artin stack of dimension $4g-4+n$. Over the open locus $\mathfrak{Bun}^{\circ}_{g,n,d}$, the universal curve*

$$\pi \colon \mathfrak{C} \to \mathfrak{Bun}^{\circ}_{g,n,d}$$

is projective.

Note that the line bundle $\omega_{\pi,\log} \otimes \mathcal{L}^{\otimes 3}$ is relatively ample by construction (see Condition 4 above).

Consider the universal curve $\pi \colon \mathfrak{C} \to \mathfrak{Bun}^{\circ}_{g,n,d}$, and the universal line bundle $\mathcal{L} \to \mathfrak{C}$. Define the vector bundle $\mathcal{V} := \mathcal{L}^{\oplus r+1}$ over $\mathfrak{C}$.

Proposition 2.12. *There is an open immersion*

$$\overline{\mathscr{M}}_{g,n}(\mathbb{P}^r, d) \hookrightarrow \operatorname{tot}(\pi_* \mathcal{V}).$$

PROOF. By Condition 4 of Definition 2.8, $\overline{\mathscr{M}}_{g,n}(\mathbb{P}^r, d)$ lies over $\mathfrak{Bun}^\circ_{g,n,d}$. By Proposition 2.3 and Lemma 2.9, $\overline{\mathscr{M}}_{g,n}(\mathbb{P}^r, d)$ is a substack of $\mathrm{tot}(\pi_*\mathcal{V})$, defined by the additional condition that the section s is nowhere vanishing. We claim that this is an open condition.

Consider the universal curve $\mathfrak{C}_{\mathrm{tot}(\pi_*\mathcal{V})} = \mathrm{tot}(\pi_*\mathcal{V}) \times_{\mathfrak{Bun}^\circ_{g,n,d}} \mathfrak{C}$ over $\mathrm{tot}(\pi_*\mathcal{V})$. On $\mathfrak{C}_{\mathrm{tot}(\pi_*\mathcal{V})}$ there is a universal vector bundle $\mathcal{V}_{\mathrm{tot}(\pi_*\mathcal{V})} \to \mathfrak{C}_{\mathrm{tot}(\pi_*\mathcal{V})}$ (pulled back from the map $\mathfrak{C}_{\mathrm{tot}(\pi_*\mathcal{V})} \to \mathfrak{C}$), together with a universal section $\tilde{s} \in \Gamma(\mathfrak{C}_{\mathrm{tot}(\pi_*\mathcal{V})}, \mathcal{V}_{\mathrm{tot}(\pi_*\mathcal{V})})$. Consider the closed substack $Z(\tilde{s}) \subset \mathfrak{C}_{\mathrm{tot}(\pi_*\mathcal{V})}$ defined by the vanishing of the universal section. Because the map $\pi\colon \mathcal{C} \to \mathfrak{Bun}^\circ_{g,n,d}$ is proper, so is the map $\pi_{\mathrm{tot}(\pi_*\mathcal{V})}\colon \mathfrak{C}_{\mathrm{tot}(\pi_*\mathcal{V})} \to \mathrm{tot}(\pi_*\mathcal{V})$. The image $\pi_{\mathrm{tot}(\pi_*\mathcal{V})}(Z(\tilde{s}))$ is therefore a closed subset of $\mathrm{tot}(\pi_*\mathcal{V})$. Its complement consists of points $(C, \mathcal{V}|_C \to C, s \in \Gamma(C, \mathcal{V}_C))$ such that s is nowhere vanishing. These are exactly the objects of Definition 2.8. We conclude that $\overline{\mathscr{M}}_{g,n}(\mathbb{P}^r, d) = \pi_{\mathrm{tot}(\pi_*\mathcal{V})}(Z(\tilde{s}))^c$ is open in $\mathrm{tot}(\pi_*\mathcal{V})$. □

2.3. The virtual class. Finally, we construct a virtual fundamental class for $\overline{\mathscr{M}}_{g,n}(\mathbb{P}^r, d)$ via a very minor modification of the procedure in Section 2.1. Let $\mathcal{V}$ be as in the previous paragraph. Note that $\pi\colon \mathfrak{C} \to \mathfrak{Bun}^\circ_{g,n,d}$ is projective. As described in Section 2.1, embed $\mathcal{V}$ into a π-acyclic vector bundle $\mathcal{A} \to \mathfrak{C}$. Define $\mathcal{B}$ as the cokernel of $\mathcal{V} \to \mathcal{A}$, and let

$$A := \pi_*(\mathcal{A}),$$
$$B := \pi_*(\mathcal{B}).$$

Then $[A \to B]$ is a two-term resolution of $\mathbb{R}\pi_*\mathcal{V}$ by vector bundles. Define

$$\tau\colon\ \mathrm{tot}(A) \to \mathfrak{Bun}^\circ_{g,n,d}$$

and $E := \tau^*(B)$ as before. Because $\mathfrak{Bun}^\circ_{g,n,d}$ is a smooth Artin stack, $\mathrm{tot}(A)$ will be as well.

For $b \in \mathfrak{Bun}^\circ_{g,n,d}$, the fiber of $\mathrm{tot}(A)$ over b is given by

$$\mathrm{tot}(A)_b = \{s \in \Gamma(C_b, \mathcal{A}|_{C_b})\}.$$

Define the open substack

$$U \subset \mathrm{tot}(A)$$

by the condition that the section s is nowhere vanishing on each fiber. Again this is an open condition. One can check that U is a Deligne–Mumford stack. In particular the automorphism group of each point of U is finite. As described in Section 2.1, the map $A \to B$ of vector bundles induces a natural section which we denote β in $\Gamma(U, E)$. The zero locus of β cuts out exactly those sections of $\mathcal{A}$ which are sections of $\mathcal{V} = \mathcal{L}^{\oplus r+1}$. As we have restricted to the locus of nowhere vanishing sections, we conclude that

$$\overline{\mathscr{M}}_{g,n}(\mathbb{P}^r, d) = Z(\beta) \subset U \subset \mathrm{tot}(A).$$

DEFINITION 2.13. Define the virtual fundamental class

$$[\overline{\mathscr{M}}_{g,n}(\mathbb{P}^r, d)]^{\mathrm{vir}} := e(E, \beta) \in A_*(\overline{\mathscr{M}}_{g,n}(\mathbb{P}^r, d)).$$

As a quick check that this definition is reasonable, we do a dimension count. The refined Euler class $e(E,\beta)$ lies in $A_k(\overline{\mathscr{M}}_{g,n}(\mathbb{P}^r,d))$, where

$$\begin{aligned} k &= \dim(\mathfrak{Bun}_{g,n,d}) + \operatorname{rank}(A) - \operatorname{rank}(B) \\ &= \dim(\mathfrak{Bun}_{g,n,d}) + \chi(V_s) \\ &= 4g - 4 + n + (r+1)(1-g+d) \\ &= (1-g)(r-3) + n + (r+1)d \\ &=: \operatorname{virdim}\left(\overline{\mathscr{M}}_{g,n}(\mathbb{P}^r,d)\right). \end{aligned}$$

This is the *expected dimension* of $\overline{\mathscr{M}}_{g,n}(\mathbb{P}^r,d)$ [**14**, Section 7.1.4]. More precisely, there is an open substack $\overline{\mathscr{M}}_{g,n}(\mathbb{P}^r,d)^\circ$ of $\overline{\mathscr{M}}_{g,n}(\mathbb{P}^r,d)$ parametrizing equivalence classes of maps $f\colon C \to \mathbb{P}^r$ such that C is smooth and $H^1(C, f^*(\mathscr{O}_{\mathbb{P}^r}(1))) = 0$. On this open locus $\overline{\mathscr{M}}_{g,n}(\mathbb{P}^r,d)^\circ$ is smooth, of dimension $(1-g)(r-3)+n+(r+1)d$.

PROPOSITION 2.14. *The virtual fundamental class of Definition 2.13 agrees with the Gromov–Witten theory virtual fundamental class for $\overline{\mathscr{M}}_{g,n}(\mathbb{P}^r,d)$ as defined by Behrend and Fantechi in* [**1,3**].

PROOF. Consider the morphism

$$\phi\colon \overline{\mathscr{M}}_{g,n}(\mathbb{P}^r,d) \to \mathfrak{M}_{g,n}$$

to the stack of pre-stable curves which forgets the map to $\mathbb{P}^r$ (but does not stabilize the curve). Following [**1, 3**], the virtual fundamental class on $\overline{\mathscr{M}}_{g,n}(\mathbb{P}^r,d)$ may be defined by using the relative perfect obstruction theory $(\mathbb{R}\pi_* f^* T\mathbb{P}^r)^\vee$ (see the introduction of [**1**] for details).

The key to the proposition is the following fact, described in the beginning of Section 6 of [**3**]. Whenever one has the fiber square:

(2.3)
$$\begin{array}{ccc} Z & \xrightarrow{\ i\ } & X \\ \downarrow & & \downarrow s \\ X & \xrightarrow{\ 0\ } & E, \end{array}$$

there is a perfect obstruction theory $F^\bullet$ given by the complex

$$s|_Z^*(E)^\vee \to i^*\Omega_X,$$

together with the natural map to the cotangent complex $L_Z^\bullet$ (see [**3**, Section 6] for details). In this special case, the virtual fundamental class $[Z,F^\bullet]$ of Behrend–Fantechi is precisely the refined Euler class $e(E,s)$ of Definition 1.4. More generally, consider the relative version of diagram (2.3) over a base stack S. Assume that X and S are smooth Deligne–Mumford stacks. Then the complex

$$s|_Z^*(E)^\vee \to i^*\Omega_{X/S},$$

defines a relative perfect obstruction theory $F_S^\bullet$ for X/S. In this case, again the virtual fundamental class $[Z,F_S^\bullet]$ is the refined Euler class $e(E,s)$.

On the other hand, as described in Sections 2.1 and 2.2, a particular case of diagram (2.3) is given by:

(2.4)
$$\begin{array}{ccc} \overline{\mathscr{M}}_{g,n}(\mathbb{P}^r,d) & \xrightarrow{\ i\ } & U \\ \downarrow & & \downarrow \beta \\ U & \xrightarrow{\ 0\ } & E. \end{array}$$

Via the morphism $\nu\colon U \to \mathfrak{M}_{g,n}$ which forgets the section of $\mathcal{A}$ and the line bundle $\mathcal{L}$, diagram (2.4) is relative to $\mathfrak{M}_{g,n}$.

By the above discussion, the class in Definition 2.13 is equal to the class

$$[\overline{\mathscr{M}}_{g,n}(\mathbb{P}^r, d), F^\bullet_{\mathfrak{M}_{g,n}}],$$

for the relative perfect obstruction theory

$$F^\bullet_{\mathfrak{M}_{g,n}} = \tau^* B^\vee \to \Omega_{U/\mathfrak{M}_{g,n}}. \tag{2.5}$$

The proposition will follow after identifying the relative perfect obstruction theory $F^\bullet_{\mathfrak{M}_{g,n}}$ with $(\mathbb{R}\pi_* f^* T\mathbb{P}^r)^\vee$.

Consider the following commutative diagram on the universal curve $\mathcal{C}$ over $\overline{\mathscr{M}}_{g,n}(\mathbb{P}^r, d)$, viewed here as the vanishing locus $Z(\beta) \subset U$:

$$\begin{array}{ccccc} \mathcal{O}_{\mathcal{C}} & \longrightarrow & \mathcal{V} & \longrightarrow & f^*(T\mathbb{P}^{r-1}) \\ \downarrow & & \downarrow & & \downarrow \\ \mathcal{O}_{\mathcal{C}} & \longrightarrow & \mathcal{A} & \longrightarrow & \mathcal{A}' \\ \downarrow & & \downarrow & & \downarrow \\ 0 & \longrightarrow & \mathcal{B} & \longrightarrow & \mathcal{B}. \end{array} \tag{2.6}$$

Here $f\colon \mathcal{C} \to \mathbb{P}^{r-1}$ is the universal map to $\mathbb{P}^{r-1}$ and the top row is obtained by pulling back the Euler sequence,

$$0 \to \mathcal{O}_{\mathbb{P}^{r-1}} \to \mathcal{O}_{\mathbb{P}^{r-1}}(1)^{\oplus r} \to T\mathbb{P}^{r-1} \to 0,$$

from $\mathbb{P}^{r-1}$. The vector bundles $\mathcal{V}$, $\mathcal{A}$, and $\mathcal{B}$ are all pulled back from the corresponding vector bundles on the universal curve over $\mathfrak{Bun}^\circ_{g,n,d}$. By abuse of notation we will denote them by the same letters. The section $\mathcal{O}_{\mathcal{C}} \to \mathcal{A}$ is defined by the tautological section and the vector bundle $\mathcal{A}'$ is the cokernel of this map. Note that all rows and columns are short exact sequences.

For a given point $\{C, \mathcal{L}|_C \to C, s \in \Gamma(C, \mathcal{A}|_C)\} \in U$, the relative tangent bundle $T\nu$ consists of deformations of the line bundle $\mathcal{L}|_C$, together with deformations of the section $s \in \Gamma(C, \mathcal{A}|_C)$. The former space is given by $R^1\pi_*(C, \mathcal{O}_C)$ and the later is $H^0(C, \mathcal{A}|_C)/\mathbb{C}$, where the quotient by $\mathbb{C}$ accounts for the fact that two sections give equivalent points of U if they agree up to a scaling.

The bundle $T\nu$ may be described globally by pushing forward the middle row of the above diagram:

$$\mathcal{O}_U \to \tau^* A \to A' \to R^1\pi_*(\mathcal{O}_{\mathcal{C}}),$$

where $A' := \pi_*(\mathcal{A}')$. This yields the short exact sequence

$$0 \to \mathrm{cok}(\mathcal{O}_U \to \tau^* A) \to A' \to R^1\pi_*(\mathcal{O}_{\mathcal{C}}) \to 0$$

Deformations of the section s up to scaling are described by exactly $\mathrm{cok}(\mathcal{O}_U \to \tau^* A)$ and we observe that $T_\nu = A'$.

Returning to (2.5), $F^\bullet_{\mathfrak{M}_{g,n}}$ may now be represented as $B^\vee \to A'^\vee$, where the map is the dual of the pushforward of $\mathcal{A}' \to \mathcal{B}$. Because the last column of (2.6) is exact, $F^\bullet_{\mathfrak{M}_{g,n}}$ is equal to $(\mathbb{R}\pi_* f^* T\mathbb{P}^r)^\vee$, and $[\overline{\mathscr{M}}_{g,n}(\mathbb{P}^r, d), F^\bullet_{\mathfrak{M}_{g,n}}]$ is therefore the virtual class of the first paragraph of the proof. □

3. Stable maps to a hypersurface

Let $w = w(x_0, \ldots, x_r)$ be a homogeneous polynomial of degree k. We assume that w defines an isolated singularity at the origin in $\mathbb{C}^{r+1}$, or equivalently, that the hypersurface $X_w = Z(w) \subset \mathbb{P}^r$ is smooth. In this section we expand the ideas of the previous section to construct a virtual class on $\overline{\mathscr{M}}_{g,n}(X_w, d)$.

Related to the moduli space of stable maps to X_w is the space of maps to $\mathbb{P}^r$ with *p-fields*, which has the advantage of being a space of sections of a vector bundle as in Section 2.1. This moduli space and various generalizations have appeared in [**5**, **13**, **16**].

DEFINITION 3.1. [**5**] Given $k \in \mathbb{Z}_{>0}$, define the space of stable maps to $\mathbb{P}^r$ with a p-field of degree k to be the moduli space lying over $\overline{\mathscr{M}}_{g,n}(\mathbb{P}^r, d)$ parametrizing families of stable maps $f : C \to \mathbb{P}^r$ of degree d from genus g curves C, together with a section

$$p \in \Gamma(C, f^*(\mathcal{O}_{\mathbb{P}^r}(-k) \otimes \omega_C)). \tag{3.1}$$

This moduli stack is denoted by $\overline{\mathscr{M}}_{g,n}(\mathbb{P}^r, d)^p$.

The stack $\overline{\mathscr{M}}_{g,n}(\mathbb{P}^r, d)^p$ is easily seen to be a Deligne–Mumford stack because $\overline{\mathscr{M}}_{g,n}(\mathbb{P}^r, d)$ is. We can also view the stack as representing families

$$(\pi : \mathcal{C} \to S, \mathcal{L} \to \mathcal{C}, (s, p) \in \Gamma(\mathcal{C}, \mathcal{L}^{\oplus r+1} \oplus \mathcal{L}^{\otimes -k} \otimes \omega_\pi))$$

where:

(1) π is a flat family of prestable curves of genus g;
(2) $\mathcal{L}$ is a line bundle on $\mathcal{C}$, of degree d on each fiber C_s;
(3) The section $s \in \Gamma(\mathcal{C}, \mathcal{L}^{\oplus r+1})$ is nowhere vanishing;
(4) The line bundle $\omega_{\pi,\log} \otimes \mathcal{L}^{\otimes 3}$ is ample.

By the same reasoning as Proposition 2.12, we see that $\overline{\mathscr{M}}_{g,n}(\mathbb{P}^r, d)^p$ is (an open subset of) a space of sections over $\mathfrak{Bun}^{\circ}_{g,n,d}$.

3.1. Smooth embedding via resolutions. We can repeat the argument of the previous section to realize $\overline{\mathscr{M}}_{g,n}(\mathbb{P}^r, d)^p$ as the zero locus of a section of a vector bundle defined over a smooth Deligne–Mumford stack. Let

$$\begin{array}{c} \mathcal{L} \\ \downarrow \\ \mathcal{C} \\ \downarrow \pi \\ S \end{array}$$

be a family of prestable curves together with a degree d line bundle over $\mathcal{C}$, pulled back from $\mathfrak{Bun}^{\circ}_{g,n,d}$ via a map $S \to \mathfrak{Bun}^{\circ}_{g,n,d}$. Define the fiber products

$$\begin{aligned} \overline{\mathscr{M}}_{g,n}(\mathbb{P}^r, d)_S &:= S \times_{\mathfrak{Bun}^{\circ}_{g,n,d}} \overline{\mathscr{M}}_{g,n}(\mathbb{P}^r, d) \\ \overline{\mathscr{M}}_{g,n}(\mathbb{P}^r, d)^p_S &:= S \times_{\mathfrak{Bun}^{\circ}_{g,n,d}} \overline{\mathscr{M}}_{g,n}(\mathbb{P}^r, d)^p \\ \overline{\mathscr{M}}_{g,n}(X_w, d)_S &:= S \times_{\mathfrak{Bun}^{\circ}_{g,n,d}} \overline{\mathscr{M}}_{g,n}(X_w, d). \end{aligned} \tag{3.2}$$

Let

$$\begin{aligned} \mathcal{V}_1 &:= \mathcal{L}^{\oplus r+1}, \\ \mathcal{V}_2 &:= \mathcal{L}^{\otimes -k} \otimes \omega_\pi, \\ \mathcal{V} &:= \mathcal{V}_1 \oplus \mathcal{V}_2. \end{aligned}$$

Choose π-acyclic vector bundles $\mathcal{A}_1$ and $\mathcal{A}_2$ together with embeddings of $\mathcal{V}_1$ and $\mathcal{V}_2$ to obtain short exact sequences

$$\begin{aligned} 0 \to &\mathcal{V}_1 \to \mathcal{A}_1 \to \mathcal{B}_1 \to 0 \\ 0 \to &\mathcal{V}_2 \to \mathcal{A}_2 \to \mathcal{B}_2 \to 0 \end{aligned}$$

where $\mathcal{B}_1$ and $\mathcal{B}_2$ are the respective cokernels. Define

$$\begin{aligned} A := A_1 \oplus A_2 &:= \pi_*(\mathcal{A}_1) \oplus \pi_*(\mathcal{A}_2) \\ B := B_1 \oplus B_2 &:= \pi_*(\mathcal{B}_1) \oplus \pi_*(\mathcal{B}_2). \end{aligned}$$

Then the two-term complex of vector bundles $A \to B$ gives a resolution of $\mathbb{R}\pi_*\mathcal{V}$ over S.

Definition 3.2. Define U to be the open substack of $\operatorname{tot}(A) = \operatorname{tot}(A_1 \oplus A_2)$, where the section $s \in \Gamma(\mathcal{C}, \mathcal{A}_1)$ is nowhere vanishing. Let $\tau \colon U \to \mathfrak{Bun}^\circ_{g,n,d}$ be the map which forgets the section of $\mathcal{A}$, and define

$$E := \tau^*(B).$$

The map $A \to B$ induces a section $\beta \in \Gamma(U, E)$. We conclude that

$$\overline{\mathscr{M}}_{g,n}(\mathbb{P}^r, d)^p_S = Z(\beta) \subset U$$

as desired.

3.2. Incorporating w. Recall that $w(x_0, \ldots, x_r)$ is a homogeneous polynomial of degree k defining a hypersurface $X_w \subset \mathbb{P}^r$. Define

$$\hat{w} = \hat{w}(x_0, \ldots, x_r, y) := y \cdot w(x_0, \ldots, x_r).$$

Proposition 3.3. *The function $\hat{w}$ induces a morphism*

$$\tilde{\alpha} \colon \operatorname{Sym}^{k+1}([A \to B]) \to \mathcal{O}_S[-1]$$

in the derived category of S.

Proof. The morphism $\tilde{\alpha}$ will be defined as a composition of morphisms in the derived category:

(3.3)

$$\begin{array}{ccccccc} \operatorname{Sym}^{k+1}([A \to B]) & \xrightarrow{\sim} & \operatorname{Sym}^{k+1}(\mathbb{R}\pi_*\mathcal{V}) & \xrightarrow{\text{nat}} & \mathbb{R}\pi_*(\operatorname{Sym}^{k+1}(\mathcal{V})) & \xrightarrow{\widetilde{\hat{w}}} & \mathbb{R}\pi_*(\omega_\pi) \\ & & & & & & \downarrow \\ & & & & & & \mathbb{R}^1\pi_*(\omega_\pi)[-1] \\ & & & \tilde{\alpha} & & & \downarrow \\ & & & & & & \mathbb{R}^0\pi_*(\mathcal{O}_S)^\vee[-1] \\ & & & & & & \downarrow \\ & & & & & \dashrightarrow & \mathcal{O}_S[-1]. \end{array}$$

In the above, the first vertical map is obtained by taking cohomology in degree one, the second vertical map is Serre duality, and the third is the trace map. The first horizontal map is due to the fact that $[A \to B]$ is equivalent to $\mathbb{R}\pi_*\mathcal{V}$. The second horizontal map

$$\text{nat}\colon \operatorname{Sym}^{k+1}(\mathbb{R}\pi_*\mathcal{V}) \to \mathbb{R}\pi_*(\operatorname{Sym}^{k+1}(\mathcal{V})) \tag{3.4}$$

is induced from the composition

$$\mathbb{L}\pi^* \operatorname{Sym}^{k+1}(\mathbb{R}\pi_*\mathcal{V}) \to \operatorname{Sym}^{k+1}(\mathbb{L}\pi^*\mathbb{R}\pi_*\mathcal{V}) \to \operatorname{Sym}^{k+1}(\mathcal{V})$$

via adjunction.

To define $\tilde{\alpha}$ it therefore remains to define $\widetilde{\hat{w}}$. This is done in Lemma 3.4 below. □

LEMMA 3.4. *The function $\hat{w}$ induces a morphism*

$$\widetilde{\hat{w}}\colon \mathbb{R}\pi_*(\operatorname{Sym}^{k+1}(\mathcal{V})) \to \mathbb{R}\pi_*(\omega_\pi)$$

in the derived category of S.

PROOF. This is explained in a general context in Section 3.1.4 of [**11**]. In this setting it can be seen most easily using local coordinates. First, let

$$M_j = c_j \prod_{i=1}^{k} x_{j_i} \cdot y$$

be a monomial of $\hat{w}$. Given a local section $(s,p) = (s_0, \dots, s_r, p)$ of $\mathcal{V}$, note that

$$M_j(s,p) := c_j \cdot \bigotimes_{i=1}^{k} s_{j_i} \otimes p$$

is a local section of ω_π. Define a map

$$\bigotimes_{i=1}^{k+1} \mathcal{V} \to \omega_\pi$$

by sending a section $(s,p)^1 \otimes \cdots \otimes (s,p)^{k+1}$ to

$$\sum_{\sigma \in S_{k+1}} \frac{c_j}{(k+1)!} \bigotimes_{i=1}^{k} s_{j_i}^{\sigma(i)} \otimes p^{\sigma(k+1)}.$$

This expression is symmetric in the factors of $\mathcal{V}$ and therefore induces a map

$$\hat{M}_j\colon \operatorname{Sym}^{k+1}(\mathcal{V}) \to \omega_\pi.$$

Summing over all monomials of $\hat{w}$ yields a map of vector bundles

$$\hat{w} := \sum_j \hat{M}_j\colon \operatorname{Sym}^{k+1}(\mathcal{V}) \to \omega_\pi. \tag{3.5}$$

Pushing forward via $\pi\colon \mathcal{C} \to S$, we obtain the desired morphism

$$\widetilde{\hat{w}} : \mathbb{R}\pi_*(\operatorname{Sym}^{k+1}(\mathcal{V})) \to \mathbb{R}\pi_*(\omega_\pi)$$

in the derived category of S. □

For simplicity of exposition, we will make the following assumption for the remainder of this section.

ASSUMPTION 3.5. Assume that the bundles A and B are such that the map $\tilde{\alpha}$ exists as a map of *complexes*.

We warn the reader that this assumption may not in fact hold in all cases. We will explain how to work around it in Section 4.

Given Assumption 3.5, $\tilde{\alpha}$ can be represented by a map of chain complexes

$$\begin{array}{ccccccc} \mathrm{Sym}^{k+1}(A) & \xrightarrow{d_{k+1}} & \mathrm{Sym}^k(A)\otimes B & \xrightarrow{d_k} & \mathrm{Sym}^{k-1}(A)\otimes\wedge^2(B) & \xrightarrow{d_{k-1}} & \cdots \\ \downarrow & & \downarrow{\scriptstyle\tilde{\alpha}} & & \downarrow & & \\ 0 & \longrightarrow & \mathcal{O}_S & \longrightarrow & 0 & \longrightarrow & \cdots. \end{array} \tag{3.6}$$

By abuse of notation we will also denote by $\tilde{\alpha}\colon \mathrm{Sym}^k(A)\otimes B\to\mathcal{O}_S$ the only non-zero vertical map.

PROPOSITION 3.6. *The differential $d\hat{w}$ induces a map of $\mathcal{O}_S$-modules over S:*

$$\widetilde{d\hat{w}}\colon \mathrm{Sym}^k(\pi_*\mathcal{V})\otimes\mathbb{R}^1\pi_*\mathcal{V}\to\mathcal{O}_S. \tag{3.7}$$

PROOF. By an analogous derivation as in (3.5), the function $\frac{\partial\hat{w}}{\partial x_i}$ defines a map of $\mathcal{O}_\mathcal{C}$-modules:

$$\mathrm{Sym}^k(\mathcal{V})\to\mathcal{L}^\vee\otimes\omega_\pi.$$

Tensoring with $\mathcal{L}$, we obtain a map:

$$\frac{\partial\hat{w}}{\partial x_i}\colon \mathrm{Sym}^k(\mathcal{V})\otimes\mathcal{L}\to\omega_\pi. \tag{3.8}$$

Similarly, one can define:

$$w=\frac{\partial\hat{w}}{\partial y}\colon \mathrm{Sym}^k(\mathcal{V})\otimes(\mathcal{L}^{-k}\otimes\omega_\pi)\to\omega_\pi.$$

For $0\le i\le r$ let $\pi_i:\mathcal{V}\to\mathcal{L}$ be the the projection onto the ith summand of $\mathcal{V}$ and let $\pi_y\colon\mathcal{V}\to\mathcal{L}^{-k}\otimes\omega_\pi$ be the projection onto the last summand $\mathcal{L}^{-k}\otimes\omega_\pi$. Then

$$d\hat{w}:=\frac{1}{k+1}\left(\sum_{i=0}^{r}\frac{\partial\hat{w}}{\partial x_i}\circ\mathrm{id}_{\mathrm{Sym}^k(\mathcal{V})}\otimes\pi_i+w\circ\mathrm{id}_{\mathrm{Sym}^k(\mathcal{V})}\otimes\pi_y\right) \tag{3.9}$$

gives a map of vector bundles $\mathrm{Sym}^k(\mathcal{V})\otimes\mathcal{V}\to\omega_\pi$. Pushing forward via $\pi\colon\mathcal{C}\to S$, we obtain a morphism

$$\widetilde{d\hat{w}}\colon \mathbb{R}\pi_*(\mathrm{Sym}^k(\mathcal{V})\otimes\mathcal{V})\to\mathbb{R}\pi_*(\omega_\pi)$$

in the derived category of S.

Composing with the natural map (defined in an analogous manner to (3.4))

$$\mathrm{nat}\colon \mathrm{Sym}^k(\mathbb{R}\pi_*\mathcal{V})\otimes\mathbb{R}\pi_*\mathcal{V}\to\mathbb{R}\pi_*(\mathrm{Sym}^k(\mathcal{V}))\otimes\mathbb{R}\pi_*\mathcal{V}\to\mathbb{R}\pi_*(\mathrm{Sym}^k(\mathcal{V})\otimes\mathcal{V})$$

and the vertical maps in (3.3), we obtain a morphism

$$\mathrm{Sym}^k(\mathbb{R}\pi_*\mathcal{V})\otimes\mathbb{R}\pi_*\mathcal{V}\to\mathcal{O}_S[-1].$$

By composing this with the morphism

$$\mathrm{Sym}^k(\pi_*\mathcal{V})\otimes\mathbb{R}\pi_*\mathcal{V}\to\mathrm{Sym}^k(\mathbb{R}\pi_*\mathcal{V})\otimes\mathbb{R}\pi_*\mathcal{V}$$

and applying H^1 to the composition we obtain a map of $\mathcal{O}_S$-modules

$$H^1\left(\mathrm{Sym}^k(\pi_*\mathcal{V})\otimes\mathbb{R}\pi_*\mathcal{V}\right)\to\mathcal{O}_S. \tag{3.10}$$

There is a canonical map

$$\mathrm{Sym}^k(\pi_*\mathcal{V})\otimes\mathbb{R}^1\pi_*\mathcal{V}\to H^1\left(\mathrm{Sym}^k(\pi_*\mathcal{V})\otimes\mathbb{R}\pi_*\mathcal{V}\right). \tag{3.11}$$

The composition of (3.10) with (3.11) gives the desired map. □

PROPOSITION 3.7. [**11**, Lemma 3.6.3] *The following diagram commutes*

$$\begin{array}{ccc} \mathrm{Sym}^k(\pi_*\mathcal{V})\otimes B & \xrightarrow{\qquad\qquad\tilde{\alpha}\qquad\qquad} & \mathcal{O}_S \\ & {\scriptstyle \mathrm{id}\otimes H^1(-)}\searrow \qquad \nearrow{\scriptstyle \widetilde{d\hat{w}}} & \\ & \mathrm{Sym}^k(\pi_*\mathcal{V})\otimes\mathbb{R}^1\pi_*\mathcal{V} & \end{array}$$

where $H^1(-)$ is the map $B\to B/A=\mathbb{R}^1\pi_*\mathcal{V}$.

PROOF. The proof rests on Euler's homogeneous function theorem, which states that

$$\hat{w}=\frac{1}{k+1}\left(\sum_{i=0}^{r}x_i\cdot\frac{\partial}{\partial x_i}\hat{w}+y\cdot w\right).$$

A direct calculation on local sections then shows that the following diagram commutes

$$\begin{array}{ccc} \mathrm{Sym}^k\mathcal{V}\otimes\mathcal{V} & \xrightarrow{m} & \mathrm{Sym}^{k+1}(\mathcal{V}) \\ & {\scriptstyle d\hat{w}}\searrow & \downarrow{\scriptstyle \hat{w}} \\ & & \omega_\pi, \end{array} \tag{3.12}$$

where $d\hat{w}$ and $\hat{w}$ are defined in (3.9) and (3.5) and m is the map $(v_1\otimes\cdots\otimes v_k)\otimes v\mapsto v_1\otimes\cdots\otimes v_k\otimes v$.

Consider the following commutative diagram

$$\begin{array}{ccc} \mathrm{Sym}^k(\pi_*\mathcal{V})\otimes B[-1] & & \\ \downarrow & & \\ \mathrm{Sym}^k(\pi_*\mathcal{V})\otimes\mathbb{R}\pi_*\mathcal{V} & & \\ \downarrow & & \\ \mathrm{Sym}^k(\mathbb{R}\pi_*\mathcal{V})\otimes\mathbb{R}\pi_*\mathcal{V} & \searrow & \\ \downarrow & & \\ \mathbb{R}\pi_*(\mathrm{Sym}^k(\mathcal{V}))\otimes\mathbb{R}\pi_*\mathcal{V} & & \mathrm{Sym}^{k+1}(\mathbb{R}\pi_*\mathcal{V}) \\ \downarrow & & \downarrow \\ \mathbb{R}\pi_*(\mathrm{Sym}^k(\mathcal{V})\otimes\mathcal{V}) & \longrightarrow & \mathbb{R}\pi_*(\mathrm{Sym}^{k+1}(\mathcal{V})) \\ & \searrow & \downarrow \\ & & \mathbb{R}\pi_*(\omega_\pi) \\ & & \downarrow \\ & & \mathcal{O}_S[-1]. \end{array} \tag{3.13}$$

Commutativity of the lower triangle is a result of (3.12). Commutativity of the upper trapezoid follows from the commutativity of the diagram below together with adjunction and the projection formula.

(3.14)

$$
\begin{array}{ccccc}
 & & \mathbb{L}\pi^* \operatorname{Sym}^k(\mathbb{R}\pi_*\mathcal{V}) \otimes \mathbb{L}\pi^*\mathbb{R}\pi_*\mathcal{V} & & \\
 & & \downarrow & & \\
 & & \operatorname{Sym}^k(\mathbb{L}\pi^*\mathbb{R}\pi_*\mathcal{V}) \otimes \mathbb{L}\pi^*\mathbb{R}\pi_*\mathcal{V} & & \\
 & \swarrow & \downarrow & \searrow & \\
\operatorname{Sym}^k(\mathcal{V}) \otimes \mathbb{L}\pi^*\mathbb{R}\pi_*\mathcal{V} & & & & \operatorname{Sym}^{k+1}(\mathbb{L}\pi^*\mathbb{R}\pi_*\mathcal{V}) \\
 & \searrow & \downarrow & & \downarrow \\
 & & \operatorname{Sym}^k(\mathcal{V}) \otimes \mathcal{V} & \longrightarrow & \operatorname{Sym}^{k+1}(\mathcal{V}).
\end{array}
$$

$\square$

Definition 3.8. [**5**, Definition 3.3] Let $\mathcal{F}$ be a coherent sheaf on a stack S. A *cosection* of $\mathcal{F}$ is a homomorphism of $\mathcal{O}_S$-modules

$$\sigma\colon \mathcal{F} \to \mathcal{O}_S.$$

Given a cosection $\sigma\colon \mathcal{F} \to \mathcal{O}_S$, we denote by $\{\sigma \equiv 0\}$ the locus

$$\{\sigma \equiv 0\} = \left\{ s \in S \;\middle|\; \sigma|_s\colon \mathcal{F} \otimes_{\mathcal{O}_S} \mathbf{k}(s) \to \mathbf{k}(s) \text{ vanishes} \right\}.$$

The map of vector bundles $\tilde{\alpha}\colon \operatorname{Sym}^k(A) \otimes B \to \mathcal{O}_S$ defines a cosection

$$\alpha\colon E = \tau^*(B) \to \mathcal{O}_{\operatorname{tot}(A)}$$

as follows. Given a local section b of E, define

$$\alpha(b) := \tau^*(\tilde{\alpha})\left(\frac{a^k}{k!} \otimes b\right) \in \mathcal{O}_{\operatorname{tot}(A)},$$

where $a \in \Gamma(\operatorname{tot}(A), \tau^*(A))$ is the tautological section. Here we view $\frac{a^k}{k!}$ as the tautological section of $\operatorname{Sym}^k \tau^*(A)$. Over U we now have the following maps of $\mathcal{O}_U$-modules,

$$
\begin{array}{ccccc}
\mathcal{O}_U & \xrightarrow{\beta} & E & \xrightarrow{\alpha} & \mathcal{O}_U \\
 & & \downarrow & & \\
 & & U. & &
\end{array}
$$

The map $\widetilde{d\hat{w}}$ defines a cosection of $\tau^*\left(\mathbb{R}^1\pi_*\mathcal{V}\right)$ over $\overline{\mathscr{M}}_{g,n}(\mathbb{P}^r, d)^p_S$ which, by abuse of notation, we also denote by $d\hat{w}$:

$$d\hat{w}\colon \tau^*\left(\mathbb{R}^1\pi_*\mathcal{V}\right) \to \mathcal{O}_{\overline{\mathscr{M}}_{g,n}(\mathbb{P}^r,d)^p_S} \tag{3.15}$$

The cosection $d\hat{w}$ can be written in coordinates as:

$$(\dot{s}, \dot{p}) \mapsto \frac{1}{k+1}\left(\frac{1}{k!}\sum_{i=0}^{r} \frac{\partial \hat{w}}{\partial x_i}(s_0, \ldots, s_r, p)\dot{s}_i + \frac{1}{k!} w(s_0, \ldots, s_r)\dot{p}\right).$$

where $(s,p) = (s_0,\ldots,s_r,p)$ is the tautological section of $\pi_*\mathcal{V}$ on $\mathrm{tot}(\pi_*\mathcal{V})$ and $(\dot{s},\dot{p})$ is a section of $\tau^*\left(\mathbb{R}^1\pi_*\mathcal{V}\right)$. The expression on the right is then an element of $\mathbb{R}^1(\omega_\pi) \cong \mathcal{O}_{\mathrm{tot}(\mathcal{V})}$. (see [**5**, Section 3.3]).

Since w defines an isolated singularity at the origin, the locus

$$\left\{(x_0,\ldots,x_r) \;\middle|\; w(x_0,\ldots,x_r) = 0, \frac{\partial w}{\partial x_i}(x_0,\ldots,x_r) = 0\right\}$$

is the origin. Consequently, the cosection $d\hat{w}$ is identically zero only when $p = 0$ and $w(s_0,\ldots,s_r) = f^*(w) = 0$. But $\overline{\mathscr{M}}_{g,n}(X_w,d)_S \subset \overline{\mathscr{M}}_{g,n}(\mathbb{P}^r,d)_S \subset \overline{\mathscr{M}}_{g,n}(\mathbb{P}^r,d)^p_S$ is exactly the locus where p and $f^*(w)$ vanish. We conclude that:

$$\{d\hat{w} \equiv 0\} = \overline{\mathscr{M}}_{g,n}(X_w,d)_S.$$

COROLLARY 3.9. *The closed substack $Z(\beta)\cap\{\alpha \equiv 0\}$ is equal to $\overline{\mathscr{M}}_{g,n}(X_w,d)_S \subset U$.*

PROOF. Consider the cosection $\alpha\colon E \to \mathcal{O}_U$ of E on U. After restricting to $Z(\beta)$ and identifying this locus with $\overline{\mathscr{M}}_{g,n}(\mathbb{P}^r,d)^p_S$, Proposition 3.7 implies that on $Z(\beta)$, the cosection α factors as a surjection followed by $d\hat{w}$. Thus the locus $Z(\beta)\cap\{\alpha \equiv 0\}$ equals $\{d\hat{w} \equiv 0\} = \overline{\mathscr{M}}_{g,n}(X_w,d)_S$. □

3.3. $\mathbb{Z}_2$-localized Chern character. In Section 1 we saw that given a vector bundle with a section, one could construct a refined Euler class supported on the vanishing locus of the section. In this section we describe how one can generalize this construction to also incorporate the data of a cosection. In this case the refined Euler class will be supported on the intersection of the vanishing locus of the section and the vanishing locus of the cosection. This is desirable because, as we have seen, the vanishing locus of just the section may not be compact, whereas the intersection of the vanishing locus of the section and the vanishing locus of the cosection may be compact. Having a virtual class supported on a compact space allows us to define numerical invariants, by integrating over the virtual class.

The generalization of the refined Euler class to incorporate a cosection was first described in [**29**]. The key is a modification of *MacPherson's graph construction* [**18, 26**].

Given a smooth variety X and a vector bundle E of finite rank, the following identity holds [**18**, Example 3.2.5]

$$e(E) = \mathrm{ch}(\wedge^\bullet E^\vee)\,\mathrm{Td}(E) \tag{3.16}$$

where ch is the Chern character, Td is the Todd class (viewed as an element of $A_*(X)$), and $\wedge^\bullet E^\vee$ is the class in K-theory given by the alternating sum

$$\oplus_{k=0}^{\mathrm{rk}(E)}(-1)^k \wedge^k E^\vee.$$

This identity can be refined to include a section $\sigma \in \Gamma(X,E)$. Define the Koszul complex

$$K := \wedge^{\mathrm{rk}(E)}E^\vee \to \cdots \to \wedge^2 E^\vee \to E^\vee \to \mathcal{O}_X,$$

where the differential is given by contraction with respect to σ, $d(-) := -\lrcorner\sigma$. Since σ is non-vanishing outside of $Z = Z(\sigma)$, a standard linear algebra exercise shows this complex is exact outside of $Z(\sigma)$, i.e. K is supported on Z. In this setting, MacPherson's graph construction describes a limiting procedure similar to that in

Definition 1.2 (see [**18**, Section 18.1] and [**26**] for details). This defines a *localized Chern character*:

$$\mathrm{ch}_Z^X(K)\colon A_*(X) \to A_*(Z).$$

The following identity holds

$$e(E,\sigma) = \mathrm{ch}_Z^X(K)(\mathrm{Td}(E)), \tag{3.17}$$

where $e(E,\sigma)$ is the refined Euler characteristic of Definition 1.3.

An insight of Polishchuk–Vaintrob in [**29**] was to adapt the definition of the localized Chern character to the case of two-periodic complexes. Given an infinite two-periodic complex K of vector bundles, let Z denote the support of K. In [**29**] a $\mathbb{Z}_2$*-localized Chern character* is defined:

$$^{\mathbb{Z}_2}\,\mathrm{ch}_Z^X(K)\colon A_*(X) \to A_*(Z).$$

The construction is a $\mathbb{Z}_2$-graded version of the original construction.

Definition 3.10. Assume we have a vector bundle $E \to X$ with a section $\sigma \in \Gamma(X,E)$ and a cosection $\rho \in \hom(E, \mathcal{O}_X)$ such that $\rho(\sigma) = 0$. Define the vector bundles

$$F_0 := \oplus_k \wedge^{2k} E^\vee$$
$$F_{-1} := \oplus_k \wedge^{2k+1} E^\vee,$$

and define a differential between them $d(-) := - \wedge \rho + - \lrcorner \sigma$. One can check that $d_{i\pm 1} \circ d_i = \mathrm{id}_{F_i} \cdot \rho(\sigma) = 0$. Denote by $\{\rho, \sigma\}$ the two-periodic complex of vector bundles:

$$\{\rho,\sigma\} := \cdots \to F_{-1} \xrightarrow{d_{-1}} F_0 \xrightarrow{d_0} F_{-1} \xrightarrow{d_{-1}} F_0 \to \cdots.$$

It is shown in [**30**, Lemma 1.4.1] that the support of $\{\rho,\sigma\}$ is $Z(\sigma) \cap \{\rho \equiv 0\}$. The complex $\{\rho,\sigma\}$ may be viewed as a two-periodic generalization of the Koszul complex to include a cosection; we will to refer it as a *two-periodic Koszul complex.*

3.4. The virtual class. Let us now return to the setting of Section 3.2. Given Assumption 3.5, we have a vector bundle $E \to U$, together with a section $\beta \in \Gamma(U,E)$ and a cosection $\alpha \in \hom(E, \mathcal{O}_U)$.

Proposition 3.11. [**30**, Lemma 4.2.6] *The function $\alpha(\beta)$ is identically zero.*

Proof. Recall from Definition 3.2, that U is an open subset of $\mathrm{tot}(A)$, $A \xrightarrow{\delta} B$ is a two term complex of vector bundles on S, E is the pullback of B via the map $\tau\colon \mathrm{tot}(A) \to S$, and $\beta \in \Gamma(\mathrm{tot}(A), E)$ is the section induced by δ.

First note that the section β is equal to $\tau^*\delta(a)$, where $a \in \Gamma(\mathrm{tot}(A), \tau^*(A))$ is the tautological section. The composition $\alpha(\beta)$ is given by $\tau^*\tilde{\alpha}\left(\frac{a^k}{k!} \otimes \tau^*\delta(a)\right)$, where $\tilde{\alpha}$ is defined in (3.6). On the other hand, $\frac{a^k}{k!} \otimes \tau^*\delta(a)$ is equal to $\tau^* d_{k+1}\left(\frac{a^{k+1}}{(k+1)!}\right)$. By the commutativity of the first square of (3.6),

$$\alpha(\beta) = \tau^*(\tilde{\alpha} \circ d_{k+1})\left(\frac{a^{k+1}}{(k+1)!}\right) = 0.$$

□

By Proposition 3.11, we can construct the two-periodic Koszul complex $\{\alpha,\beta\}$ as above. The support of $\{\alpha,\beta\}$ is given by $Z(\beta) \cap \{\alpha \equiv 0\}$ which, by Corollary 3.9, is $\overline{\mathscr{M}}_{g,n}(X_w, d)_S$. In analogy with (3.17), we make the following definition.

DEFINITION 3.12. Let $\mathscr{M}_{X_w} := \overline{\mathscr{M}}_{g,n}(X_w, d)_S$. Define the virtual class

$$\left[\overline{\mathscr{M}}_{g,n}(\mathbb{P}^r, d)_S^p\right]^{\mathrm{vir}} := {}^{\mathbb{Z}_2}\operatorname{ch}^U_{\mathscr{M}_{X_w}}(\{\alpha, \beta\})(\mathrm{Td}(E)) \in A_*(\overline{\mathscr{M}}_{g,n}(X_w, d)_S).$$

REMARK 3.13. If Assumption 3.5 held for $S = \mathfrak{Bun}^\circ_{g,n,d}$, we would obtain a virtual class

$$\left[\overline{\mathscr{M}}_{g,n}(\mathbb{P}^r, d)^p\right]^{\mathrm{vir}} \in A_*(\overline{\mathscr{M}}_{g,n}(X_w, d)).$$

Unfortunately Assumption 3.5 may not hold in this case. In Section 4 we explain how to overcome this difficulty.

4. Technical assumptions and the "two-step procedure"

The construction of a virtual class on $\overline{\mathscr{M}}_{g,n}(X_w, d)$ using the method of Section 3.4 would require Assumption 3.5 holding for $\mathfrak{Bun}^\circ_{g,n,d}$. This is not known, and likely not true. In this section we describe a "two-step procedure" to circumvent the problem. For details and a general formulation see Section 4 of [**11**].

4.1. Step 1: a projective embedding. Fix $g, n, d \geq 0$ such that $2g-2+n > 0$. We first embed $\overline{\mathscr{M}}_{g,n}(\mathbb{P}^r, d)$ into (the smooth locus of) a larger space of stable maps.

Consider the stabilization map:

$$\mathrm{st}\colon \mathfrak{Bun}^\circ_{g,n,d} \to \overline{\mathscr{M}}_{g,n},$$

which forgets the line bundle $\mathcal{L}$ and stabilizes the curve. Let $\tilde{\mathrm{st}}\colon \mathfrak{C} \to \mathcal{C}_{\mathrm{st}}$ denote the map between the universal curves over $\mathfrak{Bun}^\circ_{g,n,d} \to \overline{\mathscr{M}}_{g,n}$. The coarse underlying space $\underline{\mathcal{C}}_{\mathrm{st}}$ associated to the Deligne–Mumford stack $\mathcal{C}_{\mathrm{st}}$ is projective, and so can be embedded into $\mathbb{P}^{N-1}$. We obtain the following diagram:

$$\begin{array}{ccccc} \mathcal{L} & & & & \mathcal{O}_{\mathbb{P}^{N-1}}(1) \\ \downarrow & & & & \downarrow \\ \mathfrak{C} & \xrightarrow{\tilde{\mathrm{st}}} & \mathcal{C}_{\mathrm{st}} & \xrightarrow{\rho} & \underline{\mathcal{C}}_{\mathrm{st}} \subset \mathbb{P}^{N-1} \\ \downarrow{\scriptstyle \pi} & & \downarrow & & \\ \mathfrak{Bun}^\circ_{g,n,d} & \xrightarrow{\mathrm{st}} & \overline{\mathscr{M}}_{g,n}. & & \end{array}$$

Let $\mathcal{N} \to \mathfrak{C}$ denote the line bundle $\mathcal{N} = \tilde{\mathrm{st}}^* \circ \rho^*(\mathcal{O}_{\mathbb{P}^{N-1}}(1))$. One can choose an appropriate projective embedding $\underline{\mathcal{C}}_{\mathrm{st}} \subset \mathbb{P}^{N-1}$ so that $\mathcal{L} \otimes \mathcal{N}$ is π-acyclic. Consider the Euler sequence

$$0 \to \mathcal{O}_{\mathbb{P}^{N-1}} \to \mathcal{O}_{\mathbb{P}^{N-1}}(1)^{\oplus N} \to T\mathbb{P}^{N-1} \to 0.$$

Pulling this back to $\mathfrak{C}$ and tensoring with $\mathcal{V}_1 = \mathcal{L}^{\oplus r+1}$, we obtain

$$0 \to \mathcal{V}_1 \to \mathcal{M}^{\oplus M} \to \mathcal{Q} \to 0$$

where $\mathcal{M} := \mathcal{L} \otimes \mathcal{N}$, $M = (r+1)N$ and $\mathcal{Q} = \mathcal{V}_1 \otimes \tilde{\mathrm{st}}^* \circ \rho^*(T\mathbb{P}^{N-1})$. Pushing forward via π we obtain the long exact sequence on $\mathfrak{Bun}^\circ_{g,n,d}$:

$$\begin{aligned} 0 \to &\pi_*(\mathcal{V}_1) \to \pi_*\left(\mathcal{M}^{\oplus M}\right) \to \pi_*(\mathcal{Q}) \\ \to &\mathbb{R}^1\pi_*(\mathcal{V}_1) \to 0 \end{aligned} \tag{4.1}$$

where the last terms are zero because $\mathcal{M}$ was constructed to be π-acyclic. Letting

$$A_1' := \pi_* \left(\mathcal{M}^{\oplus M}\right),$$
$$B_1' := \pi_*(\mathcal{Q}),$$

we see that $[A_1' \to B_1']$ is a two term resolution of $\mathbb{R}\pi_*(\mathcal{V}_1)$ by vector bundles. We also obtain an embedding $\mathrm{tot}(\pi_*\mathcal{V}_1)$ in $\mathrm{tot}(A_1')$. Let

$$\tau\colon \mathrm{tot}(A_1') \to \mathfrak{Bun}^\circ_{g,n,d}$$

denote the map forgetting the sections. Then $\mathrm{tot}(\pi_*\mathcal{V}_1)$ is the zero locus of the section $\beta_1 \in \Gamma(\mathrm{tot}(A_1'), \tau^*(B_1'))$ induced by the map $A_1' \to B_1'$.

The section of $\mathcal{N}^{\oplus N}$ obtained by pulling back the first terms of the Euler sequence is nowhere vanishing. Therefore nowhere vanishing sections of $\mathcal{V}_1$ are mapped to nowhere vanishing sections of $\mathcal{M}^{\oplus M}$. Define U_1 to be the open substack of $\mathrm{tot}(A_1')$ consisting of sections of $\mathcal{M}^{\oplus M}$ which are nowhere vanishing. Note that these sections correspond to a map to $\mathbb{P}^{M-1}$ of some degree e. This realizes U_1 as an open subset of $\overline{\mathscr{M}}_{g,n}(\mathbb{P}^{M-1}, e)$. By construction U_1 is a smooth Deligne–Mumford stack, lying in the smooth locus of $\overline{\mathscr{M}}_{g,n}(\mathbb{P}^{M-1}, e)$. Let $\overline{U}_1$ denote the closure of U_1 in $\overline{\mathscr{M}}_{g,n}(\mathbb{P}^{M-1}, e)$. The zero locus of $\beta_1|_{U_1}$ is equal to $\overline{\mathscr{M}}_{g,n}(\mathbb{P}^r, d)$. We have

$$\overline{\mathscr{M}}_{g,n}(\mathbb{P}^r, d) \hookrightarrow U_1 \subset \overline{U}_1 \subset \overline{\mathscr{M}}_{g,n}(\mathbb{P}^{M-1}, e).$$

4.2. Step 2: an admissible resolution. The next step is to construct a second resolution, this time of $\tau^*(\mathbb{R}\pi_*\mathcal{V})$ over U_1. Because $\overline{U}_1$ has projective coarse moduli space, this can be done in such a way that the map $\tilde{\alpha}$ is realized at the level of complexes.

Let $\pi_{U_1}\colon \mathcal{C}_{U_1} \to U_1$ and $\pi_{\overline{U}_1}\colon \mathcal{C}_{\overline{U}_1} \to \overline{U}_1$ denote the universal curves over U_1 and $\overline{U}_1$ respectively. Let $\mathcal{L}_{U_1}$, $\mathcal{N}_{U_1}$, and $\mathcal{M}_{U_1}$ denote the pullback of $\mathcal{L}$, $\mathcal{N}$, and $\mathcal{M}$ from $\mathfrak{C}$ to $\mathcal{C}_{U_1}$. These line bundles naturally extend to line bundles $\mathcal{L}_{\overline{U}_1}$, $\mathcal{N}_{\overline{U}_1}$, and $\mathcal{M}_{\overline{U}_1}$ over $\overline{U}_1$ as follows. Since $\overline{U}_1 \subset \overline{\mathscr{M}}_{g,n}(\mathbb{P}^{M-1}, e)$, there exists a universal map $f\colon \mathcal{C}_{\overline{U}_1} \to \mathbb{P}^{M-1}$. Define $\mathcal{M}_{\overline{U}_1}$ to be the pullback of $\mathcal{O}_{\mathbb{P}^{M-1}}(1)$. On the other hand by forgetting f and stabilizing the curve, we get a map $\overline{U}_1 \subset \overline{\mathscr{M}}_{g,n}(\mathbb{P}^{M-1}, e) \to \overline{\mathscr{M}}_{g,n}$, and consequently a map

$$\mathcal{C}_{\overline{U}_1} \to \mathcal{C}_{\mathrm{st}} \to \underline{\mathcal{C}}_{\mathrm{st}} \hookrightarrow \mathbb{P}^{N-1}.$$

Define $\mathcal{N}_{\overline{U}_1}$ to be the pullback of $\mathcal{O}_{\mathbb{P}^{N-1}}(1)$. Finally, define

$$\mathcal{L}_{\overline{U}_1} := \mathcal{M}_{\overline{U}_1} \otimes \mathcal{N}^\vee_{\overline{U}_1}.$$

The line bundle $\mathcal{L}_{\overline{U}_1} \to \mathcal{C}_{\overline{U}_1}$ is a degree d line bundle over the family $\mathcal{C}_{\overline{U}_1} \to \overline{U}_1$ of pre-stable curves, therefore there is an induced map $\overline{U}_1 \to \mathfrak{Bun}_{g,n,d}$, which, when restricted to $U_1 \subset \overline{U}_1$, recovers the forgetful map $U_1 \to \mathfrak{Bun}^\circ_{g,n,d} \subset \mathfrak{Bun}_{g,n,d}$.

Let

$$\mathcal{V}_{\overline{U}_1,1} := \mathcal{L}_{\overline{U}_1}^{\oplus r+1},$$
$$\mathcal{V}_{\overline{U}_1,2} := \mathcal{L}_{\overline{U}_1}^{\otimes -k} \otimes \omega_{\pi_{\overline{U}_1},\log},$$

and define

$$\mathcal{V}_{\overline{U}_1} := \mathcal{V}_{\overline{U}_1,1} \oplus \mathcal{V}_{\overline{U}_1,2}.$$

Let $\mathcal{V}_{U_1,1}, \mathcal{V}_{U_1,2}$, and $\mathcal{V}_{U_1}$ denote the corresponding restrictions to $\mathcal{C}_{U_1}$.

Using the fact that $\overline{U}_1$ has projective coarse underlying space, we can resolve $\mathcal{V}_{\overline{U}_1}$ by a two term complex of vector bundles for which the map $\tilde{\alpha}$ is realized at the level of complexes.

PROPOSITION 4.1. [**11**, Proposition 3.5.2] *On $\overline{U}_1$, there exist resolutions*

$$[A''_{\overline{U}_1,1} \to B''_{\overline{U}_1,1}] \sim \mathbb{R}\pi_*(\mathcal{V}_{\overline{U}_1,1}),$$
$$[A''_{\overline{U}_1,2} \to B''_{\overline{U}_1,2}] \sim \mathbb{R}\pi_*(\mathcal{V}_{\overline{U}_1,2})$$

such that, if we let

$$[A''_{\overline{U}_1} \to B''_{\overline{U}_1}] := [A''_{\overline{U}_1,1} \oplus A''_{\overline{U}_1,2} \to B''_{\overline{U}_1,1} \oplus B''_{\overline{U}_1,2}],$$

there exists a map

$$\tilde{\alpha}\colon \mathrm{Sym}^k(A''_{\overline{U}_1}) \otimes B''_{\overline{U}_1} \to \mathcal{O}_{\overline{U}_1}$$

fitting into the diagram (3.6) *on $\overline{U}_1$ and realizing the map from Proposition* 3.3 *at the level of complexes.*

Let

$$A''_{U_1,1} \to B''_{U_1,1},$$
$$A''_{U_1,2} \to B''_{U_1,2},$$

denote the restrictions of

$$A''_{\overline{U}_1,1} \to B''_{\overline{U}_1,1},$$
$$A''_{\overline{U}_1,2} \to B''_{\overline{U}_1,2},$$

to U_1. Let

$$A'_{U_1,1} := \tau^*(A'_1),$$
$$B'_{U_1,1} := \tau^*(B'_1).$$

In the derived category $D(U_1)$ we have the equivalence

$$[A''_{U_1,1} \to B''_{U_1,1}] \sim [A'_{U_1,1} \to B'_{U_1,1}].$$

By standard arguments (see Lemma 3.6.5 of [**11**]) there exists another resolution $[A_{U_1,1} \to B_{U_1,1}]$ of $\mathbb{R}\pi_*(\mathcal{V}_{U_1,1})$ and a roof diagram

$$(4.2)\qquad \begin{array}{ccccccc} & & A_{U_1,1} & \xrightarrow{d_1} & B_{U_1,1} & & \\ & \swarrow & & \times & & \searrow & \\ A''_{U_1,1} & \xrightarrow{d''_1} & B''_{U_1,1} & & A'_{U_1,1} & \xrightarrow{d'_1} & B'_{U_1,1} \end{array}$$

where the diagonal maps of two-term complexes are quasi-isomorphisms. Consider the resolution of $\mathbb{R}\pi_*(\mathcal{V}_{U_1})$ given by

$$[A_{U_1} \to B_{U_1}] := [A_{U_1,1} \oplus A''_{U_1,2} \to B_{U_1,1} \oplus B''_{U_1,2}].$$

By composing the map $\tilde{\alpha}$ of Proposition 4.1 with the left diagonal map of (4.2), we obtain the map (which by abuse of notation we will also denote as $\tilde{\alpha}$)

$$(4.3)\qquad \tilde{\alpha}\colon \mathrm{Sym}^k(A_{U_1}) \otimes B_{U_1} \to \mathcal{O}_{U_1}.$$

This map fits into the diagram (3.6) over U_1 and and realizes the map from Proposition 3.3 at the level of complexes.

Consider the morphism

$$\tilde{\tau}\colon \operatorname{tot}(A_{U_1}) \to U_1$$

forgetting the section of A_{U_1}. Define $E := \tilde{\tau}^*(B_{U_1})$. The map $[A_{U_1} \to B_{U_1}]$ defines a section

$$\beta \in \Gamma(\operatorname{tot}(A_{U_1}), E)$$

and the map $\tilde{\alpha}$ of (4.3) defines a cosection

$$\alpha \in \hom(E, \mathcal{O}_{\operatorname{tot}(A_{U_1})}).$$

The function $\alpha(\beta)$ is equal to zero by Proposition 3.11. We can therefore define the two-periodic complex $\{\alpha, \beta\}$ as in Definition 3.10.

4.3. The cut-down procedure. The space $\operatorname{tot}(A_{U_1})$ lies over $U_1 \subset \operatorname{tot}(A_1')$, so the relative dimension of $\operatorname{tot}(A_{U_1}) \to \mathfrak{Bun}^\circ_{g,n,d}$ is $\operatorname{rank}(A_{U_1}) + \operatorname{rank}(A_1')$. If we view $E = \tilde{\tau}^*(B_{U_1})$ as an obstruction bundle, the relative virtual dimension over $\mathfrak{Bun}^\circ_{g,n,d}$ is

$$\operatorname{rank}(A_{U_1}) - \operatorname{rank}(B_{U_1}) + \operatorname{rank}(A_1') = \chi(\mathbb{R}\pi_*\mathcal{V}) + \operatorname{rank}(A_1')$$

which is too large by $\operatorname{rank}(A_1')$. This overcounting is due to the fact that we resolved $\mathbb{R}\pi_*(\mathcal{V}_1)$ twice.[1] We must correct for this redundancy by choosing an appropriate closed subset of codimension equal to $\operatorname{rank}(A_1')$.

We have tautological sections $\operatorname{taut} \in \Gamma(\operatorname{tot}(A_{U_1}), \tilde{\tau}^*(A_{U_1,1}))$ and $\operatorname{taut}' \in \Gamma(U_1, A'_{U_1,1})$. Let $f_1\colon A_{U_1,1} \to A'_{U_1,1}$ denote the surjective map from (4.2). Consider the section

$$\xi := \tilde{\tau}^*(f_1) \circ \operatorname{taut} - \tilde{\tau}^*(\operatorname{taut}') \in \Gamma(\operatorname{tot}(A_{U_1}), \tilde{\tau}^*(A'_{U_1,1})).$$

DEFINITION 4.2. Define the substack of $\operatorname{tot}(A_{U_1})$

$$\square := \{\xi = 0\}.$$

By definition of ξ, $\square$ consists of triples (a_1', a_1, a_2) where $a_1' \in U_1 \subset \operatorname{tot}(A_1')$, $a_1 \in \operatorname{tot}(A_{U_1,1})|_{a_1'}$, and $a_2 \in \operatorname{tot}(A_{U_1,2})|_{a_1'}$ such that $f_1(a_1) = a_1'$. Because f_1 is surjective, $\square$ will be smooth of relative dimension $\operatorname{rank}(A_{U_1})$ over $\mathfrak{Bun}^\circ_{g,n,d}$.

Let

$$\iota\colon \square \hookrightarrow \operatorname{tot}(A_{U_1})$$

denote the inclusion.

DEFINITION 4.3. Define the *Fundamental two-periodic Koszul complex*

$$K_{g,n,d} := \iota^*(\{\alpha, \beta\}).$$

The terms of $K_{g,n,d}$ consist of the wedge powers of $\iota^*(E)$.

PROPOSITION 4.4. *The support of $K_{g,n,d}$ is contained in $\overline{\mathscr{M}}_{g,n}(X_w, d)$.*

PROOF. By [**30**, Lemma 1.4.1],

$$\operatorname{supp}(K_{g,n,d}) = Z(\iota^*(\beta)) \cap \{\iota^*(\alpha) \equiv 0\}.$$

The locus $Z(\iota^*(\beta))$ is equal to $\overline{\mathscr{M}}_{g,n}(\mathbb{P}^r, d)^p$ after restricting to $\square$ in $\operatorname{tot}(A_{U_1})$. By Corollary 3.9, the intersection of this with $\{\iota^*(\alpha) \equiv 0\}$ is equal to

$$\overline{\mathscr{M}}_{g,n}(\mathbb{P}^r, d)^p \cap \{d\hat{w} \equiv 0\} = \overline{\mathscr{M}}_{g,n}(X_w, d).$$

$\square$

[1]More precisely, $[A_1' \to B_1']$ resolved $\mathbb{R}\pi_*(\mathcal{V}_1)$, and $[A_{U_1,1} \to B_{U_1,1}]$ resolved $\tau^*(\mathbb{R}\pi_*(\mathcal{V}_1))$.

DEFINITION 4.5. Let $\mathscr{M}_{X_w} = \overline{\mathscr{M}}_{g,n}(X_w, d)$. Define the virtual class

$$\left[\overline{\mathscr{M}}_{g,n}(\mathbb{P}^r, d)^p\right]^{\text{vir}} := {}^{\mathbb{Z}_2}\operatorname{ch}^{\square}_{\mathscr{M}_{X_w}}(K_{g,n,d})(\operatorname{Td}(\iota^*(E))) \in A_*(\overline{\mathscr{M}}_{g,n}(X_w, d)).$$

Definition 4.5 gives a new construction of the virtual fundamental class on the space of stable maps to a hypersurface, using a generalization of the refined Euler class to two-periodic complexes. We state without proof two comparison results to show that this class agrees with previous constructions of the virtual fundamental class.

PROPOSITION 4.6. [**11**, Proposition 6.1.7] *The virtual class* $\left[\overline{\mathscr{M}}_{g,n}(\mathbb{P}^r, d)^p\right]^{\text{vir}}$ *agrees with the* cosection-localized *virtual fundamental class defined by Chang–Li in* [**5**].

The proof uses results of [**6**] together with the compatibility of $\tilde{\alpha}$ and $\widetilde{d\hat{w}}$ described in Proposition 3.7.

On the other hand, the cosection-localized virtual fundamental class has been shown to agree with the original virtual fundamental class of Behrend-Fantechi.

PROPOSITION 4.7. [**5,7,22**] *The cosection-localized virtual fundamental of* [**5**] *is equal to the Behrend–Fantechi virtual fundamental class up to a sign of* $(-1)^{\chi(\mathcal{L}^{\otimes k})}$.

In [**5**, Theorem 1.1], Chang–J. Li prove that the degree of the virtual classes agree in the case of stable maps to a quintic three-fold with $n = 0$. In [**22**, Theorem 3.2] and [**7**, Theorem 1.1], Kim–Oh and Chang–M.-L. Li prove that the virtual classes themselves agree.

5. Conclusions and further directions

In these notes we have used two-periodic Koszul complexes and the localized top Chern class to give a new construction of the virtual fundamental class for Gromov–Witten theory of a projective hypersurface. This method was first developed by Polishchuk–Vaintrob in [**29**], where these ideas were employed to give a mathematical definition of the Witten top Chern class on the moduli space of spin curves.

In fact the methods described above apply in greater generality than has been treated here. The general context in which one might use such techniques is that of a *gauged linear sigma model* (GLSM). This includes both Gromov–Witten theory of a complete intersection as well as FJRW theory of a singularity as special cases. For a detailed description of the mathematical theory of the GLSM, as well as a construction of corresponding enumerative invariants in certain cases, see [**16**].

Although less familiar to mathematicians than Gromov–Witten theory, one place in which GLSMs arise naturally is through mirror symmetry and wall crossing correspondences with Gromov–Witten theory. The most well-known example of this is the Landau–Ginzburg/Calabi–Yau correspondence of [**8,9,24**], relating Gromov–Witten theory and FJRW theory of the quintic. An analogous result involving more exotic GLSMs, known as *hybrid models*, appears in [**13**].

In contrast to Gromov–Witten theory of hypersurfaces, for FJRW theory and general hybrid model GLSMs, a more general construction than the two-periodic Koszul complex is needed to define the corresponding enumerative theory. In this level of generality, two-periodic complexes are replaced by so-called matrix factorizations. Such a construction was accomplished first in FJRW theory in [**30**]. It was generalized to hybrid model GLSMs in [**11**].

Acknowledgments

These notes are based on a series of lectures given at the 2018 Crossing the Walls in Enumerative Geometry Summer Workshop, and those lectures are based upon the paper [**11**]. I thank the organizers of the workshop, T. Jarvis, N. Priddis, and Y. Ruan, as well as D. Coffman, the project manager who took care of much of the work required to make the workshop such a success. I am grateful to my collaborators on [**11**], I. Ciocan-Fontanine, D. Favero, J. Guéré, and B. Kim. I thank the University of Minnesota's Department of Mathematics for the use of their facilities during several extended collaborative meetings during the writing of [**11**]. I am grateful to T. Jarvis and Y. Ruan for many discussions explaining their beautiful construction of the GLSM. Finally, I am grateful to the referee for providing feedback and suggesting many improvements to the first draft of this manuscript.

References

[1] K. Behrend, *Gromov-Witten invariants in algebraic geometry*, Invent. Math. **127** (1997), no. 3, 601–617, DOI 10.1007/s002220050132. MR1431140

[2] K. Behrend, *Algebraic Gromov-Witten invariants*, New trends in algebraic geometry (Warwick, 1996), London Math. Soc. Lecture Note Ser., vol. 264, Cambridge Univ. Press, Cambridge, 1999, pp. 19–70, DOI 10.1017/CBO9780511721540.004. MR1714820

[3] K. Behrend and B. Fantechi, *The intrinsic normal cone*, Invent. Math. **128** (1997), no. 1, 45–88, DOI 10.1007/s002220050136. MR1437495

[4] Raoul Bott and Loring W. Tu, *Differential forms in algebraic topology*, Graduate Texts in Mathematics, vol. 82, Springer-Verlag, New York-Berlin, 1982. MR658304

[5] Huai-Liang Chang and Jun Li, *Gromov-Witten invariants of stable maps with fields*, Int. Math. Res. Not. IMRN **18** (2012), 4163–4217, DOI 10.1093/imrn/rnr186. MR2975379

[6] Huai-Liang Chang, Jun Li, and Wei-Ping Li, *Witten's top Chern class via cosection localization*, Invent. Math. **200** (2015), no. 3, 1015–1063, DOI 10.1007/s00222-014-0549-5. MR3348143

[7] Huai-Liang Chang and Mu-Lin Li, *Invariants of stable quasimaps with fields*, Trans. Amer. Math. Soc. **373** (2020), no. 5, 3669–3691, DOI 10.1090/tran/8011. MR4082252

[8] Alessandro Chiodo, Hiroshi Iritani, and Yongbin Ruan, *Landau-Ginzburg/Calabi-Yau correspondence, global mirror symmetry and Orlov equivalence*, Publ. Math. Inst. Hautes Études Sci. **119** (2014), 127–216, DOI 10.1007/s10240-013-0056-z. MR3210178

[9] Alessandro Chiodo and Yongbin Ruan, *Landau-Ginzburg/Calabi-Yau correspondence for quintic three-folds via symplectic transformations*, Invent. Math. **182** (2010), no. 1, 117–165, DOI 10.1007/s00222-010-0260-0. MR2672282

[10] Ionuţ Ciocan-Fontanine and Bumsig Kim, *Moduli stacks of stable toric quasimaps*, Adv. Math. **225** (2010), no. 6, 3022–3051, DOI 10.1016/j.aim.2010.05.023. MR2729000

[11] Ionuţ Ciocan-Fontanine, David Favero, Jérémy Guéré, Bumsig Kim, and Mark Shoemaker, *Fundamental factorization of a GLSM, part I: Construction*, preprint: `arXiv:1802.05247`, 2018.

[12] Ionuţ Ciocan-Fontanine, Bumsig Kim, and Davesh Maulik, *Stable quasimaps to GIT quotients*, J. Geom. Phys. **75** (2014), 17–47, DOI 10.1016/j.geomphys.2013.08.019. MR3126932

[13] Emily Clader, *Landau-Ginzburg/Calabi-Yau correspondence for the complete intersections $X_{3,3}$ and $X_{2,2,2,2}$*, Adv. Math. **307** (2017), 1–52, DOI 10.1016/j.aim.2016.11.010. MR3590512

[14] David A. Cox and Sheldon Katz, *Mirror symmetry and algebraic geometry*, Mathematical Surveys and Monographs, vol. 68, American Mathematical Society, Providence, RI, 1999. MR1677117

[15] Huijun Fan, Tyler Jarvis, and Yongbin Ruan, *The Witten equation, mirror symmetry, and quantum singularity theory*, Ann. of Math. (2) **178** (2013), no. 1, 1–106, DOI 10.4007/annals.2013.178.1.1. MR3043578

[16] Huijun Fan, Tyler J. Jarvis, and Yongbin Ruan, *A mathematical theory of the gauged linear sigma model*, Geom. Topol. **22** (2018), no. 1, 235–303, DOI 10.2140/gt.2018.22.235. MR3720344

[17] Kenji Fukaya and Kaoru Ono, *Arnold conjecture and Gromov-Witten invariant*, Topology **38** (1999), no. 5, 933–1048, DOI 10.1016/S0040-9383(98)00042-1. MR1688434

[18] William Fulton, *Intersection theory*, 2nd ed., Ergebnisse der Mathematik und ihrer Grenzgebiete. 3. Folge. A Series of Modern Surveys in Mathematics [Results in Mathematics and Related Areas. 3rd Series. A Series of Modern Surveys in Mathematics], vol. 2, Springer-Verlag, Berlin, 1998. MR1644323

[19] W. Fulton and R. Pandharipande, *Notes on stable maps and quantum cohomology*, Algebraic geometry—Santa Cruz 1995, Proc. Sympos. Pure Math., vol. 62, Amer. Math. Soc., Providence, RI, 1997, pp. 45–96, DOI 10.1090/pspum/062.2/1492534. MR1492534

[20] A. Grothendieck, *Éléments de géométrie algébrique. III. Étude cohomologique des faisceaux cohérents. II* (French), Inst. Hautes Études Sci. Publ. Math. **17** (1963), 91. MR163911

[21] Robin Hartshorne, *Algebraic geometry*, Springer-Verlag, New York-Heidelberg, 1977. Graduate Texts in Mathematics, No. 52. MR0463157

[22] Bumsig Kim and Jeongseok Oh, *Localized chern characters for 2-periodic complexes*, preprint `arXiv:1804.03774`, 2018.

[23] Joachim Kock and Israel Vainsencher, *An invitation to quantum cohomology*, Progress in Mathematics, vol. 249, Birkhäuser Boston, Inc., Boston, MA, 2007. Kontsevich's formula for rational plane curves. MR2262630

[24] Yuan-Pin Lee, Nathan Priddis, and Mark Shoemaker, *A proof of the Landau-Ginzburg/Calabi-Yau correspondence via the crepant transformation conjecture* (English, with English and French summaries), Ann. Sci. Éc. Norm. Supér. (4) **49** (2016), no. 6, 1403–1443, DOI 10.24033/asens.2312. MR3592361

[25] Jun Li and Gang Tian, *Virtual moduli cycles and Gromov-Witten invariants of algebraic varieties*, J. Amer. Math. Soc. **11** (1998), no. 1, 119–174, DOI 10.1090/S0894-0347-98-00250-1. MR1467172

[26] R. D. MacPherson, *Chern classes for singular algebraic varieties*, Ann. of Math. (2) **100** (1974), 423–432, DOI 10.2307/1971080. MR361141

[27] Brian Osserman, *Notes on cohomology and base change*, `https://www.math.ucdavis.edu/~osserman/math/cohom-base-change.pdf`.

[28] R. Pandharipande and R. P. Thomas, *13/2 ways of counting curves*, Moduli spaces, London Math. Soc. Lecture Note Ser., vol. 411, Cambridge Univ. Press, Cambridge, 2014, pp. 282–333. MR3221298

[29] Alexander Polishchuk and Arkady Vaintrob, *Algebraic construction of Witten's top Chern class*, Advances in algebraic geometry motivated by physics (Lowell, MA, 2000), Contemp. Math., vol. 276, Amer. Math. Soc., Providence, RI, 2001, pp. 229–249, DOI 10.1090/conm/276/04523. MR1837120

[30] Alexander Polishchuk and Arkady Vaintrob, *Matrix factorizations and cohomological field theories*, J. Reine Angew. Math. **714** (2016), 1–122, DOI 10.1515/crelle-2014-0024. MR3491884

[31] Yongbin Ruan, *Virtual neighborhoods and pseudo-holomorphic curves*, Proceedings of 6th Gökova Geometry-Topology Conference, Turkish J. Math. **23** (1999), no. 1, 161–231. MR1701645

[32] Edward Witten, *Phases of $N = 2$ theories in two dimensions*, Mirror symmetry, II, AMS/IP Stud. Adv. Math., vol. 1, Amer. Math. Soc., Providence, RI, 1997, pp. 143–211. MR1416338

Department of Mathematics, Colorado State University, 1874 Campus Delivery, Fort Collins, Colorado 80523-1874

Email address: `mark.shoemaker@colostate.edu`

Contemporary Mathematics
Volume **763**, 2021
https://doi.org/10.1090/conm/763/15325

Localized Chern characters for 2-periodic complexes and virtual cycles

Jeongseok Oh

ABSTRACT. This survey note is based on the talk at the May 2018 Snowbird workshop 'Crossing the Walls in Enumerative Geometry.' We explain a definition of localized Chern characters for 2-periodic complexes, their properties and applications to Gromov-Witten theory. To do so, we explain how the localized Chern character for a 2-periodic complexe gives rise to a virtual fundamental class.

1. Introduction

Let Q be a smooth projective variety over the complex numbers $\mathbb{C}$. The moduli space of stable maps $\overline{\mathcal{M}}_{g,k}(Q,d)$ is a compactification of a moduli space $\mathcal{M}_{g,k}(Q,d)$ of maps of degree $d \in H_2(Q;\mathbb{Z})$ from genus g, smooth curves with distinct k marked points to Q. More precisely, $\overline{\mathcal{M}}_{g,k}(Q,d)$ is a proper Deligne-Mumford (DM for short) stack over $\mathbb{C}$ that contains $\mathcal{M}_{g,k}(Q,d)$ as an open substack. By an argument of a deformation theory, the virtual dimension of $\overline{\mathcal{M}}_{g,k}(Q,d)$ is

$$\mathrm{vdim} := c_1(T_Q) \cap d + (1-g)(\dim Q - 3) + k,$$

where c_1 stands for the first Chern class. A starting point for the Gromov-Witten theory for Q is to get a cycle (or a homology) class of the moduli space $\overline{\mathcal{M}}_{g,k}(Q,d)$, which is related to the deformation theory of $\overline{\mathcal{M}}_{g,k}(Q,d)$ in a certain way (this implies that the degree of such a cycle must be vdim; we will see it briefly later in Section 3.1 and 4.1). By cap products with this cycle, or push-forwards of such products by proper morphisms, we obtain the Gromov-Witten potentials for Q, a cohomological field theory of the Gromov-Witten theory for Q, etc. This cycle is constructed in [**2**] and it is denoted by

$$[\overline{\mathcal{M}}_{g,k}(Q,d)]^{vir} \in A_{\mathrm{vdim}}(\overline{\mathcal{M}}_{g,k}(Q,d))_{\mathbb{Q}} \text{ (or } H_{2\mathrm{vdim}}(\overline{\mathcal{M}}_{g,k}(Q,d);\mathbb{Q})),$$

where $A(-)$ denotes the Chow group (see [**9, 15**] for the definition of Chow groups) and $A(-)_{\mathbb{Q}}$ denotes $A(-) \otimes_{\mathbb{Z}} \mathbb{Q}$.

Although Gromov-Witten theory is always well-defined for a smooth projective Q (in the sense that $[\overline{\mathcal{M}}_{g,k}(Q,d)]^{vir}$ is naturally defined), it is not easy in general to study it further. When Q is a (locally) complete intersection in a GIT quotient $V /\!\!/ G = [V^{ss}/G]$ of a vector space V over $\mathbb{C}$ by a reductive algebraic group G, we

2020 *Mathematics Subject Classification.* Primary 14C17; Secondary 14N35.
Key words and phrases. Algebraic geometry, Intersection theory.

have another framework to study Gromov-Witten theory for Q, which is called a gauged linear sigma model. This model defines a moduli space $\overline{\mathcal{M}}_{g,k}(V /\!\!/ G, d)^p$ for each g, k, and $d \in H_2(V /\!\!/ G; \mathbb{Z})$ that is a DM stack (but not proper in general). This moduli space is parametrized by an object in $\overline{\mathcal{M}}_{g,k}(V /\!\!/ G, d)$ with some extra data determined by numerical information of an embedding $Q \subset V /\!\!/ G$; we will see the details in Section 4.1. Note that the extra data forms a vector space at each point in $\overline{\mathcal{M}}_{g,k}(V /\!\!/ G, d)$ and the rank defines an upper semi-continuous function on $\overline{\mathcal{M}}_{g,k}(V /\!\!/ G, d)$. Hence, $\overline{\mathcal{M}}_{g,k}(V /\!\!/ G, d)^p$ is decomposed by vector bundles on a stratification of $\overline{\mathcal{M}}_{g,k}(V /\!\!/ G, d)$. By this reason, $\overline{\mathcal{M}}_{g,k}(V /\!\!/ G, d)^p$ in general has more symmetries than $\overline{\mathcal{M}}_{g,k}(Q, d)$. Also note that the extra data is designed to satisfy that the virtual dimension of $\overline{\mathcal{M}}_{g,k}(V /\!\!/ G, d)^p$ (expected by a deformation theory) is vdim. The class

$$[\overline{\mathcal{M}}_{g,k}(V /\!\!/ G, d)^p]^{vir} \in A_{\mathrm{vdim}}(\overline{\mathcal{M}}_{g,k}(V /\!\!/ G, d)^p)_{\mathbb{Q}},$$

defined by a method in [**2**], can be localized to the class

$$[\overline{\mathcal{M}}_{g,k}(V /\!\!/ G, d)^p]^{vir}_{loc} \in A_{\mathrm{vdim}}(\overline{\mathcal{M}}_{g,k}(Q, d))_{\mathbb{Q}},$$

defined by a cosection localization method (see [**11**] for the construction). In other words, we obtain an equivalence of cycles

$$i_*[\overline{\mathcal{M}}_{g,k}(V /\!\!/ G, d)^p]^{vir}_{loc} = [\overline{\mathcal{M}}_{g,k}(V /\!\!/ G, d)^p]^{vir},$$

where $i : \overline{\mathcal{M}}_{g,k}(Q, d) \hookrightarrow \overline{\mathcal{M}}_{g,k}(V /\!\!/ G, d) \hookrightarrow \overline{\mathcal{M}}_{g,k}(V /\!\!/ G, d)^p$, is an embedding defined by a composition of an embedding induced by $Q \subset V /\!\!/ G$ and an embedding defined by a zero section $\overline{\mathcal{M}}_{g,k}(V /\!\!/ G, d) \hookrightarrow \overline{\mathcal{M}}_{g,k}(V /\!\!/ G, d)^p$. In [**12**], it is proven that we have an equivalence of cycles

$$[\overline{\mathcal{M}}_{g,k}(Q, d)]^{vir} = \pm[\overline{\mathcal{M}}_{g,k}(V /\!\!/ G, d)^p]^{vir}_{loc}, \tag{1.1}$$

(see Theorem 4.1 and Remark 4.2 for the clear statement). Here, the homology class d is in $H_2(V /\!\!/ G; \mathbb{Z})$ and $\overline{\mathcal{M}}_{g,k}(Q, d)$ on the left-hand side is the union of $\overline{\mathcal{M}}_{g,k}(Q, \beta)$ over all classes β satisfying $\iota_*\beta = d$, where $\iota : Q \hookrightarrow V/\!\!/G$ is an embedding. It tells us that a gauged linear sigma model has at least an information of Gromov-Witten theory for complete intersections, while their moduli spaces have more symmetries.

This short survey article is about a discussion of (1.1) for a simple case: when Q is a complete intersection in a projective space. A brief idea of the proof is introduced in Section 4. The key machinery of the proof is to express virtual classes in terms of localized Chern characters and use their properties. In Section 2, we discuss a definition of localized Chern characters for 2-periodic complexes and their properties. In Section 3, we explain what virtual fundamental classes are and how they can be written in terms of localized Chern characters.

2. Localized Chern characters for 2-periodic complexes

2.1. Review of localized Chern characters. Let $\mathbf{k}$ be any base field. Let Y be a finite type DM stack over $\mathbf{k}$. Let $E_\bullet$ be a 2-periodic complex of vector bundles on Y

$$E_\bullet : \cdots \to E_+ \xrightarrow{d_+} E_- \xrightarrow{d_-} E_+ \xrightarrow{d_+} E_- \to \cdots, \quad d_+ \circ d_- = d_- \circ d_+ = 0,$$

where E_+ is on even degree and E_- is on odd degree. We say $E_\bullet$ is *exact off a closed substack* $X \subset Y$ if $E_\bullet$ is exact on $Y - X$. We say $E_\bullet$ is *strictly exact off* X if $E_\bullet$ is exact off X and $\ker d_+$, $\ker d_-$ are vector bundles on $Y - X$.

EXAMPLE 2.1. Let E be a vector bundle on a finite type DM stack Y. Let $\alpha : E \to \mathcal{O}_Y$ and $\beta : \mathcal{O}_Y \to E$ be a cosection and a section, respectively, of E such that $\alpha \circ \beta = 0$. The complex

$$\cdots \to \oplus_k \wedge^{2k} E^\vee \xrightarrow{\wedge\alpha^\vee + i_\beta} \oplus_k \wedge^{2k-1} E^\vee \xrightarrow{\wedge\alpha^\vee + i_\beta} \oplus_k \wedge^{2k} E^\vee \to \cdots$$

is strictly exact off $Z(\alpha, \beta) := Z(\alpha) \cap Z(\beta)$, where $Z(\alpha)$ is a closed substack of Y defined by an ideal sheaf $\mathrm{im}\alpha \subset \mathcal{O}_Y$, and $Z(\beta) \subset Y$ is a zero locus of the section β. Here, $\wedge\cdot$ and $i.$ stand for the wedge and interior product, respectively. We will denote this 2-periodic complex by $\{\alpha, \beta\}$. This is called the *Koszul 2-periodic complex associated to* α *and* β. To write virtual classes in terms of localized Chern characters, we will use Koszul 2-periodic complexes (see Theorem 3.6).

By generalizing the idea of Baum, Fulton and MacPherson in [**1**], Polishchuk and Vaintrob define an additive (but not homogeneous) homomorphism

$$\mathrm{ch}^Y_X(E_\bullet) : A_*(Y) \to A_*(X)_\mathbb{Q}$$

for a 2-periodic complex $E_\bullet$ strictly exact off X (see [**13**]).

Here is the brief explanation of the construction of $\mathrm{ch}^Y_X(E_\bullet)$. Let $r_\pm$ be $\mathrm{rank}E_\pm$. Let $G := Gr(E_+ \oplus E_-, r_+) \times_Y Gr(E_+ \oplus E_-, r_-)$. Here, $Gr(E_+ \oplus E_-, r)$ is the Grassmannian stack parametrizing rank r sub-bundles in $E_+ \oplus E_-$. On a Grassmannian space $Gr(E_+ \oplus E_-, r)$, there is a rank r tautological bundle. Since G is the fiber product of $Gr(E_+ \oplus E_-, r_\pm)$, there are two tautological bundles ξ_+, and ξ_-. Let ξ be $\xi_+ - \xi_-$, an element in a K-group of vector bundles. Consider the locally closed embeddings (we consider the second one only when $Y \neq X$, which implies $r_+ = r_-$)

$$\phi : Y \times \mathbb{A}^1 \to G \times \mathbb{P}^1, \ (y, t) \mapsto (\Gamma(td_+(y)), \Gamma(td_-(y)), t),$$
$$\psi : Y - X =: U \to G|_U, \ y \mapsto ((\ker d_+(y), \ker d_-(y)), (\ker d_+(y), \ker d_-(y))),$$

where $\Gamma(f) \subset A \oplus B$ for a homomorphism $f : A \to B$ is a graph space, an image of $(id, f) : A \to A \oplus B$. Let $[Y_1]$ be the class in $Z_{\dim Y}(G)$ defined to be $\infty^![\overline{\phi(Y \times \mathbb{A}^1)}]$, where

$$\infty^! : Z_{\dim Y+1}(G \times \mathbb{P}^1) \to Z_{\dim Y}(G), \ [W] \mapsto (G \times \{\infty\}) \cap [W]$$

is a specialization morphism. The overline here means taking a closure. We then have $\infty^!_U[\overline{\phi(Y \times \mathbb{A}^1)}|_{G|_U \times \mathbb{P}^1}] = [\psi(U)]$ if U is non-empty, where

$$\infty^!|_U : Z_{\dim Y+1}(G|_U \times \mathbb{P}^1) \to Z_{\dim Y}(G|_U), \ [W] \mapsto (G|_U \times \{\infty\}) \cap [W]$$

is also a specialization morphism. Let $[Y_2] := [\overline{\psi(U)}] \in Z_{\dim Y}(G)$. By construction, $[Y_1] - [Y_2]$ can be considered as an element in $Z_{\dim Y}(G|_X)$. Let

$$\mathrm{ch}^Y_X(E_\bullet)([Y]) := p_*(\mathrm{ch}(\xi|_{G|_X}) \cap ([Y_1] - [Y_2])),$$

where $p : G|_X \to X$ is the projection. For any class in $[W] \in Z_*(Y)$, we can define $\mathrm{ch}^Y_X(E_\bullet)([W])$ using the same idea, i.e. embedding of $W \times \mathbb{A}^1$ into $G \times \mathbb{P}^1$. Note that $\mathrm{ch}^Y_X(E_\bullet)$ is well-defined on $A_*(Y)$ because $[Y_2] \in Z_{\dim Y}(\Delta)$ (if it exists) where Δ is a diagonal substack of G. On Δ, we see that $\xi = 0$. The homomorphism $\mathrm{ch}^Y_X(E_\bullet)$ is called the *localized Chern character for a 2-periodic complex* $E_\bullet$.

A localized Chern character for a 2-periodic complex $E_\bullet$ is related to a usual Chern character by the following identity: for $[W] \in A_*(Y)$, we have

$$i_*(\mathrm{ch}^Y_X(E_\bullet)([W])) = (\mathrm{ch}(E_+) - \mathrm{ch}(E_-)) \cap [W],$$

where $i : X \to Y$ is an embedding.

EXAMPLE 2.2. Let $Y := \mathbb{A}^1_{\mathbf{k}}$ and $E_\pm$ be trivial bundles $\mathbb{A}^1_{\mathbf{k}} \times \mathbb{A}^1_{\mathbf{k}}$ of rank one. Let $d_+ : E_+ \to E_-$ be a morphism defined by $(a, s) \mapsto (a, as)$ and let $d_- := 0$. The 2-periodic complex $E_\bullet$ is strictly exact off $X := \{0\} \subset Y$. Note that $E_\bullet$ is a Koszul complex $\{d_+, 0\}$ by regarding d_+ as a cosection of a trivial bundle. In this case, we have $G \cong Y \times \mathbb{P}^1 \times \mathbb{P}^1$. The morphism $\phi : Y \times \mathbb{A}^1 \to G \times \mathbb{P}^1 \cong Y \times \mathbb{P}^1 \times \mathbb{P}^1 \times \mathbb{P}^1$ is

$$(a, t) \mapsto (a, [1; at], [0; 1], t).$$

The restriction to $t = \infty$ of the closure of an image of ϕ is

$$(Y \times \{[0; 1]\} \times \{[0; 1]\}) \cup (\{0\} \times \mathbb{P}^1 \times \{[0; 1]\}) \subset G \cong Y \times \mathbb{P}^1 \times \mathbb{P}^1 \tag{2.1}$$

and it is Y_1. The first part of the decomposition (2.1) $(Y \times \{[0; 1]\} \times \{[0; 1]\})$ is Y_2. Hence, $Y_1 - Y_2$ is $\{0\} \times \mathbb{P}^1 \times \{[0; 1]\}) \cong \mathbb{P}^1$. Thus, a localized Chern character

$$\mathrm{ch}^Y_X(E_\bullet) : A_*(Y) = A_1(Y) \cong \mathbb{Z} \to A_*(X)_{\mathbb{Q}} = A_0(X)_{\mathbb{Q}} \cong \mathbb{Q}$$

sends a generator $[Y]$ to $\int_{\mathbb{P}^1} \mathrm{ch}(\mathcal{O}_{\mathbb{P}^1}(-1)) - \mathrm{ch}(\mathcal{O}_{\mathbb{P}^1}) = \int_{\mathbb{P}^1} c_1(\mathcal{O}_{\mathbb{P}^1}(-1)) = -1 \in \mathbb{Q}$. This example shows that even if $\mathrm{ch}(E_+) - \mathrm{ch}(E_-)$ is trivial, $\mathrm{ch}^Y_X(E_\bullet)$ can be non-trivial.

2.2. Properties of localized Chern characters. It is known that the usual Chern characters of vector bundles are bivariant classes; see [**8**, Section 17]. Also, they define a ring homomorphism from a K-group of vector bundles (with a tensor product) to a bivariant Chow ring (of an identity morphism). Localized Chern characters for 2-periodic complexes also have similar properties: see bivariant property, additive and weak multiplicative properties below. These properties have not only theoretical importance but also usefulness for the proofs of Theorem 3.6 and Theorem 4.1. For instance, the last property, called $\mathbb{A}^1$-homotopy below, is a very useful tool for the comparison of localized Chern characters. They are listed in [**13**, chapter 2].

Bivariant property. Consider the fiber product diagram of stacks:

$$\begin{array}{ccccc} X' & \hookrightarrow & Y' & \longrightarrow & Z' \\ \downarrow{\scriptstyle p'} & & \downarrow{\scriptstyle p} & & \downarrow{\scriptstyle q} \\ X & \hookrightarrow & Y & \longrightarrow & Z. \end{array}$$

Let $E_\bullet$ be a 2-periodic complex of vector bundles on Y strictly exact off X. Let W and W' be closed substacks of Y and Y', respectively.

(1) Suppose that p is proper. Then we have

$$p'_*(\mathrm{ch}^{Y'}_{X'}(p^*E_\bullet)([W'])) = \mathrm{ch}^Y_X(E_\bullet)(p_*[W']).$$

(2) Suppose that p is flat. Then we have

$$p'^*(\mathrm{ch}^Y_X(E_\bullet)([W])) = \mathrm{ch}^{Y'}_{X'}(p^*E_\bullet)(p^*[W]).$$

(3) Suppose that q is a locally complete intersection morphism. Then we have

$$q^!(\mathrm{ch}^Y_X(E_\bullet)([W])) = \mathrm{ch}^{Y'}_{X'}(p^*E_\bullet)(q^![W]).$$

Additive property. For a short exact sequence of 2-periodic complexes of vector bundles on Y strictly exact off X,

$$0 \to E^1_\bullet \to E^2_\bullet \to E^3_\bullet \to 0,$$

we have $\mathrm{ch}^Y_X(E^2_\bullet) = \mathrm{ch}^Y_X(E^1_\bullet) + \mathrm{ch}^Y_X(E^3_\bullet)$.

Weak multiplicative property. Let $X \subset Y$, $E_\bullet$ be as above. Let $p : F \to Y$ be a vector bundle. On the total space F, the diagonal morphism $F \to F \times_Y F$ gives a *tautological section* $t_F : \mathcal{O}_F \to p^*F$. With the zero cosection $0 : p^*F \to \mathcal{O}_F$, it defines a Koszul 2-periodic complex $\{0, t_F\}$ on F strictly exact off Y. Then, we have

$$\mathrm{ch}^F_X(p^*E_\bullet \otimes^2 \{0, t_F\}) = \mathrm{ch}^Y_X(E_\bullet) \circ \mathrm{ch}^F_Y(\{0, t_F\}).$$

REMARK 2.3. For 2-periodic complexes of vector bundles on Y, namely $E^1_\bullet$ and $E^2_\bullet$, strictly exact off $X_1 \subset Y$ and $X_2 \subset Y$, respectively, its tensor product

$$E^1_\bullet \otimes^2 E^2_\bullet$$

is well-defined and a 2-periodic complex of vector bundles on Y strictly exact off $X_1 \cap X_2$. See [**14**] for the definition.

$\mathbb{A}^1$-homotopy property. Let $\tilde{Y} \to \mathbb{A}^1$ be a stack over $\mathbb{A}^1$ and let $\mathcal{E}_\bullet$ be a 2-periodic complex of vector bundles on $\tilde{Y}$. Suppose that there exists a stack X such that $X \times \mathbb{A}^1$ is embedded into $\tilde{Y}$, whose composition with $\tilde{Y} \to \mathbb{A}^1$ is a projection to $\mathbb{A}^1$, and $\mathcal{E}_\bullet$ is strictly exact off $X \times \mathbb{A}^1$. Let $[\tilde{W}] \in A_*(\tilde{Y})$. Suppose

$$\lambda^!(\mathrm{ch}^{\tilde{Y}}_{X \times \mathbb{A}^1}(\mathcal{E}_\bullet)([\tilde{W}])) = \mathrm{ch}^{\tilde{Y}|_\lambda}_X(\mathcal{E}_\bullet|_\lambda)(\lambda^![\tilde{W}]) = 0$$

for $\lambda \in \mathbb{A}^1 - \{0\}$, where $\lambda : \{\lambda\} \to \mathbb{A}^1$ is a regular embedding. Then we have

$$0^!(\mathrm{ch}^{\tilde{Y}}_{X \times \mathbb{A}^1}(\mathcal{E}_\bullet)([\tilde{W}])) = \mathrm{ch}^{\tilde{Y}|_0}_X(\mathcal{E}_\bullet|_0)(0^![\tilde{W}]) = 0.$$

3. Virtual fundamental classes

3.1. Review of virtual fundamental classes. Let M be a finite type DM stack over $\mathbf{k}$. Let $[A \to F]$ be a 2-term complex of vector bundles. Its dual $[F^\vee \to A^\vee]$ is called a *perfect obstruction theory* if there is a finite cover $M = \cup_\alpha M_\alpha$ of M and if for each α, there is a data $(Y_\alpha, E_\alpha, s_\alpha)$ where Y_α is a smooth stack, $E_\alpha \to Y_\alpha$ is a vector bundle, and $s_\alpha : \mathcal{O}_{Y_\alpha} \to E_\alpha$ is a section such that

(1) $s_\alpha^{-1}(0) \cong M_\alpha$,

(2) $[A \to F]|_{M_\alpha} \sim_{qis} [T_{Y_\alpha}|_{M_\alpha} \xrightarrow{ds_\alpha|_{E_\alpha|M_\alpha}} E_\alpha|_{M_\alpha}]$, where $\sim_{qis}$ denotes a quasi-isomorphism in $D^b(M_\alpha)$, a derived category of bounded complexes of coherent sheaves on M_α.

Suppose that $[A \to F]$ is a dual of a perfect obstruction theory on M. A class $[C_M] \in A_{\mathrm{rank}A}(F)_\mathbb{Q}$ is constructed in [**2**].

REMARK 3.1. Behrend and Fantechi constructed an Artin stack $\mathfrak{C}_M$ of pure dimension 0 called the *intrinsic normal cone of M*, which does not depend on perfect obstruction theories for a finite type DM stack M, and then they showed that $[\mathfrak{C}_M] \in A_0([F/A])$ if $[A \to F]$ is a dual of a perfect obstruction theory on M. A DM stack C_M of pure dimension rank$A$ is defined by the following fiber product:

$$\begin{array}{ccc} C_M & \hookrightarrow & F \\ \downarrow & & \downarrow \\ \mathfrak{C}_M & \hookrightarrow & [F/A]. \end{array}$$

The Gysin map $0_F^! : A_*(F) \to A_{*-\mathrm{rank}F}(M)$ of a zero section $0_F : M \to F$ defines a *virtual fundamental class*

$$[M]^{vir} := 0_F^![C_M] \in A_{\mathrm{rank}A-\mathrm{rank}F}(M)_{\mathbb{Q}}.$$

Here, rank$A$ – rank$F$ is called the *virtual dimension.*

EXAMPLE 3.2. Let M and F be as above. Note that a Gysin map $0_F^!$ is a bivariant class in $A^{\mathrm{rank}F}(M \to F)_{\mathbb{Q}}$, where $M \to F$ is given by a zero section (denoted by 0_F). Let $s : M \to F$ be any section of F. Since $0_F^!$ is a bivariant class, by (a part of) the definition, we have a homomorphism

$$0_F^! : A_*(M) \to A_{*-\mathrm{rank}F}(s^{-1}(0))_{\mathbb{Q}}. \tag{3.1}$$

This homomorphism is called the *refined Euler class* and denoted by $e^r(F)\cap$.

Assume that M is smooth. Then, $[T_M \xrightarrow{ds} F]|_{s^{-1}(0)}$ becomes a perfect obstruction theory for $s^{-1}(0)$. By considering a deformation to the normal cone $M^0_{s^{-1}(0)/M}$, one can show that

$$[s^{-1}(0)]^{vir} = 0_F^![C_{s^{-1}(0)/M}] = e^r(F) \cap [M] =: e^r(F).$$

Let $i : s^{-1}(0) \hookrightarrow M$ be an embedding. It is known that

$$i_*e^r(F) = e(F) \cap [M] =: e(F) \in A_{\dim M - \mathrm{rank}F}(M)_{\mathbb{Q}},$$

where $e(-)$ stands for an Euler class.

EXAMPLE 3.3. Let Q be a smooth projective variety over $\mathbb{C}$ and let C be a projective scheme over $\mathbb{C}$. Let $f : C \to Q$ be a morphism of schemes. The deformation and obstruction spaces of f are $H^0(C, f^*T_Q)$ and $H^1(C, f^*T_Q)$, respectively, where T_Q is a tangent bundle of Q. Hence, on the moduli space of stable maps $\overline{\mathcal{M}}_{g,k}(Q,d)$ (recall that each point here is a curve C with a morphism $f : C \to Q$), the family of deformation and obstruction spaces are written by $\mathbb{R}^0\pi_*f^*T_Q$ and $\mathbb{R}^1\pi_*f^*T_Q$, respectively, where $\pi : \mathfrak{C} \to \overline{\mathcal{M}}_{g,k}(Q,d)$ is a universal curve and $f : \mathfrak{C} \to Q$ is a universal morphism. They are kernel and cokernel of the two-term complex $\mathbb{R}\pi_*f^*T_Q$. Note that it has a locally free, two-term representation because $\overline{\mathcal{M}}_{g,k}(Q,d)$ is known to be projective. This is a perfect obstruction theory for $\overline{\mathcal{M}}_{g,k}(Q,d)$ (relative to $\mathfrak{M}_{g,k}$ the moduli space of genus g prestable curves with k-marked points. Since $\mathfrak{M}_{g,k}$ is a smooth Artin stack, we get an absolute perfect obstruction theory; this process is briefly explained in Section 4.1). Conversely, any perfect obstruction theory is related to defomation and obstruction spaces (see [**2**]).

EXAMPLE 3.4. A smooth projective variety Q is *convex* if $H^1(C, f^*T_Q) = 0$ for any genus 0 curve C and a morphism $f : C \to Q$. For a convex Q, a (relative) perfect obstruction theory $\mathbb{R}\pi_*f^*T_Q$ for $\overline{\mathcal{M}}_{0,k}(Q,d)$ is a locally free sheaf (on 0-degree). Note that an induced absolute perfect obstruction theory is also a locally free sheaf on degree zero. In this case, we have $C_{\overline{\mathcal{M}}_{0,k}(Q,d)} = \overline{\mathcal{M}}_{0,k}(Q,d)$. This implies

$$[\overline{\mathcal{M}}_{0,k}(Q,d)]^{vir} = [\overline{\mathcal{M}}_{0,k}(Q,d)],$$

i.e., the virtual fundamental class is the fundamental class.

EXAMPLE 3.5. A vector bundle U on Q is *convex* if $H^1(C, f^*U) = 0$ for any genus 0 curve C and a morphism $f : C \to Q$. Let $P := s^{-1}(0)$ be a zero locus of a section s of a convex bundle U on Q (it is a locally complete intersection in a space Q). Since

$$\mathbb{R}\pi_* f^*T_P \to \phi^*\mathbb{R}\pi_* f^*T_Q \to \phi^*\mathbb{R}\pi_* f^*U$$

is an exact triangle, where $\phi : \overline{\mathcal{M}}_{0,k}(P,d) \hookrightarrow \overline{\mathcal{M}}_{0,k}(Q,d)$ is an embedding induced by $P \hookrightarrow Q$, the ((−1)-shifting of the) cone of $\phi^*\mathbb{R}\pi_* f^*T_Q \to \phi^*\mathbb{R}\pi_* f^*U$ is a (relative) perfect obstruction theory for $\overline{\mathcal{M}}_{0,k}(P,d)$.

Assume further that Q is convex. Let $F := \mathbb{R}\pi_* f^*U$ be a locally free sheaf. By Example 3.2, we have

$$[\overline{\mathcal{M}}_{0,k}(P,d)]^{vir} = e^r(F) \text{ and } \phi_*[\overline{\mathcal{M}}_{0,k}(P,d)]^{vir} = e(F).$$

Let M and $[A \xrightarrow{d} F]$ be as above. Suppose further that there is a cosection $\sigma : F \to \mathcal{O}_M$ such that $\sigma \circ d = 0$. Let $Y := (p^*\sigma \circ t_F)^{-1}(0)$ and $X := Z(\sigma)$ where $p : F \to M$ is a projection. A definition of a closed substack $Z(\sigma) \subset M$ is introduced in Example 2.1. We have a sequence of closed immersions

$$X \subset M \subset Y \subset F,$$

where the second embedding $M \subset Y$ is given by a zero section of F. Kiem and Li proved that $C_M \subset Y$ in [**11**]. In the same paper, they constructed a cosection localized Gysin map

$$0^!_{F,\sigma} : A_*(Y) \to A_{*-\mathrm{rank}F}(X)_{\mathbb{Q}},$$

satisfying that

$$\begin{array}{ccc} A_*(Y) & \longrightarrow & A_*(F) \\ \Big\downarrow 0^!_{F,\sigma} & & \Big\downarrow 0^!_F \\ A_*(X)_{\mathbb{Q}} & \longrightarrow & A_*(M)_{\mathbb{Q}} \end{array} \tag{3.2}$$

is commutative. Note that the horizontal morphisms are proper pushforwards by closed immersions. They define a *cosection localized virtual fundamental class*

$$[M]^{vir}_\sigma := 0^!_{F,\sigma}[C_M] \in A_{\mathrm{rank}A-\mathrm{rank}F}(X)_{\mathbb{Q}}.$$

Hence, we have $[M]^{vir}_{\sigma=0} = [M]^{vir}$ by (3.2).

3.2. Virtual classes via localized Chern characters. Let M, $[A \xrightarrow{d} F]$, $p : F \to M$, $\sigma : F \to \mathcal{O}_M$, Y, and X be as above. The following chain on F

$$\cdots \to \oplus_k \wedge^{2k} p^*F^\vee \xrightarrow{\wedge p^*\sigma^\vee + i_{t_F}} \oplus_k \wedge^{2k-1} p^*F^\vee \xrightarrow{\wedge p^*\sigma^\vee + i_{t_F}} \oplus_k \wedge^{2k} p^*F^\vee \to \cdots$$

can be a complex only on Y. For simple notation, let us denote this complex on Y (not on F) by $\{p^*\sigma, t_F\}$ and call the Koszul complex as well. A 2-periodic complex $\{p^*\sigma, t_F\}$ on Y is exact off X. Thus, we have a homomorphism

$$\mathrm{ch}^Y_X(\{p^*\sigma, t_F\}) : A_*(Y) \to A_*(X)_{\mathbb{Q}}.$$

In [**12**], it is proven that

THEOREM 3.6 ([**12**]).

$$\operatorname{td}(F|_X)\cdot \operatorname{ch}_X^Y(\{p^*\sigma, t_F\}) = 0^!_{F,\sigma} : A_*(Y) \to A_*(X)_{\mathbb{Q}}. \tag{3.3}$$

For $\sigma = 0$, Theorem 3.6 is proven in [**13**, Proposition 2.3(vi)]. In [**12**], the result in [**13**] for the $\sigma = 0$ case is used for proving (3.3). Other parts of the proof are done by the $\mathbb{A}^1$-homotopy property. Kim and the author proved the so called *splitting principle* by the $\mathbb{A}^1$-homotopy property, which is the key to prove (3.3).

Splitting principle. Let Y be a DM stack. Suppose that the following are given:

(1) There is a short exact sequence of vector bundles on Y,

$$0 \to K \xrightarrow{f} E \xrightarrow{q} Q \to 0.$$

(2) There are sections $\beta : \mathcal{O}_Y \to E$ and $\beta_K : \mathcal{O}_Y \to K$.
(3) There are cosections $\alpha : E \to \mathcal{O}_Y$ and $\alpha_Q : Q \to \mathcal{O}_Y$.

Suppose that the following are true:

(1) $\alpha \circ \beta = 0$.
(2) $\alpha = \alpha_Q \circ q$.
(3) $Z(\alpha, \beta) = Z(\alpha_Q) \cap Z(\beta_K)(=: X)$.
(4) $\alpha_Q^\vee$ and β_K are regular sections.

Then we have

$$\operatorname{ch}_X^Y(\{\alpha, \beta\}) = (-1)^{\operatorname{rank} Q}\operatorname{td}(E|_X)^{-1} i^!_{Z(\alpha_Q,\beta_K)} \circ i^!_{Z(\beta_K)}, \tag{3.4}$$

where $i^!_{Z(\alpha_Q,\beta_K)}$ and $i^!_{Z(\beta_K)}$ are Gysin pullbacks with respect to regular immersions of $Z(\alpha_Q, \beta_K)$ and $Z(\beta_K)$ into $Z(\beta_K)$ and Y, respectively. Note that the assumptions (3) and (4) can be weakened to get a similar statement of (3.4) (see [**12**, Lemma 2.4]).

REMARK 3.7. The key idea of the proof of the splitting principle is the construction of a vector bundle P on $Y \times \mathbb{A}^1$ of the same rank with E, a section β_P, and a cosection α_P of P such that its restriction to $\lambda \in (\mathbb{A}^1\backslash\{0\})$ is E with β and α, and its restriction to $\lambda = 0$ is $K \oplus Q$ with $\beta_K \oplus 0$ and $0 \oplus \alpha_Q$. This can be done by the following: let P be a cokernel of

$$\bar{K} \xrightarrow{(\lambda id_K, f)} \bar{K} \oplus \bar{E},$$

where $\bar{K}$ and $\bar{E}$ are pullbacks of K and E to $Y \times \mathbb{A}^1$, respectively. A section $((1-\lambda)\beta_K, \lambda\beta) : \mathcal{O}_{Y\times\mathbb{A}^1} \to \bar{K} \oplus \bar{E}$ defines a section β_P on P by a composition. A cosection $(0, \alpha) : \bar{K} \oplus \bar{E} \to \mathcal{O}_{Y\times\mathbb{A}^1}$ can be factored to define a cosection α_P on P. Note that $\alpha_P \circ \beta_P = 0$. The Gysin pullback of $\operatorname{ch}_{X\times\mathbb{A}^1}^{Y\times\mathbb{A}^1}(\{\alpha_P, \beta_P\})$ by a regular morphism $\{\lambda\} \hookrightarrow \mathbb{A}^1$ proves the splitting principle by $\mathbb{A}^1$-homotopy property.

REMARK 3.8. A proof of (3.3) is as follows: let $p : M' := Bl_{Z(\sigma)}M \to M$ be a blow up of M along $Z(\sigma)$ and let D be an exceptional divisor,

$$\begin{array}{ccc} D & \hookrightarrow & M' \\ \downarrow & & \downarrow{\scriptstyle p} \\ Z(\sigma) & \hookrightarrow & M \end{array}$$

Let K be a vector bundle on M' defined by the kernel of $p^*F \to \mathcal{O}_{M'}(-D)$. Let $Y' := Y \times_M M'$ be a fiber product and let $q : Y' \to M'$ be a natural morphism. We have the following:

(1) a short exact sequence of vector bundles on Y'

$$0 \to q^*K \to q^*p^*F \to q^*\mathcal{O}_{M'}(-D) \to 0,$$

(2) a section r^*t_F, where $r : Y' \to Y$ is a projection,
(3) a cosection $r^*p^*\sigma$ of q^*p^*F and a cosection q^*s_D of $q^*\mathcal{O}_{M'}(-D)$, where $s_D : \mathcal{O}_{M'}(-D) \to \mathcal{O}_{M'}$ is a homomorphism defined by D.

Note that the total space of K is embedded in Y'. The restriction of the above data to the total space of K, with a tautological section t_K gives rise to the splitting principle. It proves (3.3) (see [**12**, section 2.4.2]).

Consequently, we have

$$[M]^{vir}_{\sigma} = \mathrm{td}(F|_X) \cdot \mathrm{ch}^Y_X(\{p^*\sigma, t_F\})[C_M]. \tag{3.5}$$

4. Application to Gromov-Witten theory

4.1. Comparison of virtual cycles in Gromov-Witten theory. We assume that our base field $\mathbf{k}$ from now is $\mathbb{C}$.

Let l_1, ..., l_c be positive integers and let s_1, ..., s_c be sections of the line bundles $\mathcal{O}_{\mathbb{P}^n}(l_1)$, ..., $\mathcal{O}_{\mathbb{P}^n}(l_c)$ on $\mathbb{P}^n$, $n+1 \geqslant c$. Let E be a direct sum of line bundles $\mathcal{O}_{\mathbb{P}^n}(-l_1) \oplus \cdots \oplus \mathcal{O}_{\mathbb{P}^n}(-l_c)$. Suppose that the section $s = (s_1, ..., s_l)$ of $E^\vee$ is regular. Let $Q := Z(s) \subset \mathbb{P}^n$. Note that s induces a function $w_s : \mathrm{tot}E \to \mathbb{A}^1$ defined by a pairing, where $\mathrm{tot}E$ denotes the total space of E. Since s is regular, we have $Z(s) = Z(dw_s)$.

Consider $\overline{\mathcal{M}}_{g,k}(Q,d)$, a moduli space of genus g, degree d stable maps to Q with k-marked points. It carries a natural perfect obstruction theory relative to a smooth Artin stack $\mathfrak{Bun}^{g,k}_{\mathbb{C}^*}$, parametrizing a genus g prestable curve C with k-marked points and a $\mathbb{C}^*$-bundle (or equivalently, line bundle) L on C. Let $\mathfrak{C}$, $\mathcal{L}$ be the universal curve on $\overline{\mathcal{M}}_{g,k}(Q,d)$, the universal line bundle on $\mathfrak{C}$, respectively. Let $\pi : \mathfrak{C} \to \overline{\mathcal{M}}_{g,k}(Q,d)$ be a projection. A relative perfect obstruction theory is given by the dual of a cone of

$$\mathbb{R}\pi_* f^* ds : \mathbb{R}\pi_*(\mathcal{L}^{\oplus n+1}) \to \mathbb{R}\pi_*(\oplus_i \mathcal{L}^{\otimes l_i})$$

in $D^b(\overline{\mathcal{M}}_{g,k}(Q,d))$, where $ds : \mathcal{O}_{\mathbb{P}^1}(1)^{\oplus n+1} \to T_{\mathbb{P}^1} \xrightarrow{ds} T_{E^\vee}|_{\mathbb{P}^1} \to E^\vee$ is a composition and $f : \mathfrak{C} \to Q$ is the universal morphism. Note that $f^*\mathcal{O}_{\mathbb{P}^1}(1) \cong \mathcal{L}$ and that $Cone(\mathbb{R}\pi_* f^* ds)$ is a two-term complex of locally free sheaves (see [**7**] for details). Since $\mathfrak{Bun}^{g,k}_{\mathbb{C}^*}$ is a smooth, a cone of $Cone(\mathbb{R}\pi_* f^* ds)[-1] \to T_{\mathfrak{Bun}^{g,k}_{\mathbb{C}^*}}|_{\overline{\mathcal{M}}_{g,k}(Q,d)}$ gives rise to an absolute perfect obstruction theory. It carries a virtual fundamental class

$$[\overline{\mathcal{M}}_{g,k}(Q,d)]^{vir} \in A_{(n+1-\sum_i l_i)d+(n-c-2)(1-g)+k}(\overline{\mathcal{M}}_{g,k}(Q,d))_{\mathbb{Q}}.$$

Assume that $(n+1-\sum_i l_i)d + (n-c-2)(1-g) + k \geqslant 0$.

Let us consider $\overline{\mathcal{M}}_{g,k}(\mathbb{P}^n,d)^p$, a moduli space of genus g, degree d stable maps to $\mathbb{P}^n$ with k-marked points and p-fields. More precisely,

$$\overline{\mathcal{M}}_{g,k}(\mathbb{P}^n,d)^p := \{(C,\mathbf{x},L,u,p) | (C,\mathbf{x},L,u) \in \overline{\mathcal{M}}_{g,k}(\mathbb{P}^n,d), p \in \Gamma(C, \oplus_i L^{\otimes -l_i} \otimes \omega_C)\},$$

where ω_C is a dualizing sheaf of C. By abusing notation, let $\pi : \mathfrak{C} \to \overline{\mathcal{M}}_{g,k}(\mathbb{P}^n,d)^p$, $\mathcal{L}$ be the universal curve, universal line bundle on $\mathfrak{C}$, respectively. A dual of a relative perfect obstruction theory to $\mathfrak{Bun}^{g,k}_{\mathbb{C}^*}$ is given by

$$\mathbb{R}\pi_*(\mathcal{L}^{\oplus n+1}) \oplus \mathbb{R}\pi_*(\oplus_i \mathcal{L}^{\otimes -l_i} \otimes \omega_{\mathfrak{C}}),$$

which is a two-term complex of locally free sheaves in $D^b(\overline{\mathcal{M}}_{g,k}(\mathbb{P}^n,d)^p)$. We can define a cosection in the following way: for $(u,p) := (C,\mathbf{x},L,u,p) \in \overline{\mathcal{M}}_{g,k}(\mathbb{P}^n,d)^p$, define a linear homomorphism

$$\begin{aligned}\mathbb{R}^1\pi_*(\mathcal{L}^{\oplus n+1}\bigoplus\oplus_i\mathcal{L}^{\otimes -l_i}\otimes\omega_{\mathfrak{C}})|_{(u,p)} &\cong H^1(L^{\oplus n+1}\bigoplus\oplus_i L^{\otimes -l_i}\otimes\omega_C)\\ &\to H^1(\omega_C)\cong H^0(\mathcal{O}_C)\cong\mathbb{C}\\ (u',p') &\mapsto s(u)\cdot p' + \sum_j \frac{\partial s}{\partial u_j}|_{u=u'}\cdot p\,.\end{aligned}$$

It gives rise to a cosection

$$\sigma' : \mathbb{R}^1\pi_*(\mathcal{L}^{\oplus n+1}\bigoplus\oplus_i\mathcal{L}^{\otimes -l_i}\otimes\omega_{\mathfrak{C}}) \to \mathcal{O}_{\overline{\mathcal{M}}_{g,k}(\mathbb{P}^n,d)^p}.$$

Let $[A \xrightarrow{d} F]$ be a two-term complex presentation of $\mathbb{R}\pi_*(\mathcal{L}^{\oplus n+1}\bigoplus\oplus_i\mathcal{L}^{\otimes -l_i}\otimes\omega_{\mathfrak{C}})$ by locally free sheaves. Let $\sigma : F \to \mathcal{O}_{\overline{\mathcal{M}}_{g,k}(\mathbb{P}^n,d)^p}$ be a composition

$$\sigma : F \to F/d(A) \cong \mathbb{R}^1\pi_*(\mathcal{L}^{\oplus n+1}\bigoplus\oplus_i\mathcal{L}^{\otimes -l_i}\otimes\omega_{\mathfrak{C}}) \xrightarrow{\sigma'} \mathcal{O}_{\overline{\mathcal{M}}_{g,k}(\mathbb{P}^n,d)^p}.$$

It induces that $\sigma\circ d = 0$ and $Z(\sigma) = \overline{\mathcal{M}}_{g,k}(Q,d) \subset \overline{\mathcal{M}}_{g,k}(\mathbb{P}^n,d) \subset \overline{\mathcal{M}}_{g,k}(\mathbb{P}^n,d)^p$, where the second immersion is by taking $p=0$. Thus, it carries a cosection localized virtual fundamental class

$$[\overline{\mathcal{M}}_{g,k}(\mathbb{P}^n,d)^p]^{vir}_\sigma \in A_{(n+1-\sum_i l_i)d+(n-c-2)(1-g)+k}(\overline{\mathcal{M}}_{g,k}(Q,d))_{\mathbb{Q}}.$$

In [**12**], we have the following equivalence of cycles:

THEOREM 4.1 ([**12**]). *The following equality holds*

$$[\overline{\mathcal{M}}_{g,k}(Q,d)]^{vir} = (-1)^{c-cg+d\sum_i l_i}[\overline{\mathcal{M}}_{g,k}(\mathbb{P}^n,d)^p]^{vir}_\sigma$$

in $A_{(n+1-\sum_i l_i)d+(n-c-2)(1-g)+k}(\overline{\mathcal{M}}_{g,k}(Q,d))_{\mathbb{Q}}$.

REMARK 4.2. In [**12**], Theorem 4.1 is proven with a more general set-up. To be precise, let V be a vector space over $\mathbb{C}$ and let G be a linear reductive group acting on V. Let $\theta : G \to \mathbb{C}^*$ be a character such that every semi-stable point in V with respect to θ is stable, that is $V^s(\theta) = V^{ss}(\theta)$. Let $E := [(V^s(\theta)\times V')/G]$ be a vector bundle on $V /\!\!/_\theta G$, where V' is a suitable G-representation space.

Let $w \in ((\mathrm{Sym}^{\geqslant 1}V^\vee)\otimes V'^\vee)^G$ be a G-invariant polynominal, where Sym denotes the (graded) symmetric algebra. It induces a section $s : \mathcal{O}_{V/\!\!/_\theta G} \to E^\vee$ and a function $w_s : \mathrm{tot}E \to \mathbb{A}^1$. Suppose that the zero locus of $dw_s \in \Omega^1_{\mathrm{tot}E}$ (or equivalently, the critical locus of w_s) lies on the zero section $V /\!\!/_\theta G$ as a smooth, closed subscheme of codimension rankV'. It implies that $Z(s) = Z(dw_s)$. Let Q denote $Z(s) = Z(dw_s) \subset V /\!\!/_\theta G$.

For $\varepsilon \in \mathbb{Q}_+$, a moduli space $Q^\varepsilon_{g,k}(Q,d)$ of genus g, degree d ε-stable quasimaps to Q with k-marked points is constructed and it carries a virtual fundamental class

$$[Q^\varepsilon_{g,k}(Q,d)]^{vir} \in A_{c_1(Q).d+(\dim Q-3)(1-g)+k}(Q^\varepsilon_{g,k}(Q,d))_{\mathbb{Q}}$$

(see [**6**] for details). Also, a moduli space $Q^\varepsilon_{g,k}(V /\!\!/_\theta G,d)^p$ of genus g, degree d ε-stable quasimaps to $V /\!\!/_\theta G$ with k-marked points and p-fields is constructed, and it carries a cosection localized virtual fundamental class

$$[Q^\varepsilon_{g,k}(V /\!\!/_\theta G,d)^p]^{vir}_{loc} \in A_{c_1(Q).d+(\dim Q-3)(1-g)+k}(Q^\varepsilon_{g,k}(Q,d))_{\mathbb{Q}}$$

(see [**3**, **12**] for details). The main theorem in [**12**] states that

$$[Q^{\varepsilon}_{g,k}(Q,d)]^{vir} = (-1)^{\chi(\mathbb{R}\pi_*\mathcal{E}^\vee)}[Q^{\varepsilon}_{g,k}(V /\!\!/_\theta G, d)^p]^{vir}_{loc},$$

where $\pi : \mathfrak{C} \to Q^{\varepsilon}_{g,k}(Q,d)$ is a universal curve; $\mathfrak{P} \to \mathfrak{C}$ is a universal G-bundle; $\mathcal{E} := (\mathfrak{P} \times_{\mathfrak{C}} V')/G$; and $\chi(-)$ denotes an Euler characteristic.

REMARK 4.3. In [**5**], Chang and M-l. Li proved Theorem 4.1 for $c = 1$ and for ε-stable quasimaps. They mentioned that their method can work for all c.

REMARK 4.4. In [**3**], Chang and J. Li proved the following: let Q be a quintic threefold in $\mathbb{P}^4$. Then, for $d > 0$,

$$\deg[\overline{\mathcal{M}}_{g,0}(Q,d)]^{vir} = (-1)^{1-g+5d}\deg[\overline{\mathcal{M}}_{g,0}(\mathbb{P}^4,d)^p]^{vir}_{\sigma} \tag{4.1}$$

in $H_0(\overline{\mathcal{M}}_{g,0}(Q,d);\mathbb{Q}) \cong \mathbb{Q}$ where $\deg : A_0(\overline{\mathcal{M}}_{g,0}(Q,d))_{\mathbb{Q}} \to H_0(\overline{\mathcal{M}}_{g,0}(Q,d);\mathbb{Q})$ is a cycle map. In [**4**], Chang and J. Li proved Zinger's theorem in [**16**, **17**] algebraically using (4.1). Zinger's theorem played a crucial role for studying genus one mirror symmetry for quintic threefolds (see [**16**]).

In [**10**], Guo, Janda and Ruan studied genus 2 (and 3) Gromov-Witten invariants for quintic threefold Q using the following fact:

$$\deg[Q^{\varepsilon}_{g,0}(Q,d)]^{vir} = (-1)^{1-g+5d}\deg[Q^{\varepsilon}_{g,0}(\mathbb{P}^4,d)^p]^{vir}_{\sigma},$$

which is implied by Theorem 4.1.

As explained in Section 1, Theorem 4.1 gave a justification of how the gauged linear sigma model can be a framework for studying Gromov-Witten theory which is an implement of studying Landau-Ginzburg & Calabi-Yau correspondence, etc.

4.2. Idea of Theorem 4.1. The remaining of the paper is a technical part of [**12**, Section 3]. Here, we introduce only a short strategy. On (a certain open substack of) $\mathfrak{Bun}^{g,k}_{\mathbb{C}^*}$, we can find 2-term presentations of locally free sheaves

$$\mathbb{R}\pi_*(\mathcal{L}^{\oplus n+1}) = [A \to B], \quad \mathbb{R}\pi_*(\oplus_i \mathcal{L}^{\otimes l_i}) = [Q \to P^\vee].$$

Hence it induces $\mathbb{R}\pi_*(\oplus_i \mathcal{L}^{\otimes -l_i} \otimes \omega_C) = [P \to Q^\vee]$.

For simple notation, let $\mathfrak{B} := \mathfrak{Bun}^{g,k}_{\mathbb{C}^*}$. For a vector bundle S, we denote $|S| = \mathrm{tot} S$ the total space of S.

STEP 1. Consider a (smooth) space $U := |A| \times_{\mathfrak{B}} |P| \times_{\mathfrak{B}} |Q| \times \mathbb{A}^1$ and a vector bundle $F' := B \oplus Q^\vee \oplus Q|_U$ on U. We can construct a section $\beta : \mathcal{O}_U \to F'$ such that

$$Z(\beta)|_\lambda \cong \begin{cases} \overline{\mathcal{M}}_{g,k}(\mathbb{P}^n,d)^p & \text{if } 0 \neq \lambda \in \mathbb{A}^1 \\ \overline{\mathcal{M}}_{g,k}(Q,d) \times_{\mathfrak{B}} |P| \times_{\mathfrak{B}} |Q| & \text{if } 0 = \lambda, \end{cases}$$

(We omit the details of β). Let $M := Z(\beta)$, $F := F'|_M$, and $Y := C_M U \subset |F|$.

STEP 2. We can construct a cosection $\sigma : F \to \mathcal{O}_M$ such that $p^*\sigma \circ t_F = 0$ on Y, where $p : |F| \to M$ is a projection and $Z(\sigma) \cong \overline{\mathcal{M}}_{g,k}(Q,d) \times \mathbb{A}^1$, (We omit the details of σ). Let $X := Z(\sigma)$. Consider the following class

$$\mathrm{td}(F|_X) \cdot \mathrm{ch}^Y_X(\{p^*\sigma, t_F\})[Y] \in A_*(X)_{\mathbb{Q}}.$$

The construction in [**12**] implies

$$\lambda^!(\mathrm{td}(F|_X) \cdot \mathrm{ch}^Y_X(\{p^*\sigma, t_F\})[Y]) = [\overline{\mathcal{M}}_{g,k}(\mathbb{P}^n,d)^p]^{vir}_{\sigma} \in A_*(\overline{\mathcal{M}}_{g,k}(Q,d))_{\mathbb{Q}} \tag{4.2}$$

directly for $0 \neq \lambda \in \mathbb{A}^1$. Note that we use (3.5) for proving (4.2). By a bivariant property, we have

$$\begin{aligned} &0^!(\mathrm{td}(F|_X) \cdot \mathrm{ch}_X^Y(\{p^*\sigma, t_F\})[Y]) \\ &= \mathrm{td}(F|_{\overline{\mathcal{M}}_{g,k}(Q,d)}) \cdot \mathrm{ch}_{\overline{\mathcal{M}}_{g,k}(Q,d)}^{Y|_0}(\{p^*\sigma, t_F\}|_0)(0^![Y]). \end{aligned}$$

In [**7**, Lemma 3.6], it is proven that $0^![Y] = [C_{\overline{\mathcal{M}}_{g,k}(Q,d)}|A| \times_{\mathfrak{B}} |P| \times_{\mathfrak{B}} |Q|] \in A_*(Y|_0)$. Let $Y_0 := C_{\overline{\mathcal{M}}_{g,k}(Q,d)}|A| \times_{\mathfrak{B}} |P| \times_{\mathfrak{B}} |Q|$.

Step 3. In [**7**], the authors constructed a short exact sequence of vector bundles on $M_0 := \overline{\mathcal{M}}_{g,k}(Q,d)$,

$$0 \to K \to F|_{M_0} \to (P^\vee \oplus Q^\vee)|_{M_0} \to 0,$$

such that the cosection $\sigma|_{M_0}$ factors through

$$pair : (P^\vee \oplus Q^\vee)|_{M_0} \to \mathcal{O}_{M_0},$$

the pairing cosection and $C_{\overline{\mathcal{M}}_{g,k}(Q,d)}|A| \subset |K|$. By the splitting principle, we have

$$\mathrm{ch}_{\overline{\mathcal{M}}_{g,k}(Q,d)}^{Y_0}(\{p^*\sigma, t_F\}|_{Y_0})([Y_0]) = \mathrm{ch}_{\overline{\mathcal{M}}_{g,k}(Q,d)}^{Y_0}(\{pair, 0\}|_{Y_0} \otimes^2 \{0, t_K\}|_{Y_0})([Y_0]).$$

By the weak multiplicative property, the construction, and (3.5), we obtain

$$\begin{aligned} (4.3)\quad &0^!(\mathrm{td}(F|_X) \cdot \mathrm{ch}_X^Y(\{p^*\sigma, t_F\})[Y]) \\ &= \mathrm{td}(F|_{\overline{\mathcal{M}}_{g,k}(Q,d)}) \cdot \mathrm{ch}_{\overline{\mathcal{M}}_{g,k}(Q,d)}^{Y_0}(\{pair, 0\}|_{Y_0} \otimes^2 \{0, t_K\}|_{Y_0})([Y_0]) \\ &= \mathrm{td}(K|_{\overline{\mathcal{M}}_{g,k}(Q,d)}) \cdot \mathrm{td}((P^\vee \oplus Q^\vee)|_{\overline{\mathcal{M}}_{g,k}(Q,d)}) \\ &\quad \cdot \mathrm{ch}_{\overline{\mathcal{M}}_{g,k}(Q,d)}^{C_{\overline{\mathcal{M}}_{g,k}(Q,d)}|A|}(\{0, t_K\}) \circ \mathrm{ch}_{C_{\overline{\mathcal{M}}_{g,k}(Q,d)}|A|}^{Y_0}(\{pair, 0\}|_{Y_0})([Y_0]) \\ &= (-1)^{c-cg+d\sum_i l_i} \mathrm{td}(K|_{\overline{\mathcal{M}}_{g,k}(Q,d)}) \cdot \mathrm{ch}_{\overline{\mathcal{M}}_{g,k}(Q,d)}^{C_{\overline{\mathcal{M}}_{g,k}(Q,d)}|A|}(\{0, t_K\})([C_{\overline{\mathcal{M}}_{g,k}(Q,d)}|A|]) \\ &= (-1)^{c-cg+d\sum_i l_i} [\overline{\mathcal{M}}_{g,k}(Q,d)]^{vir}. \end{aligned}$$

By (4.2), (4.3), and the $\mathbb{A}^1$-homotopy property, we have Theorem 4.1.

Acknowledgment

The author would like to thank the anonymous referee for precious advice for better shaping of the paper.

References

[1] Paul Baum, William Fulton, and Robert MacPherson, *Riemann-Roch for singular varieties*, Inst. Hautes Études Sci. Publ. Math. **45** (1975), 101–145. MR412190

[2] K. Behrend and B. Fantechi, *The intrinsic normal cone*, Invent. Math. **128** (1997), no. 1, 45–88, DOI 10.1007/s002220050136. MR1437495

[3] Huai-Liang Chang and Jun Li, *Gromov-Witten invariants of stable maps with fields*, Int. Math. Res. Not. IMRN **18** (2012), 4163–4217, DOI 10.1093/imrn/rnr186. MR2975379

[4] Huai-Liang Chang and Jun Li, *An algebraic proof of the hyperplane property of the genus one GW-invariants of quintics*, J. Differential Geom. **100** (2015), no. 2, 251–299. MR3343833

[5] Huai-Liang Chang and Mu-Lin Li, *Invariants of stable quasimaps with fields*, Trans. Amer. Math. Soc. **373** (2020), no. 5, 3669–3691, DOI 10.1090/tran/8011. MR4082252

[6] Daewoong Cheong, Ionuţ Ciocan-Fontanine, and Bumsig Kim, *Orbifold quasimap theory*, Math. Ann. **363** (2015), no. 3-4, 777–816, DOI 10.1007/s00208-015-1186-z. MR3412343

[7] Ionuţ Ciocan-Fontanine and Bumsig Kim, *Quasimap wall-crossings and mirror symmetry*, Publ. Math. Inst. Hautes Études Sci. **131** (2020), 201–260, DOI 10.1007/s10240-020-00114-0. MR4106795

[8] William Fulton, *Intersection theory*, 2nd ed., Ergebnisse der Mathematik und ihrer Grenzgebiete. 3. Folge. A Series of Modern Surveys in Mathematics [Results in Mathematics and Related Areas. 3rd Series. A Series of Modern Surveys in Mathematics], vol. 2, Springer-Verlag, Berlin, 1998. MR1644323

[9] Henri Gillet, *Intersection theory on algebraic stacks and Q-varieties*, Proceedings of the Luminy conference on algebraic K-theory (Luminy, 1983), J. Pure Appl. Algebra **34** (1984), no. 2-3, 193–240, DOI 10.1016/0022-4049(84)90036-7. MR772058

[10] Longting Wu, *A remark on Gromov-Witten invariants of quintic threefold*, Adv. Math. **326** (2018), 241–313, DOI 10.1016/j.aim.2017.12.016. MR3758430

[11] Young-Hoon Kiem and Jun Li, *Localizing virtual cycles by cosections*, J. Amer. Math. Soc. **26** (2013), no. 4, 1025–1050, DOI 10.1090/S0894-0347-2013-00768-7. MR3073883

[12] B. Kim and J. Oh, *Localized Chern characters for 2-periodic complexes*, arXiv:1804.03774

[13] Alexander Polishchuk and Arkady Vaintrob, *Algebraic construction of Witten's top Chern class*, Advances in algebraic geometry motivated by physics (Lowell, MA, 2000), Contemp. Math., vol. 276, Amer. Math. Soc., Providence, RI, 2001, pp. 229–249, DOI 10.1090/conm/276/04523. MR1837120

[14] Alexander Polishchuk and Arkady Vaintrob, *Matrix factorizations and cohomological field theories*, J. Reine Angew. Math. **714** (2016), 1–122, DOI 10.1515/crelle-2014-0024. MR3491884

[15] Angelo Vistoli, *Intersection theory on algebraic stacks and on their moduli spaces*, Invent. Math. **97** (1989), no. 3, 613–670, DOI 10.1007/BF01388892. MR1005008

[16] Aleksey Zinger, *Standard versus reduced genus-one Gromov-Witten invariants*, Geom. Topol. **12** (2008), no. 2, 1203–1241, DOI 10.2140/gt.2008.12.1203. MR2403808

[17] Aleksey Zinger, *Reduced genus-one Gromov-Witten invariants*, J. Differential Geom. **83** (2009), no. 2, 407–460. MR2577474

School of Mathematics, KIAS, Seoul 02455, Korea
Email address: jeongseok@kias.re.kr

Contemporary Mathematics
Volume **763**, 2021
https://doi.org/10.1090/conm/763/15324

Singularity theory and mirror symmetry

Todor Milanov

ABSTRACT. The main goal of these lectures is to outline K. Saito's theory of primitive forms in singularity theory and use it to state mirror symmetry for Calabi–Yau orbifolds of a Fermat type. We also present an efficient method to compute explicitly the analytic continuation of Gromov–Witten invariants with respect to the Novikov variables.

1. Givental's higher genus reconstruction

Following Dubrovin [**5**], we recall the notion of a Frobenius manifold. Then we proceed by defining the so-called *total ancestor potential* of a semi-simple Frobenius manifold (see [**6, 7**]).

1.1. Frobenius manifolds. Suppose that B is a complex manifold and $\mathcal{T}_B$ is the sheaf of holomorphic vector fields on B. The manifold B is equipped with the following structures:

(F1) A non-degenerate symmetric bilinear pairing

$$(\,\cdot\,,\,\cdot\,):\mathcal{T}_B\otimes\mathcal{T}_B\to\mathcal{O}_B.$$

(F2) A Frobenius multiplication: commutative associative multiplication

$$\cdot\bullet\cdot:\mathcal{T}_B\otimes\mathcal{T}_B\to\mathcal{T}_B,$$

such that $(v_1\bullet w,v_2)=(v_1,w\bullet v_2)\ \forall v_1,v_2,w\in\mathcal{T}_B$.

(F3) A unit vector field: global vector field $e\in\mathcal{T}_B(B)$ such that

$$e\bullet v=v,\quad \nabla^{\mathrm{L.C.}}_v e=0,\quad \forall v\in\mathcal{T}_B,$$

where $\nabla^{\mathrm{L.C.}}$ is the Levi–Civita connection of the pairing $(\cdot,\cdot)$.

(F4) An Euler vector field $E\in\mathcal{T}_B(B)$ such that

$$E(v_1,v_2)-([E,v_1],v_2)-(v_1,[E,v_2])=(2-D)(v_1,v_2)$$

for some constant $D\in\mathbb{C}$.

2010 *Mathematics Subject Classification.* Primary 14N35, 32S30.

This work was partially supported by the JSPS Grant-In-Aid (Kiban C) 17K05193 and by the World Premier International Research Center Initiative (WPI Initiative), MEXT, Japan.

Given the data (F1)-(F4), we define the so called *Dubrovin's connection* on the vector bundle $TB \times \mathbb{C}^* \to B \times \mathbb{C}^*$

$$\nabla_v := \nabla_v^{\mathrm{L.C.}} - z^{-1} v\bullet, \quad v \in \mathcal{T}_B,$$
$$\nabla_{\partial/\partial z} := \frac{\partial}{\partial z} - z^{-1}\theta + z^{-2} E\bullet,$$

where z is the standard coordinate on $\mathbb{C}^* = \mathbb{C} \setminus \{0\}$, where $v\bullet$ is an endomorphism of $\mathcal{T}_B$ defined by the Frobenius multiplication by the vector field v, and where $\theta : \mathcal{T}_B \to \mathcal{T}_B$ is an $\mathcal{O}_B$-modules morphism defined by

$$\theta(v) := \nabla_v^{\mathrm{L.C.}}(E) - \Big(1 - \frac{D}{2}\Big) v.$$

DEFINITION 1.1. The data $((\cdot,\cdot), \bullet, e, E)$, satisfying the properties $(F1) - (F4)$, is said to be a *Frobenius structure* of conformal dimension D if the corresponding Dubrovin connection is flat.

The flatness of the Dubrovin connection implies that the Levi–Civita connection $\nabla^{\mathrm{L.C.}}$ of the pairing $(\cdot,\cdot)$ is flat. Suppose that B is contractible and let us fix a reference point $t^\circ \in B$. Let us trivialize the tangent bundle $TB = B \times H$, $H := T_{t^\circ} B$, by using the parallel transport with respect to $\nabla^{\mathrm{L.C.}}$. The vector space H is naturally identified with the space of flat vector fields. Let us fix a basis $\{\phi_a\}_{a=1}^N$ of H. The dual basis $\{\phi^a\}_{a=1}^N \subset H^*$ consist of global holomorphic 1-forms. The axioms of a Frobenius manifold imply that ϕ^a are closed and hence, $\phi^a = dt_a$ for some holomorphic functions t_a. Although $t = (t_1, \ldots, t_N)$ might fail to be global coordinates on B, they are always local holomorphic coordinates. Moreover, by definition the coordinate vector fields $\partial/\partial t_a = \phi_a$.

1.2. Givental's R-matrix.

Suppose that B is a Frobenius manifold.

DEFINITION 1.2. A point $t_0 \in B$ is called *semi-simple* if there exists an open neighborhood B_0 of t_0 in B and coordinates $u(t) = (u_1(t), \ldots, u_N(t))$ on B_0, called *canonical coordinates*, such that

$$\frac{\partial}{\partial u_i} \bullet \frac{\partial}{\partial u_j} = \delta_{i,j} \frac{\partial}{\partial u_j}, \quad \Big(\frac{\partial}{\partial u_i}, \frac{\partial}{\partial u_j}\Big) = \frac{\delta_{i,j}}{\Delta_j},$$

where $\Delta_j \in \mathcal{O}_B(B_0)$ and $\Delta_j(t) \neq 0$ for all $t \in B_0$.

The subset $\mathcal{K}$ of all points $t \in B$ that are not semi-simple is called *caustic.* It is either an empty subset, the entire manifold B, or an analytic hypersurface. In these lecture we will be interested in the case where $\mathcal{K} \neq B$.

Suppose that $t_0 \in B$ is a semi-simple point and that B_0 is a neighborhood of t_0 that has canonical coordinates $u(t) = (u_1(t), \ldots, u_N(t))$. Let us construct a trivialization of the tangent bundle TB_0 using the canonical coordinates.

$$\Psi : B_0 \times \mathbb{C}^N \to TB_0 = B_0 \times H, \quad (t, v) \mapsto (t, \Psi(t) v),$$

where

$$\Psi(t) e_i = \sqrt{\Delta_i}\, \frac{\partial}{\partial u_i} = \sum_{a=1}^N \sqrt{\Delta_i}\, \frac{\partial t_a}{\partial u_i}\, \phi_a,$$

where $\{e_i\}_{i=1}^N$ is the standard basis of $\mathbb{C}^N$.

PROPOSITION 1.3. *There exists a unique series*

$$R(t,z) = 1 + \sum_{k=1}^{\infty} R_k(t) z^k, \quad R_k(t) \in \mathrm{End}(\mathbb{C}^N),$$

such that

$$\nabla\Big(\Psi(t)R(t,z)e^{U(t,z)}\Big) = 0,$$

where $U(t)$ is a diagonal matrix with diagonal entries $u_1(t), \ldots, u_N(t)$. The series satisfies the symplectic condition $R(t,z)R(t,-z)^{\mathrm{T}} = 1$, *where* ^{T} *is the transpose with respect to the Frobenius pairing.*

1.3. Quantization formalism. Given a vector space H equipped with a non-degenerate pairing $(\cdot,\cdot)$ and a vector $e \in H$, we construct the vector space $\mathcal{H} := H(\!(z)\!)$ of formal Laurent series in z with coefficients in H. The form

$$\Omega(f_1(z), f_2(z)) = \mathrm{Res}_{z=0}(f_1(-z), f_2(z))dz$$

turns $\mathcal{H}$ into a symplectic vector space.

Let $\{\phi^a\}_{a=1}^N \subset H$ be the basis dual to $\{\phi_a\}_{a=1}^N$ with respect to the pairing $(\cdot,\cdot)$. Then the following linear functions

$$p_{k,a}(f) := \Omega(f, \phi_a z^k), \quad q_{k,a}(f) := \Omega(\phi^a(-z)^{-k-1}, f)$$

are Darboux coordinates on $\mathcal{H}$. Finally, let us introduce also the *Fock space*

$$\mathbb{C}_\hbar[\![q_{k,a} + \delta_{1,a}c_a \ (1 \le a \le N, k \ge 0)]\!],$$

where $\mathbb{C}_\hbar$ is the field of formal Laurent series in $\hbar^{\frac{1}{2}}$ and the constants c_a are the coordinates of e in the basis $\{\phi_a\}$, that is, $e = \sum_{a=1}^N c_a\phi_a$. The quantization formalism consists of two projective representations on the Fock space. The first one is defined on the Poisson Lie algebra of quadratic Hamiltonians on $\mathcal{H}$ and the second one on the Lie algebra of infinitesimal symplectic transformations of $\mathcal{H}$.

1) Suppose that $h \in \mathbb{C}[\![p_{k,a}, q_{k,a}]\!]$ is an arbitrary formal series. Then we define

$$\widehat{h} :=: h(\widehat{p}, \widehat{q}) : ,$$

where

$$\widehat{p}_{k,a} = \sqrt{\hbar}\frac{\partial}{\partial q_{k,a}}, \quad \widehat{q}_{k,a} = \frac{1}{\sqrt{\hbar}} q_{k,a}$$

and : : is the normal ordering, i.e., where differentiation is applied first. Note that $\widehat{h}$ is a formal differential operator. In general, it might fail to act on the Fock space.

2) Suppose now that $A \in \mathrm{End}(\mathcal{H})$ is an infinitesimal symplectic transformation, i.e.,

$$\Omega(Af_1, f_2) + \Omega(f_1, Af_2) = 0, \quad \forall f_1, f_2 \in \mathcal{H}.$$

Then the map $f \mapsto Af$ can be viewed as a linear Hamiltonian vector field on $\mathcal{H}$ with Hamiltonian

$$h_A(f) = \frac{1}{2}\Omega(Af, f).$$

By definition, $f = \sum_{k,a}(q_{k,a}\phi_a z^k + p_{k,a}\phi^a(-z)^{-k-1})$, so h_A is a quadratic function in p and q. We define $\widehat{A} := \widehat{h}_A$, in other words,

$$\widehat{A} = \sum_{k,l\geq 0}\sum_{1\leq i,j\leq N}\Big(\frac{1}{2\hbar}\Omega(A\phi_i z^k, \phi_j z^l)q_{k,i}q_{l,j} + \Omega(A\phi_i z^k, \phi^j(-z)^{-l-1})q_{k,i}\frac{\partial}{\partial q_{l,j}}$$
$$+\frac{\hbar}{2}\Omega(A\phi^i(-z)^{-k-1}, \phi^j(-z)^{-l-1})\frac{\partial^2}{\partial q_{k,i}\partial q_{l,j}}\Big).$$

1.4. The total ancestor potential. By definition, the Witten-Kontsevich tau-function is a generating function for certain intersection numbers on the moduli space $\overline{\mathcal{M}}_{g,n}$ of stable Riemann surfaces of genus-g equipped with n marked points and having, at most, nodal singularities. More precisely,

$$\mathcal{D}_{\rm pt}(\hbar, \mathbf{t}) = \exp\Big(\sum_{g,n=0}^{\infty}\frac{\hbar^{g-1}}{n!}\int_{\overline{\mathcal{M}}_{g,n}}\mathbf{t}(\psi_1)\cdots\mathbf{t}(\psi_n)\Big),$$

where $\mathbf{t}(\psi) = \sum_{k=0}^{\infty} t_k\psi^k$, where $\hbar$ and $t_k (k\geq 0)$ are formal variables, and where ψ_i $(1\leq i\leq n)$ is the 1st Chern class of the orbifold line bundle on $\overline{\mathcal{M}}_{g,n}$ formed by the cotangent lines at the ith marked point. It was conjectured by Witten (see [**20**]) and proved by Kontsevich (see [**10**]) that $\mathcal{D}_{\rm pt}$ is a tau-function of the KdV hierarchy.

If $t\in B\setminus\mathcal{K}$ is a semisimple point, then we define

$$\mathcal{A}_t(\hbar, \mathbf{q}) := \Big(\Psi(t)R(t,z)\Psi(t)^{-1}\Big)^{\widehat{}}\ \prod_{i=1}^{N}\mathcal{A}_t^{(i)}(\hbar, \mathbf{q}), \tag{1.1}$$

where the notation is as follows: the operator series $\Psi(t)R(t,z)\Psi(t)^{-1}$ is a symplectic transformation of the form e^A, where A is an infinitesimal symplectic transformation. Its quantization is by definition $e^{\widehat{A}}$. The formal power series $\mathcal{A}_t^{(i)}(\hbar, \mathbf{q})$ are obtained from the Witten-Kontsevich tau-function via the substitutions

$$\hbar\mapsto\hbar\Delta_i, \quad t_k\mapsto\delta_{k,1}+q_k(u_i),$$

where $\mathbf{q} = (q_0, q_1, \dots)$ is a sequence of formal vector variables taking values in H. The coordinates of q_k with respect to the basis $\{\phi_a\}$ will be denoted by $q_{k,a}$. Then

$$q_k(u_i) = \sum_{a=1}^{N} q_{k,a}\frac{\partial u_i}{\partial t_a}.$$

The formal power series (1.1) are called the *total ancestor potential* of the corresponding semi-simple Frobenius structure. It is easy to check that $\mathcal{A}_t(\hbar, \mathbf{q})$ is an element of the Fock space.

2. Opposite subspaces in singularity theory

The first examples of Frobenius manifolds can be found in the work of K. Saito (see [**18**]). The notion of a Frobenius manifold did not exist yet. Saito studied the periods of certain holomorphic forms that he called *primitive*. In particular, he introduced a certain flat structure on the base of miniversal unfolding of a holomorphic function with an isolated critical point. It turns out that Saito's flat structure coincides with the flat structure of a Frobenius manifold. In this lecture we would like to outline Saito's theory.

2.1. Singularity theory. Suppose that $f \in \mathcal{O}_{\mathbb{C}^n,0}$ is the germ of a holomorphic function with an isolated critical point at 0 with critical value $f(0) = 0$. Let us denote by $x = (x_1, \ldots, x_n)$ the standard coordinate system on $\mathbb{C}^n$. The algebra

$$H = \mathcal{O}_{\mathbb{C}^n,0}/(\partial_{x_1} f, \ldots, \partial_{x_n} f)$$

is known as the *local algebra* of f. The names *Milnor ring*, *Jacobi algebra*, and *Turina algebra* are also used in the literature. The Rückert Nullstellensatz theorem implies that H is a finite dimensional vector space over $\mathbb{C}$. Its dimension N is called the *Milnor number*. Let us fix a set of holomorphic germs $\phi_i \in \mathcal{O}_{\mathbb{C}^n,0}$ $(1 \le i \le N)$ that represent a basis of H. Put

$$F(x, s) = f(x) + \sum_{i=1}^{N} s_i \phi_i(x),$$

where $s = (s_1, \ldots, s_N)$ are deformation parameters.

Let us make a careful choice of the domain X for the function F (see [**3**]). Put $|x| = \sqrt{|x_1|^2 + \cdots + |x_n|^2}$ for the standard Euclidean norm. First, we choose a small number $r > 0$ such that the germs f and ϕ_i $(1 \le i \le N)$ can be represented by holomorphic functions in the ball $|x| \le r$, the function $f(x)$ does not have other critical points in $|x| < r$ besides $x = 0$, and the hypersurface defined by $f(x) = 0$ intersects transversely the spheres $|x| = r'$ for all $0 < r' \le r$. Then we choose a small open disk $\Delta = \{\lambda \in \mathbb{C} \mid |\lambda| < \delta\}$ and a small open ball $B = \{s \in \mathbb{C}^N \mid |s| < \eta\}$ such that for all $(s, \lambda) \in B \times \Delta$, the hypersurfaces

$$X_{s,\lambda} := \{x \in \mathbb{C}^n \mid |x| \le r,\ F(x, s) = \lambda\}$$

intersect transversely the sphere $|x| = r$. Let

$$X := \{(x, s) \in \mathbb{C}^n \times B \mid |x| < r,\ F(x, s) \in \Delta\}.$$

Note that X is biholomorphic to the hypersurface

$$\{(x, s, \lambda) \mid |x| < r,\ |s| < \eta,\ |\lambda| < \delta,\ F(x, s) = \lambda\}.$$

Since a product of open balls is a Stein manifold, and a non-singular hypersurface of a Stein manifold is also Stein, we get that X is a Stein domain, that is, $H^i(X, \mathcal{F}) = 0$ for $i > 0$ for all coherent sheaves $\mathcal{F}$ on X.

Let us denote by $F_{x_i} = \partial F/\partial x_i$ $(1 \le i \le n)$ and recall the corresponding *Koszul complex*, i.e., complex of free $\mathcal{O}_X$-modules $(\wedge^p(\mathcal{O}_X^n), \partial_p)$ concentrated in degrees $p \in [0, n]$, where the differential in degree p is defined by

$$\partial_p(e_{i_1} \wedge \cdots \wedge e_{i_p}) := \sum_{s=1}^{p} (-1)^{s-1} F_{x_{i_s}} e_{i_1} \wedge \cdots \wedge \widehat{e_{i_s}} \wedge \cdots \wedge e_{i_p}.$$

Put $\mathcal{O}_C := \mathcal{O}_X/(F_{x_1}, \ldots, F_{x_n})$. The support of the sheaf $\mathcal{O}_C$ coincides with the *relative critical set* $C \subset X$ defined by the equations $F_{x_1} = \cdots = F_{x_n} = 0$. Let $p : X \to B$ be the map induced by the projection $\mathbb{C}^n \times B \to B$. The map

$$\mathcal{T}_B \to p_* \mathcal{O}_C, \quad \frac{\partial}{\partial s_i} \mapsto \frac{\partial F}{\partial s_i} \pmod{\partial_{x_1} F, \ldots, \partial_{x_n} F}$$

is known as the *Kodaira–Spencer map*.

PROPOSITION 2.1. *If the numbers r, δ, η in the definition of the domain X are sufficiently small then the following two conditions are satisfied:*

a) The Koszul complex yields a resolution of $\mathcal{O}_C$, that is,

$$0 \to \wedge^n(\mathcal{O}_X^n) \to \wedge^{n-1}(\mathcal{O}_X^n) \to \cdots \to \wedge^1(\mathcal{O}_X^n) \to \mathcal{O}_X \to \mathcal{O}_C \to 0$$

is an exact sequence of $\mathcal{O}_X$-modules.

b) The Kodaira–Spencer map is an isomorphism.

PROOF. Part a) is a standard fact about complete intersections. Let us prove part b). Since $0 \in X$ is an isolated point of $(p|_C)^{-1}(0)$, by choosing the numbers (r, δ, η) so small, we can decrease X and arrange that $p|_C : C \to B$ is a finite map. Then $p_*\mathcal{O}_C$ is a coherent sheaf. Therefore, we need only to check that the germ of the Kodaira–Spencer map at $0 \in B$ is an isomorphism.

Note that the stalk of $p_*\mathcal{O}_C$ at 0 is $\mathcal{N} := \mathcal{O}_{X,0}/(F_{x_1}, \ldots, F_{x_n})$. We claim that $\mathcal{N}$ is a flat $\mathcal{O}_{B,0}$-module. Put $A = \mathcal{O}_{X,0}$, $R = \mathcal{O}_{B,0}$ and $\mathfrak{m} = (s_1, \ldots, s_N)R$. Let $F_i \in A$ be the germ of F_{x_i}. Note that under the quotient map

$$A \to \overline{A} := A/\mathfrak{m}A = \mathcal{O}_{\mathbb{C}^n,0},$$

F_i is mapped to $f_i := f_{x_i}$. We recall [**1**] Corollary 5.7 and get that $\mathcal{N} = A/(F_1, \ldots, F_n)$ is a flat R-module if and only if every relation between $(f_1, \ldots, f_n)$ can be lifted to a relation between $F_1, \ldots, F_n$. Suppose that $a_1 f_1 + \cdots + a_n f_n = 0$ $(a_i \in \mathcal{O}_{\mathbb{C}^n,0})$ is a given relation. The Koszul complex corresponding to the sequence $(f_1, \ldots, f_n)$ is exact because the corresponding zero locus is a complete intersection. Therefore, there exists a skew-symmetric matrix $(b_{ij})_{1\leq i,j\leq n}$ with entries in $\mathcal{O}_{\mathbb{C}^n,0}$ such that $a_i = \sum_j b_{ij} f_j$. Let us choose a lift $\widetilde{b}_{ij} \in A$ for each entry and define $\widetilde{a}_i := \sum_j \widetilde{b}_{ij} F_j$. Then $\widetilde{a}_1 F_1 + \cdots + \widetilde{a}_n F_n = 0$ is a lift of the given relation.

Put $\mathcal{M} := \mathcal{T}_{B,0}$ and let $\varphi : \mathcal{M} \to \mathcal{N}$ be the germ of the Kodaira-Spencer map. Let us prove that φ is surjective. The injectivity is similar. Put $\mathcal{M}' = \varphi(\mathcal{M})$ and let us tensor the exact sequence

$$0 \to \mathcal{M}' \to \mathcal{N} \to \mathcal{N}' = \mathcal{N}/\mathcal{M}' \to 0$$

by $\otimes_R R/\mathfrak{m}$ and look at the corresponding long exact sequence

$$0 \to \mathrm{Tor}_1^R(\mathcal{N}', R/\mathfrak{m}) \to \mathcal{M}' \otimes_R R/\mathfrak{m} \to \mathcal{N} \otimes_R R/\mathfrak{m} \to \mathcal{N}' \otimes_R R/\mathfrak{m} \to 0,$$

where we used that $\mathcal{N}$ is a flat R-module, so $\mathrm{Tor}_1^R(\mathcal{N}, R/\mathfrak{m}) = 0$. Note that the construction of F is such that the Kodaira–Spencer map induces an isomorphism between the vector spaces $\mathcal{M} \otimes_R R/\mathfrak{m} = T_0 B$ and $\mathcal{N} \otimes_R R/\mathfrak{m} = H$ (the local ring of f). Therefore, $\mathcal{M}' \otimes_R R/\mathfrak{m} \cong \mathcal{N} \otimes_R R/\mathfrak{m}$. The above long exact sequence yields $\mathrm{Tor}_1^R(\mathcal{N}', R/m) = 0$ and $\mathcal{N}' \otimes_R R/\mathfrak{m} = 0$. The vanishing of the Tor-group implies (local criteria for flatness) that $\mathcal{N}'$ is a flat R-module $\Rightarrow 0 = \mathcal{N}' \otimes_R (R/\mathfrak{m}) = \mathcal{N}'/\mathfrak{m}\mathcal{N}'$, that is, $\mathcal{N}' = \mathfrak{m}\mathcal{N}'$. In particular, since $\mathfrak{m}$ is contained in the maximal ideal $\mathfrak{m}_A$ of the local ring A, we get that $\mathcal{N}' = \mathfrak{m}_A\mathcal{N}'$. It remains only to recall Nakayama's lemma, so $\mathcal{N}' = 0$, that is, $\varphi(\mathcal{M}) = \mathcal{M}' = \mathcal{N}$. □

From now on we fix the numbers (r, δ, η) in the definition of the domain X such that both the transversality conditions described above and the statements in Proposition 2.1 hold.

2.2. Twisted de Rham cohomology. Let $p : X \to B$ be the map induced by the projection $\mathbb{C}^n \times B \to B$. Let $(\Omega^i_{X/B}, d_{X/B})$ be the relative holomorphic de Rham complex, i.e.,

$$\Omega^i_{X/B} := \Omega^i_X / p^*\Omega^1_B \wedge \Omega^{i-1}_X,$$

where $\Omega_X^\bullet$ is the sheaf of holomorphic differential forms on X and the relative de Rham differential $d_{X/B}$ is induced from the de Rham differential d_X on $\Omega_X^\bullet$. Let us define the following sheaf on B:

$$\mathcal{F} := p_*\Omega^n_{X/B}[\![z]\!]/(zd_{X/B} + d_{X/B}F\wedge)p_*\Omega^{n-1}_{X/B}[\![z]\!]. \tag{2.1}$$

The sheaf $\mathcal{F}$ is equipped with a *Gauss–Manin* connection defined as follows: Let U be an open Stein subset of B. Then one can prove that

$$\mathcal{F}(U) = \Omega^n_{X/B}(p^{-1}U)[\![z]\!]/(zd_{X/B} + d_{X/B}F\wedge)\Omega^{n-1}_{X/B}(p^{-1}U)[\![z]\!].$$

If $\varphi \in \Omega^n_{X/B}(p^{-1}U)[\![z]\!]$, then we denote by $[\varphi]$ the equivalence class of φ in $\mathcal{F}(U)$. Let us denote by

$$\mathrm{rel} : \Omega^1_X \to \Omega^1_{X/B} = \Omega^1_X/p^*\Omega_B$$

the canonical quotient map. Finally, every vector field $v \in \mathcal{T}_B(U)$ has a lift $\widetilde{v} \in \mathcal{T}_X(p^{-1}U)$ such that $v(g) = \widetilde{v}(g \circ p)$ for every $g \in \mathcal{O}_B(U)$.

$$\nabla_v[\varphi] := \Big[\mathrm{rel} \circ \iota_{\widetilde{v}}\Big((d_X + z^{-1}d_X F\wedge)\varphi\Big)\Big],$$

$$\nabla_{\partial/\partial z}[\varphi] := \Big[\frac{\partial\varphi}{\partial z} - z^{-2}F\varphi - \frac{n}{2}z^{-1}\varphi\Big],$$

where $v \in \mathcal{T}_B(U)$.

Intuitively, the Gauss–Manin connection can be understood as follows: Suppose that we have a differential form

$$\varphi = g(x, s, z)dx_1 \cdots dx_n, \quad g \in \mathcal{O}_X(p^{-1}U)[\![z]\!].$$

The oscillatory form $e^{F/z}\varphi$ is exact with respect to the relative de Rham differential $zd_{X/B}$ if and only if φ is exact with respect to the twisted differential $zd_{X/B} + d_{X/B}F\wedge$. Therefore, we can think of the twisted de Rham cohomology class $[\varphi]$ as a formal oscillatory integral

$$[\varphi] = \frac{1}{(-2\pi z)^{n/2}} \int e^{F(x,s)/z} g(x, s, z)dx_1 \cdots dx_n. \tag{2.2}$$

In this interpretation, the Gauss–Manin derivatives $\nabla_{\partial/\partial s_i}$ and $\nabla_{\partial/\partial z}$ correspond to formally differentiating the integral with respect to the parameters s_i and z.

PROPOSITION 2.2. *Suppose that*

$$\omega = \sum_{k=0}^{\infty} g_k(x, s)z^k dx_1 \cdots dx_n \quad \in \quad \Omega^n_{X/B}(X)[\![z]\!]$$

is such that $g_0(x, s) \neq 0$ *for all* $(x, s) \in X$. *Then the map*

$$P_\omega : \mathcal{T}_B[\![z]\!] \to \mathcal{F}, \quad \frac{\partial}{\partial s_i} \mapsto z\nabla_{\frac{\partial}{\partial s_i}}[\omega]$$

is an isomorphism of $\mathcal{O}_B[\![z]\!]$*-modules.*

The proof of Proposition 2.2 is obtained after a direct computation using that we fixed a domain X in which Proposition 2.1 holds.

PROPOSITION 2.3. *There exists a unique* $\mathcal{O}_B$*-bilinear pairing*

$$K_F : \mathcal{F} \otimes_{\mathcal{O}_B} \mathcal{F} \to \mathcal{O}_B[\![z]\!]$$

satisfying the following conditions

(i) $K_F(\omega_1, \omega_2) = K_F(\omega_2, \omega_1)^*$, *where* * *is the involution* $z \mapsto -z$.

(ii) $p(z)K_F(\omega_1,\omega_2) = K_F(p(z)\omega_1,\omega_2) = K_F(\omega_1,p(-z)\omega_2)$.
(iii) *If* $\xi = \partial/\partial s_a$ *or* $\xi = z\partial/\partial z$, *then*

$$\xi K_F(\omega_1,\omega_2) = K_F(\nabla_\xi\omega_1,\omega_2) + K_F(\omega_1,\nabla_\xi\omega_2).$$

(iv) *Put* $K_F(\omega_1,\omega_2) = \sum_{m=0}^{\infty} K_F^{(m)}(\omega_1,\omega_2)z^m$. *Then*

$$K_F^{(0)}(\omega_1,\omega_2) = \frac{1}{(2\pi\mathbf{i})^n}\int_{|F_{x_1}|=\cdots=|F_{x_n}|=\epsilon} \frac{\phi_1(x,s)\phi_2(x,s)}{F_{x_1}\cdots F_{x_n}} dx_1\cdots dx_n,$$

where

$$\omega_i = (\phi_i(x,s)+O(z))dx_1\cdots dx_n, \quad i=1,2.$$

2.3. Opposite subspaces. Let $\widehat{\mathbb{H}}(f,\mathbb{C})$ be the fiber of the sheaf $\mathcal{F}(\!(z)\!)$ at $0\in B$. There is a natural restriction map from the fiber of the sheaf $p_*\Omega_X$ at 0 to $\Omega^n_{\mathbb{C}^n,0}$, which induces an isomorphism

$$\widehat{\mathbb{H}}(f,\mathbb{C}) = \Omega^n_{\mathbb{C}^n,0}(\!(z)\!)/(zd+df\wedge)\Omega^{n-1}_{\mathbb{C}^n,0}(\!(z)\!).$$

Using the higher residue pairing, we define the following symplectic pairing

$$\Omega(\omega_1,\omega_2) := \mathrm{Res}_{z=0}\, K_f(\omega_1,\omega_2)dz, \quad \omega_1,\omega_2\in\widehat{\mathbb{H}}(f,\mathbb{C}),$$

where the residue is understood formally as the coefficient in front of dz/z. Note that

$$\widehat{\mathbb{H}}_+(f,\mathbb{C}) := \Omega^n_{\mathbb{C}^n,0}[\![z]\!]/(zd+df\wedge)\Omega^{n-1}_{\mathbb{C}^n,0}[\![z]\!]$$

is a Lagrangian subspace.

DEFINITION 2.4. A Lagrangian subspace $P\subset\widehat{\mathbb{H}}(f,\mathbb{C})$ is said to be

a) an *opposite subspace* if $\widehat{\mathbb{H}}(f,\mathbb{C}) = \widehat{\mathbb{H}}_+(f,\mathbb{C})\bigoplus P$ and $z^{-1}P\subset P$.
b) *homogeneous* if $\nabla_{z\partial/\partial z}P\subset P$.

To avoid cumbersome notation, let us put $\widehat{\mathbb{H}}_+ := \widehat{\mathbb{H}}_+(f,\mathbb{C})$ and $\widehat{\mathbb{H}} := \widehat{\mathbb{H}}(f,\mathbb{C})$.

PROPOSITION 2.5. *If P is a homogeneous opposite subspace then the following statements hold.*

a) The projection $\widehat{\mathbb{H}}_+\to\widehat{\mathbb{H}}_+/z\widehat{\mathbb{H}}_+$ has a unique section σ_P such that $\mathrm{Im}(\sigma_P) = \widehat{\mathbb{H}}_+\bigcap zP$.

b) Put $\Omega_f := \widehat{\mathbb{H}}_+/z\widehat{\mathbb{H}}_+ = \Omega^n_{\mathbb{C}^n,0}/df\wedge\Omega^{n-1}_{\mathbb{C}^n,0}$ and let $\widehat{\sigma}_P : \Omega_f(\!(z)\!)\to\widehat{\mathbb{H}}$ be the unique $\mathbb{C}(\!(z)\!)$-linear extension of σ_P. Then $\widehat{\sigma}_P$ is an isomorphism and

$$\widehat{\mathbb{H}}_+ = \Omega_f[\![z]\!], \quad P = \Omega_f[z^{-1}]z^{-1}.$$

c) If $\{v_i\}_{i=1}^N$ is a basis of $\widehat{\mathbb{H}}_+\bigcap zP$, then $K^{(m)}(v_i,v_j)=0$ for all $m>0$ and $1\le i,j\le N$.

d) Under $\widehat{\sigma}_P$, the Gauss–Manin connection takes the form

$$z\nabla_{\partial/\partial z} = z\frac{\partial}{\partial z} - z^{-1}\rho - \theta_P,$$

where $\rho\in\mathrm{End}(\Omega_f)$ is the operator of multiplication by f, and $\theta_P\in\mathrm{End}(\Omega_f)$ is a linear operator independent of z.

The proof is a direct consequence of the definitions.

REMARK 2.6. In general, we do not have an intrinsic description of the operator θ_P. If the homogeneous opposite subspace P is of special type, then θ_P coincides with the grading operator of the so called Kashiwara–Malgrange filtration (see Sections 3.2 and 3.3).

3. Frobenius manifolds in singularity theory

In this section we will outline the construction of a Frobenius structure corresponding to a choice of a homogeneous opposite subspace. First, we will work formally following the work of [**11**]. Then we would like to work out a criteria for the analyticity of the formal construction.

3.1. Formal Frobenius manifolds. Let us recall the holomorphic germ $f \in \mathcal{O}_{\mathbb{C}^n,0}$, its miniversal deformation F, the deformation parameters $s = (s_1, \ldots, s_N)$, and the twisted de Rham cohomology groups $\widehat{\mathbb{H}}_+ = \widehat{\mathbb{H}}_+(f,\mathbb{C})$ and $\widehat{\mathbb{H}} := \widehat{\mathbb{H}}(f,\mathbb{C})$. All this data is introduced in Section 2.

Suppose that $P \subset \widehat{\mathbb{H}}$ is a homogeneous opposite subspace. Note that multiplication by $e^{(F-f)/z}$ defines a $\mathbb{C}[\![s,z]\!]$-linear map $\widehat{\mathcal{F}}_0 \to \widehat{\mathbb{H}}[\![s]\!]$, where

$$\widehat{\mathcal{F}}_0 := \Omega^n_{\mathbb{C}^n,0}[\![s,z]\!]/(zd + dF\wedge)\Omega^{n-1}_{\mathbb{C}^n,0}[\![s,z]\!].$$

is the formal completion at $s = 0$ of the germ $\mathcal{F}_0$ of the sheaf (2.1). Note that in the above definition, d is induced from the de Rham differential on $\mathbb{C}^n$, so it is $\mathbb{C}[\![s,z]\!]$-linear. Let us choose holomorphic forms $v_i \in \Omega^n_{\mathbb{C}^n,0}[\![z]\!]$ $(1 \le i \le N)$ that represent a basis of $\sigma_P(\Omega_f) = \widehat{\mathbb{H}}_+ \cap zP$. Then we have

$$\left[e^{(F-f)/z} v_j\right]_f = \sum_{i=1}^{N} [v_i]_f E_{i,j}(s,z),$$

where for a holomorphic form $v \in \Omega^n_{\mathbb{C}^n,0}$ we denote by $[v]_f$ (resp. $[v]_F$) its equivalence class in $\widehat{\mathbb{H}}_+$ (resp. $\widehat{\mathcal{F}}_0$) and $E_{i,j}(s,z) \in \mathbb{C}(\!(z)\!)[\![s]\!]$. Let us denote by $E(s,z)$ the $N \times N$ matrix with entries $E_{i,j}(s,z)$. Note that there exists a Birkhoff factorization $E(s,z) = T(s,z)A(s,z)^{-1}$, where the entries of $T(s,z)$ (resp. $A(s,z)$) are formal power series in $\mathbb{C}[z^{-1}][\![s]\!]z^{-1}$ (resp. $\mathbb{C}[\![s,z]\!]z^{-1}$), and at $z = \infty$, the matrix $T = 1$. Such a factorization is evident for $s = 0$ and in general is obtained perturbatively by comparing the coefficients in front of the Taylor series expansion in s.

Put $\widetilde{v}_j = \sum_{i=1}^N v_i A_{i,j}(s,z)$ and let

$$[\widetilde{v}]_F = ([\widetilde{v}_1]_F, \ldots, [\widetilde{v}]_F)$$

be a vector row whose entries are in $\widehat{\mathcal{F}}_0$.

PROPOSITION 3.1. *The Gauss–Manin connection takes the form*

$$z\nabla_{\partial/\partial s_i}[\widetilde{v}]_F = [\widetilde{v}]_F\,\Gamma_i(s) \quad (1 \le i \le N)$$
$$z\nabla_{\partial/\partial z}[\widetilde{v}]_F = [\widetilde{v}]_F\,(-\rho(s)z^{-1} - \theta_P),$$

where $\Gamma_i(s)$ and $\rho(s)$ are $N \times N$ matrices representing multiplication by $\frac{\partial F}{\partial s_i}$ and F in

$$\widehat{\Omega}_F := \Omega^n_{\mathbb{C}^n,0}[\![s]\!]/dF \wedge \Omega^{n-1}_{\mathbb{C}^n,0}[\![s]\!] = \widehat{\mathcal{F}}_0/z\widehat{\mathcal{F}}_0$$

with respect to the basis $[\widetilde{v}]_F$, and θ_P is a $N \times N$ matrix independent of s.

Note that for $s=0$, the row vector $[\widetilde{v}]_F$ turns into $([v_1]_f,\ldots,[v_N]_f)$. According to Proposition 2.5, the cohomology classes $[v_i]_f$ form a $\mathbb{C}(\!(z)\!)$-basis of $\widehat{\mathbb{H}}$. Slightly abusing the notation, we denote by θ_P also the linear operators in $\mathrm{End}(\widehat{\mathbb{H}})$ and $\mathrm{End}(\Omega_f)$ represented by the matrix θ_P in respect he bases $([v_1]_f,\ldots,[v_N]_f)$ and $(\sigma_P^{-1}([v_1]_f),\ldots,\sigma_P^{-1}([v_N]_f))$. We will use the operator $\theta_P\in\mathrm{End}(\Omega_f)$ to define a grading on Ω_f. Suppose that $\omega\in\Omega_f$ is a homogeneous holomorphic volume form, that is, $\theta_P(\omega)=-\alpha\,\omega$ and $\widehat{\Omega}_f=\mathcal{O}_{\mathbb{C}^n,0}\,\omega$. Put

$$\sigma_P(\omega)=\sum_{i=1}^N c_i\,[v_i]_f$$

and

$$\widetilde{\omega}:=\sum_{i=1}^N c_i\,\widetilde{v}_i\quad\in\quad\Omega^n_{\mathbb{C}^n,0}[\![s,z]\!].$$

We have the following commutative diagram:

$$\begin{array}{ccccc}
\widehat{\mathcal{T}}_{B,0}[\![z]\!] & \xrightarrow{\ \mathrm{mod}\ z\ } & \widehat{\mathcal{T}}_{B,0} & \xrightarrow{\ \cong\ } & \mathcal{O}_{\mathbb{C}^n,0}[\![s]\!]/(\partial_{x_1}F,\ldots,\partial_{x_n}F) \\
\Big\downarrow P_{\widetilde{\omega}} & & \Big\downarrow & & \Big\downarrow \times\pi(\widetilde{\omega}) \\
\widehat{\mathcal{F}}_0 & \xrightarrow{\ \pi\ } & \widehat{\Omega}_F=\widehat{\mathcal{F}}_0/z\widehat{\mathcal{F}}_0 & \xrightarrow{\ \cong\ } & \mathcal{O}_{\mathbb{C}^n,0}[\![s]\!]\pi(\widetilde{\omega})
\end{array}$$

where $P_{\widetilde{\omega}}$ is the *period map* defined in Proposition 2.2, the natural quotient map is π, the isomorphism in the first row is the Kodaira–Spencer isomorphism, and the isomorphism in the second row is the identity map. Let $e,E\in\widehat{\mathcal{T}}_{B,0}$ and $\xi_i\in\widehat{\mathcal{T}}_{B,0}[\![z]\!]$ $(1\le i\le N)$ be such that

$$\pi(z\nabla_e[\widetilde{\omega}]_F-[\widetilde{\omega}]_F)=0,\quad \pi(z\nabla_E[\widetilde{\omega}]_F-F\,[\widetilde{\omega}]_F)=0,\quad z\nabla_{\xi_i}[\widetilde{\omega}]_F=[\widetilde{v}_i]_F.$$

Note that e and E correspond respectively to 1 and F under the Kodaira–Spencer isomorphism. Moreover, under the Kodaira–Spencer isomorphism, $\widehat{\mathcal{T}}_{B,0}$ becomes an algebra.

PROPOSITION 3.2. *a) There exists a formal coordinate system $t=(t_1,\ldots,t_N)$ $(t_i\in\mathbb{C}[\![s]\!])$ such that $\xi_i=\frac{\partial}{\partial t_i}$ $(1\le i\le N)$.*

b) The higher residue pairing $K_F(\widetilde{v}_i,\widetilde{v}_j)=K_f(v_i,v_j)$ does not depend on s.

c) Put

$$\left(\frac{\partial}{\partial t_i},\frac{\partial}{\partial t_j}\right):=K_F(\widetilde{v}_i,\widetilde{v}_j)=K_f(v_i,v_j)=K_f^{(0)}(v_i,v_j).$$

Then $(\widehat{\mathcal{T}}_{B,0},E,e,(\ ,\ ))$ is a formal Frobenius manifold.

3.2. Analyticity of the Frobenius structure. The formal Frobenius manifold of Li-Li-Saito (see Proposition 3.2) will be analytic if σ_P happens to be a *good section* in the sense of M. Saito (see Definition 3.3 in [**15**]). Let us recall the notion of a good section. Put

$$\mathbb{H}_+(f,\mathbb{C})=\Omega^n_{\mathbb{C}^n,0}/df\wedge d\Omega^{n-2}_{\mathbb{C}^n,0}.$$

This is a $\mathcal{O}_{\mathbb{C}^n,0}$-module known as the *Brieskorn lattice*. It is also a $\mathbb{C}[z]$-module (even a $\mathbb{C}\{\{z\}\}$-module) where the action of z is given by

$$z\cdot[\varphi]=-[df\wedge d^{-1}\varphi],\quad \varphi\in\Omega^n_{\mathbb{C}^n,0},$$

where $d^{-1}\varphi$ is a $(n-1)$-form $\eta \in \Omega^{n-1}_{\mathbb{C}^n,0}$ such that $d\eta = \varphi$. Since the higher cohomologies of the de Rham complex $(\Omega^{\bullet}_{\mathbb{C}^n,0}, d)$ vanish, we can always choose such a form η, and the equivalence class $[df\wedge\eta] \in \mathbb{H}_+(f,\mathbb{C})$ is independent of the choice of η. The quotient map $\Omega^n_{\mathbb{C}^n,0} \to \widehat{\mathbb{H}}_+(f,\mathbb{C})$ induces an embedding $\mathbb{H}_+(f,\mathbb{C}) \subset \widehat{\mathbb{H}}_+(f,\mathbb{C})$. In other words, the Brieskorn lattice is identified with the cohomology classes in $\widehat{\mathbb{H}}_+(f,\mathbb{C})$ that can be represented by holomorphic forms.

Let us recall also the so called *Kashiwara–Malgrange* filtration. Put

$$\widehat{\mathbb{H}}^{\alpha}_+ := \{v \in \widehat{\mathbb{H}}_+(f,\mathbb{C}) \mid (z\nabla_{\partial/\partial z} - \alpha)^r v = 0 \text{ for } r \gg 0\}.$$

It is known that $\alpha \in \mathbb{Q}$, otherwise $\widehat{\mathbb{H}}^{\alpha}_+ = 0$, and we have

$$\widehat{\mathbb{H}} = \widehat{\bigoplus_{\alpha\in\mathbb{Q}}} \widehat{\mathbb{H}}^{\alpha}_+.$$

Then $\widehat{\mathbb{H}}^{\geq\alpha}_+ := \widehat{\bigoplus}_{\beta\geq\alpha}\widehat{\mathbb{H}}^{\beta}_+$ defines a decreasing exhaustive filtration known as the Kashiwara–Malgrange filtration. Let $\Omega^{\alpha}_f := \widehat{\mathbb{H}}^{\alpha}_+/z\widehat{\mathbb{H}}^{\alpha-1}_+$ and $\Omega^{\geq\alpha}_f := \bigoplus_{\beta\geq\alpha}\Omega^{\beta}_f$ be the induced filtrations.

PROPOSITION 3.3 (M. Saito). *If P is a homogeneous opposite subspace, s.t.,*
(i) $\sigma_P(\Omega_f) \subset \mathbb{H}_+(f,\mathbb{C})$,
(ii) $\sigma_P(\Omega^{\geq\alpha}_f) \subseteq \mathbb{H}^{\geq\alpha}_+$,
then the formal Frobenius structure in Proposition 3.2 *is analytic.*

The proof of this proposition can be obtained from the results of B. Malgrange ([**12, 13**]). We refer to Section 4.3 in [**15**] for more details.

3.3. Hodge grading operators. Suppose that P is a homogeneous opposite subspace and let $\widehat{\sigma}_P : \Omega_f(\!(z)\!) \to \widehat{\mathbb{H}}(f,\mathbb{C})$ be the isomorphism from Proposition 2.5, b). The Gauss–Manin connection takes the form

$$\nabla_{z\partial/\partial z} = z\frac{\partial}{\partial z} - z^{-1}\rho - \theta_P, \quad \rho, \theta_P \in \mathrm{End}(\Omega_f).$$

DEFINITION 3.4. The linear operator θ_P is said to be a *Hodge grading operator* if it is diagonalizable and if $[\theta_P, \rho] = -\rho$.

LEMMA 3.5. *Suppose that θ_P is a Hodge grading operator. Under the isomorphism $\widehat{\sigma}_P$, the Kashiwara–Malgrange filtration takes the form*

$$\widehat{\mathbb{H}}^{\geq\alpha} = \mathrm{Span}_{\mathbb{C}}\{\varphi_i z^k \mid k - \theta_i \geq \alpha\},$$

where $\{\varphi_i\}_{i=1}^N$ is an eigenbasis of θ_P and where $\{\theta_i\}_{i=1}^N$ are the corresponding eigenvalues.

PROOF. In order to avoid cumbersome notation, let us put $\theta := \theta_P$. Under the isomorphism $\widehat{\sigma}_P$, the Gauss–Manin connection takes the form

$$z\nabla_{\partial/\partial z} = z\partial_z - z^{-1}\rho - \theta = z^{\theta}\,(z\partial_z - \rho)\,z^{-\theta}.$$

Let us write $v = \sum_{i,k} c_{i,k}\varphi_i z^k \in \Omega_f(\!(z)\!)$. We have

$$(z\nabla_{\partial/\partial z} - \alpha)^n(\varphi_i z^k) = z^{\theta}z^{-\theta_i}\,(k - \theta_i - \alpha - \rho)^n\,\varphi_i z^k.$$

This is 0 for $n \gg 0$ if and only if $k = \theta_i + \alpha$. Note that there are finitely many (k,i) for which $k = \theta_i + \alpha$, therefore if $v \in \widehat{\mathbb{H}}^{\alpha}$, then v is a finite linear combination of $\varphi_i z^k$. $\square$

Let us order the set of numbers α such that $\Omega_f^\alpha \neq 0$ in an increasing sequence and let us repeat each α as many times as $\dim_{\mathbb{C}} \Omega_f^\alpha$. We get a sequence $\alpha_1 \leq \alpha_2 \leq \cdots \leq \alpha_N$ known as the *Steenbrink's spectrum.* Note that our definition of the Gauss–Manin connection differs from the standard one by a shift corresponding to the normalization factor $(-2\pi z)^{-n/2}$ in front of the oscillatory integral (see formula (2.2)). The Steenbrink's spectrum in our settings has the following symmetry: $\alpha_i + \alpha_{N+1-i} = 0$. Lemma 3.5 implies that if θ_P is a Hodge grading operator, then the eigenvalues of θ_P coincide with the Steenbrink spectrum. In particular, they are rational numbers independent of P.

THEOREM 3.6. *If θ_P is a Hodge grading operator, then σ_P is a good section, that is, $\sigma_P(\Omega_f) \subset \mathbb{H}_+(f,\mathbb{C})$ and $\sigma_P(\Omega_f^{\geq\alpha}) \subseteq \mathbb{H}_+^{\geq\alpha}$.*

PROOF. First we prove that $\sigma_P(\Omega_f^{\geq\alpha}) \subset \widehat{\mathbb{H}}^{\geq\alpha}(f,\mathbb{C})$. Let $\{\varphi_i\}$ be an eigenbasis of θ_P with eigenvalues $\{\theta_i\}$. Put $v_i = \sigma_P(\varphi_i)$. According to Lemma 3.5 we have the following formulas for the generalized eigenspaces of the Gauss–Manin connection

$$\widehat{\mathbb{H}}_+^\alpha = \mathrm{Span}_{\mathbb{C}}\{v_i z^k \mid k \geq 0 \text{ and } \alpha = k - \theta_i\}.$$

Therefore, under the projection $\pi : \widehat{\mathbb{H}}_+ \to \Omega_f$ we have

$$\pi(\widehat{\mathbb{H}}_+^\alpha) = \mathrm{Span}_{\mathbb{C}}\{\varphi_i \mid -\theta_i = \alpha\}.$$

Therefore,

$$\Omega_f^{\geq\alpha} = \mathrm{Span}_{\mathbb{C}}\{\varphi_i \mid -\theta_i \geq \alpha\}.$$

The inclusion $\sigma_P(\Omega_f^{\geq\alpha}) \subset \widehat{\mathbb{H}}_+^{\geq\alpha}$ is a direct consequence of the above formula.

Suppose that there exists a homogeneous opposite subspace P for which θ_P is a Hodge grading operator, and the image of σ_P is in $\mathbb{H}_+(f,\mathbb{C})$. Let Q be another opposite subspace for which θ_Q is a Hodge grading operator. Let $\{v_i\} \subset \sigma_P(\Omega_f)$ and $\{w_i\} \subset \sigma_Q(\Omega_f)$ be bases corresponding to eigenbases of respectively θ_P and θ_Q. Since the eigenvalues of θ_P and θ_Q counted with multiplicities coincide, we may assume that $v_i, w_i \in \widehat{\mathbb{H}}_+^{\alpha_i}$. We get that

$$w_j = \sum_{i=1}^N v_i R_{ij}(z), \quad R_{ij} \in \mathbb{C}[\![z]\!].$$

Applying to both sides $(z\nabla_{\partial/\partial z} - \alpha_j)^r$ with $r \gg 0$ such that the LHS is 0, we get that if $R_{ij}(z) \neq 0$, then $\alpha_j - \alpha_i$ is a non-negative integer and $R_{ij}(z) \in \mathbb{C}z^{\alpha_j - \alpha_i}$. In other words, w_j is a finite linear combination of terms of the type $v_i z^k$ with $k \in \mathbb{Z}_{\geq 0}$. Since we assumed that $v_i \in \mathbb{H}_+(f,\mathbb{C})$ we get that $v_i z^k \in \mathbb{H}_+(f,\mathbb{C})$ as well, so $w_j \in \mathbb{H}_+(f,\mathbb{C})$.

The existence of a single homogeneous subspace P for which θ_P is a Hodge grading operator is a very non-trivial fact. M. Saito (see [**15**]) proved that the splittings of Steenbrink's mixed Hodge structure on the vanishing cohomology of f are in one-to-one correspondence with the set of opposite subspace P for which σ_P is a good section. This correspondence was studied further by C. Hertling in [**8**]. He proved that Steenbrink's mixed Hodge structure has a polarizing form, i.e., it is a Polarized Mixed Hodge Structure (PMHS), and introduced the notion of an *opposite filtration* for a PMHS (see Definition 10.19 in [**8**]). If we take an opposite subspace P corresponding to an opposite filtration for the PMHS, then the operator θ_P is automatically diagonalizable and the commutation relation $[\theta_P,\rho] = -\rho$ follows from Theorem 11.9, part d) in [**8**]. □

The argument in the above theorem allows us to classify all homogeneous opposite subspaces P for which θ_P is a Hodge grading operator. The classification can be stated as follows: Let us introduce the following bilinear form

$$S(a,b) := (a, e^{-\pi \mathrm{i}\rho} e^{\pi \mathrm{i}\theta} b), \quad a,b\in\Omega_f,$$

where $(\ ,\ ) = K_f^{(0)}(\ ,\ )$ is the residue pairing and $\theta\in \mathrm{End}(\Omega_f)$ is the grading operator associated with the Steenbrink spectrum, i.e.,

$$\theta(\varphi) := -\alpha\,\varphi, \quad \text{for} \quad \varphi\in\Omega_f^\alpha.$$

Let $\mathrm{Aut}(\Omega_f^{\leq\bullet}, S,\rho)$ be the group of all $g\in\mathrm{End}(\Omega_f)$ satisfying the following conditions:

(i) $g(v)\in v+\Omega_f^{<\alpha}$ for all $v\in\Omega_f^\alpha$.
(ii) $S(ga,gb) = S(a,b)$ for all $a,b\in\Omega_f$.
(iii) $[g,\rho]=0$.

The definition of the group $\mathrm{Aut}(\Omega_f^{\leq\bullet}, S,\rho)$ can be stated equivalently using Givental's symplectic space formalism associated with $\Omega_f(\!(z)\!)$. Namely, an element $g\in \mathrm{End}(\Omega_f)$ belongs to $\mathrm{Aut}(\Omega_f^{\leq\bullet}, S,\rho)$ if and only if

(i) $R_g(z) := z^\theta g z^{-\theta}$ has the form

$$R_g(z) = 1 + R_{g,1}z + R_{g,2}z^2+\cdots.$$

(ii) $R_g(z)$ is a symplectic transformation: $R_g(z)R_g(-z)^T = 1$.
(iii) $[R_g(z),\rho] = 0$.

PROPOSITION 3.7. *The group* $\mathrm{Aut}(\Omega_f^{\leq\bullet}, S,\rho)$ *acts transitively on the space of homogeneous opposite subspaces P whose grading operator θ_P is a Hodge grading operator. The action is given by the following formula:*

$$g\cdot P := \widehat{\sigma}_P\circ R_g(z)\circ\widehat{\sigma}_P^{-1}(P) = \widehat{\sigma}_P\circ R_g(z)\left(\Omega_f[z^{-1}]z^{-1}\right). \quad \square$$

REMARK 3.8. The above proposition and Lemma 10.21 in [**8**] proves that there exists a one-to-one correspondence between the set of opposite subspaces whose grading operator is a Hodge grading operator and the set of opposite filtrations to the Hertling–Steenbrink's PMHS on the vanishing cohomology of the Milnor fiber of f.

The Frobenius structures corresponding to Hodge grading operators can be used to define a set of total ancestor potentials. Since the group $\mathrm{Aut}(\Omega_f^{\leq\bullet}, S,\rho)$ acts on the opposite subspaces, it also acts on the set of total ancestor potentials. Let us describe this action. To begin with, let us recall that according to M. Saito [**15**] the smallest spectral number α_1 in the Steenbrink spectrum has multiplicity 1, that is, $\dim_{\mathbb{C}}\Omega_f^{\alpha_1} = 1$. Suppose that $\omega\in\Omega_f^{\alpha_1}\setminus\{0\}$ and that $P\subset\widehat{\mathbb{H}}_+$ is a homogeneous opposite subspace whose grading operator θ_P is a Hodge grading operator. Let $v=(v_1,\dots,v_N)$ be a row vector whose entries v_i form a basis of $\sigma_P(\Omega_f) = \mathbb{H}_+\cap zP$. The Frobenius structure of Li-Li-Saito (see Proposition 3.2) is analytic. Therefore, we can define the total ancestor potential $\mathcal{A}_{\omega,v,t}(\hbar,\mathbf{q})$ for all $t\in B\setminus\mathcal{K}$. Note that in order to define the total ancestor potential (see formula (1.1)) we have to choose a basis of flat vector fields. We make such a choice using the basis $\{v_1, \ldots, v_N\}$ and the period map $P_{\widetilde{\omega}}$ as stated in Proposition 3.2, a. According to the main result in [**14**], the total ancestor potential $\mathcal{A}_{\omega,v,t}(\hbar, \mathbf{q})$ extends analytically with respect to t for all $t \in B$. We define

$$\mathcal{A}_{\omega,v}(\hbar, \mathbf{q}) := \lim_{t\to 0} \mathcal{A}_{\omega,v,t}(\hbar, \mathbf{q}).$$

Suppose that $\omega' \in \Omega_f^{\alpha_1} - \{0\}$, $P' \subset \widehat{\mathbb{H}}(f, \mathbb{C})$ is a homogeneous opposite subspace whose grading operator is a Hodge grading operator, and that $v' := \{v_i'\}_{i=1}^N \subset \widehat{\mathbb{H}}_+(f, \mathbb{C}) \cap P'z$ is a basis. The transformation rules between the corresponding total ancestor potentials can be described by the following two cases:

(i) If $P' = P$, with $v' = vB$, and $\omega' = \omega c$ where $B \in \mathrm{GL}_N(\mathbb{C})$ and $c \in \mathbb{C}^*$, then

$$\mathcal{A}_{v',\omega'}(\hbar, \mathbf{q}) = \mathcal{A}_{v,\omega}(\hbar c^{-2}, c^{-1}B\mathbf{q}), \tag{3.1}$$

where $(c^{-1}B\mathbf{q})_{k,i} = \sum_{j=1}^N c^{-1}B_{ij}q_{k,j}$.

(ii) If $\omega' = \omega$ and $v' = vR(z)$ where $R(z) = z^\theta g z^{-\theta}$ for some $g \in \mathrm{Aut}(\Omega_f^{\leq \bullet}, S, \rho)$, then

$$\mathcal{A}_{v',\omega}(\hbar, \mathbf{q}) = (R(z)^T)^\wedge \mathcal{A}_{v,\omega}(\hbar; \mathbf{q}), \tag{3.2}$$

where $R(z)^T$ is the transpose of $R(z)$ with respect to the residue pairing.

4. Mirror symmetry for Fermat Calabi–Yau orbifolds

The goal of this section is to give a class of very interesting examples where the formalism of opposite subspaces allows us to construct the mirror model of an orbifold. The proofs are quite technical, so we refer the interested reader to [**9**] for more details.

Let d_i $(1 \leq i \leq n)$ be a set of positive integers satisfying $\frac{1}{d_1} + \cdots + \frac{1}{d_n} = 1$. Put

$$G = \{t \in (\mathbb{C}^*)^n \mid t_1^{d_1} = \cdots = t_n^{d_n}\}$$

and

$$Z = \{z \in \mathbb{C}^n \setminus \{0\} \mid z_1^{d_1} + \cdots + z_n^{d_n} = 0\}.$$

Our main interest is in the Gromov–Witten theory of the orbifold groupoid $Y = [Z/G]$. The orbit space of Y is $|Y| = \mathbb{P}^{n-1}$, so classically, Y corresponds to an orbifold structure on the projective space $\mathbb{P}^{n-1}$.

4.1. Orbifold cohomology ring.

The inertia orbifold $IY = \bigsqcup_c Y_c$, where the disjoint union is over all sequences $c = (c_1, \ldots, c_n)$ of rational numbers such that

(i) $c_i = \frac{k_i}{d_i}$, where k_i is an integer satisfying $0 \leq k_i \leq d_i$.
(ii) $c_i = 0$ for at least two i's.

The connected components $Y_c = [Z_c/G]$, where

$$Z_c = \{z \in Z \mid z_1c_1 = \cdots = z_nc_n = 0\}.$$

The *orbifold cohomology ring* or *Chen–Ruan cohomology ring* is by definiton $H := H_{\mathrm{CR}}(Y, \mathbb{C}) = \bigoplus_c H(|Y_c|, \mathbb{C})$. Note that the orbit space $|Y_c|$ is a projective space of dimension $\dim(c) :=$ (the number of zero entries in c) $- 2$. Therefore, H has the following basis

$$p^k \mathbf{1}_c, \quad 0 \leq k \leq \dim(c),$$

where $p = c_1(\mathcal{O}(1))$ is the hyper-plane class and $\mathbf{1}_c$ is the unit in $H(|Y_c|, \mathbb{C})$. The vector space H is a graded vector space. The grading in $H(|Y_c|, \mathbb{C})$ is by definiton the shift of the standard cohomology degree grading on the topological space $|Y_c|$ by a rational number called the *age* of Y_c. The above basis is homogeneous of the degree

$$\deg_{\mathrm{CR}}(p^k \mathbf{1}_c) := k + c_1 + \cdots + c_n.$$

Finally, let us point out that the dimension of H is $N = (d_1 - 1)\cdots(d_n - 1)$.

4.2. I-function. The I-function of a complete intersection in a toric orbifold is a certain hypergeometric type series defined through the combinatorics of the toric orbifold. Let us recall the precise definition for the case at hand.

Suppose that $\beta = (\beta_1, \ldots, \beta_n)$ is a sequence of non-negative integers. Put

$$I^\beta(Q, z) = e^{p \log Q/z} \sum_{d=0}^{\infty} I_d^\beta(z) Q^d,$$

where

$$I_d^\beta(z) = \frac{\Gamma(1 + d + p/z)}{\Gamma(1 + p/z)} \prod_{i=1}^{n} \frac{\Gamma(1 - c_i + (p/d_i)z^{-1})}{\Gamma(1 - c_i + k_i + (p/d_i)z^{-1})} \mathbf{1}_c z^{d - |\kappa|},$$

where the sequences $\kappa = (k_1, \ldots, k_n)$ and $c = (c_1, \ldots, c_n)$ are defined by

$$\frac{\beta_i - d}{d_i} =: -k_i + c_i, \quad 0 \le c_i < 1, \quad k_i \in \mathbb{Z},$$

$|\kappa| = k_1 + \cdots + k_n$.

4.3. The mirror map. Let us fix a sufficiently small disk Δ with center 0 such that the coefficients of the I-function

$$I^0(Q, z) = f_0(Q)\mathbf{1} + f_1(Q)z^{-1} + f_2(Q)z^{-2} + \cdots$$

are analytic for all $Q \in \Delta$. The above series is a solution to a certain Fuchsian differential equation, so the existence of Δ is guaranteed by the general theory of Fuchsian differential equations. Put

$$\tau(Q) := \frac{f_1(Q)}{f_0(Q)} = \tau_0(Q)p + \tau_{\mathrm{tw}}(Q) \quad \in \quad H^2_{\mathrm{CR}}(Y, \mathbb{C}),$$

where $\tau_0(Q) \in \log Q + \mathcal{O}(\Delta)$ and $\tau_{\mathrm{tw}}(Q) = \sum_{c:|c|=1} \tau_c(Q)\mathbf{1}_c$, $\tau_c \in \mathcal{O}(\Delta)$. The function τ defines an embedding

$$\Delta \setminus \{0\} \to \mathbb{C}^* \times H^2_{\mathrm{CR}}(Y, \mathbb{C})_{\mathrm{tw}}, \quad Q \mapsto (e^{\tau_0(Q)}, \tau_{\mathrm{tw}}(Q)),$$

where $H^2_{\mathrm{CR}}(Y, \mathbb{C})_{\mathrm{tw}} = \bigoplus_{c:|c|=1} H^2_{\mathrm{CR}}(|Y_c|, \mathbb{C})$. Note that the complex circle $\mathbb{C}^* \cong H^2(|Y|, \mathbb{C})/2\pi\mathbf{i}H^2(|Y|, \mathbb{Z})$.

Recall also the S-matrix from the orbifold Gromov–Witten theory of Y

$$S(t, q, z) = 1 + \sum_{k=1}^{\infty} S_k(t, q) z^{-k}, \quad S_k \in \mathrm{End}(H),$$

defined by

$$(S(t, q, z)\phi_i, \phi_j) := (\phi_i, \phi_j) + \sum_{d,l=0}^{\infty} \frac{q^d}{l!} \langle \phi_i \psi^k, \phi_j, t, \ldots, t \rangle_{0, l+2, d},$$

where $\{\phi_i\}_{i=1}^N$ is a basis of H, and q is a formal variable, and the correlator denotes an orbifold Gromov–Witten invariant (see [**9**] for more details).

Put

$$f_Q(x) = x_1^{d_1} + \cdots + x_n^{d_n} - \frac{1}{Q} x_1 \cdots x_n$$

and let $\{a_1, \ldots, a_r\}$ be the set of values of Q for which the function has a non-isolated critical point. Let $\Sigma = \mathbb{C} \backslash \{0, a_1, \ldots a_r\}$ be the punctured Riemann sphere. Let $\mathbb{F} \to \Sigma$ be the vector bundle on Σ whose fiber over $Q \in \Sigma$ is

$$\mathbb{F}_Q := \widehat{H}_+(f_Q, \mathbb{C}) := \Omega^n_{\mathbb{C}^n,0}[\![z]\!]/(zd + df_Q \wedge)\Omega^{n-1}_{\mathbb{C}^n,0}[\![z]\!].$$

Let us point out that f_Q is a weighted homogeneous polynomial, so the operator $\rho = 0$, that is, the grading operator of any homogeneous opposite subspace, is automatically a Hodge grading operator. The mirror map

$$\mathrm{Mir} : \mathbb{F}|_{\Delta^*} \to \Delta^* \times H[\![z]\!],$$

where $\Delta^* := \Delta \setminus \{0\} \subset \Sigma$, is defined by

$$\mathrm{Mir}\left(x_1^{\beta_1} \cdots x_n^{\beta_n} \frac{dx_1 \cdots dx_n}{Q} \right) := S(\tau(Q), 1, -z) I^\beta(Q, z).$$

Using the divisor equation from Gromov–Witten theory, we can also write

$$S(\tau(Q), 1, -z) = S(\tau_{\mathrm{tw}}(Q), q, -z) e^{-p \log q/z}, \quad q = e^{\tau_0(Q)}.$$

4.4. Mirror symmetry. The ancestor Orbifold Gromov–Witten theory of Y defines a Cohomological Field Theory (CohFT)

$$\Lambda_{g,n} : H^{\otimes n} \to H^*(\overline{\mathcal{M}}_{g,n}, \mathbb{C})$$

by the following formula

$$\Lambda_{g,n}(a_1, \ldots, a_n) = \sum_{d=0}^{\infty} q^d \mathrm{ft}_* \Big(\mathrm{ev}^*(a_1 \otimes \cdots \otimes a_n) \cap [\overline{\mathcal{M}}_{g,n}(Y, d)]^{\mathrm{virt}} \Big)$$

where $\overline{\mathcal{M}}_{g,n}(Y, d)$ is the moduli space of orbifold stable maps to Y, while $[\ ]^{\mathrm{virt}}$ denotes the *virtual fundamental cycle*, and ev and ft are respectively the evaluation map and the forgetful map

$$\begin{array}{ccc} \overline{\mathcal{M}}_{g,n}(Y, d) & \xrightarrow{\mathrm{ev}} & IY^n \, . \\ \big\downarrow{\scriptstyle \mathrm{ft}} & & \\ \overline{\mathcal{M}}_{g,n} & & \end{array}$$

Recall that the evaluation map ev is defined by evaluating the orbifold map at the marked points, and the forgetful map ft is defined by forgetting the map to the target Y, the orbifold structure on the domain curve, and by contracting unstable components.

The total ancestor potential of Y is by definition the following generating function of intersection numbers

$$\mathcal{A}^Y_{t,q}(\hbar, \mathbf{t}) = \exp\Big(\sum_{g,n=0}^{\infty} \frac{\hbar^{g-1}}{n!} \int_{\overline{\mathcal{M}}_{g,n}} \Lambda^t_{g,n}(\mathbf{t}(\psi_1), \ldots, \mathbf{t}(\psi_n)) \Big),$$

where $\mathbf{t}(\psi) = \sum_{k=0}^{\infty}\sum_{i=1}^{N} t_{k,i}\psi^k\phi_i$, $\{t_{k,i}\}$ $(1 \le i \le N,\ k \ge 0)$ are formal variables, and

$$\Lambda_{g,n}^t(a_1,\ldots,a_n) := \sum_{l=0}^{\infty}\frac{1}{l!}\Lambda_{g,n+l}(a_1,\ldots,a_n,t,\ldots,t)$$

is a deformation of the CohFT depending on the choice of a parameter $t \in H$. The total ancestor potential is identified with an element of the Fock space via the so-called *dilaton shift*

$$t_{k,i} = q_{k,i} + \delta_{k,1}c_i,$$

where $\mathbf{1} = \sum_{i=1}^{N} c_i\phi_i$.

Let us fix a basis $\{\phi_i\}_{i=1}^{N}$ of H and define

$$v_i := \mathbf{i}\ \mathrm{Mir}^{-1}(\phi_i), \quad \omega := \mathbf{i}\ \mathrm{Mir}^{-1}(1) \quad \in \quad \Gamma(\Sigma,\mathbb{F}). \tag{4.1}$$

Put

$$P = \bigoplus_{1\le i\le N}\bigoplus_{k=0}^{\infty}\mathbb{C}v_i z^{-k-1} \quad \subset \quad \widehat{\mathbb{H}}(f_Q,\mathbb{C}).$$

Then we have the following mirror symmetry theorem.

THEOREM 4.1 ([**9**]). *The subspace P is a homogeneous opposite subspace for which θ_P is a Hodge grading operator and*

$$\mathcal{A}_{\omega,v}(\hbar,\mathbf{q}) = \mathcal{A}^{Y}_{\tau_{\mathrm{tw}},q}(\hbar,\mathbf{q}),$$

where $q = e^{\tau_0(Q)}$.

5. The quartic example

The main goal in this section is to show on an example that the basis $v = (v_1,\ldots,v_N)$ used in our mirror Theorem 4.1 can be computed explicitly in terms of hypergeometric functions. Furthermore, this allows us to extend analytically the total ancestor potential for all $Q \in \Sigma$ and to obtain explicit formulas for the corresponding monodromy transformations.

5.1. Good basis via Birkhoff factorization.

DEFINITION 5.1. A set of vectors in $\widehat{\mathbb{H}}_+(f,\mathbb{C})$ is said to be a *good basis* if they form the basis of $\sigma_P(\Omega_f)\cap\widehat{\mathbb{H}}_+(f,\mathbb{C})$ for some homogeneous opposite subspace P such that $\sigma_P(\Omega_f^{\ge\alpha}) \subset \widehat{\mathbb{H}}^{\ge\alpha}$ (i.e. the section σ_P is compatible with the Kashiwara–Malgrange filtration).

Let $\phi = (\phi_1,\ldots,\phi_N)$ and $\varphi = (x^{\beta_1}dx/Q,\ldots,x^{\beta_N}dx/Q)$ be row vectors whose entries form bases respectively of H as a $\mathbb{C}$-vector space and of $\widehat{\mathbb{H}}_+(f_Q,\mathbb{C})$ as a $\mathbb{C}[\![z]\!]$-module. Here, $\beta_i = (\beta_{i,1},\ldots,\beta_{i,n})$ are multi-indices and we abbreviated

$$x^{\beta_i} := x_1^{\beta_{i,1}}\cdots x_n^{\beta_{i,n}}, \quad dx := dx_1\wedge\cdots\wedge dx_n.$$

By definition the mirror map has the form

$$\mathrm{Mir}(\varphi) = \phi\,S(\tau,1,-z)\,I(Q,z),$$

where, slightly abusing the notation we denoted by $S(\tau,1,-z)$, the matrix representing the operator series $S(\tau,1,-z)$ in the basis ϕ and $I(Q,z)$ is a $N\times N$ matrix whose (i,j) entry is the coefficient in front of ϕ_i in $I^{\beta_j}(Q,Z)$.

Let $v = (v_1,\ldots,v_N)$ be the row vector whose entries are given by the mirror map (see formula (4.1)). These vectors provide a *good basis* for the twisted de

Rham cohomology group $\widehat{\mathbb{H}}_+(f_Q,\mathbb{C})$. By definition, $v=\varphi\,A(Q,z)$ where $A(Q,z)$ is a $N\times N$ matrix whose entries are in $\mathbb{C}[\![z]\!]$, and

$$\mathbf{i}\,\phi=\mathrm{Mir}(\varphi)\,A(Q,z)=\phi\,S(\tau,1,-z)\,I(Q,z)\,A(Q,z).$$

Therefore, we get

$$I(Q,z)^{-1}=-\mathbf{i}\,A(Q,z)S(\tau,1,-z). \tag{5.1}$$

Note that $I(Q,z)$ is a matrix that is given explicitly in terms of the I-functions, i.e., it is a matrix whose entries are some explicitly given hypergeometric series. The above formula (5.1) allows us to find both matrices A and Q by computing the Birkhoff factorization of the matrix $I(Q,z)^{-1}$. Moreover, using degree reasons, one can prove that $A(Q,z)$ is polynomial in z. Therefore, the Birkhoff factorization can be performed not only perturbatively but, in fact, we can get explicit closed formulas in terms of the generalized hypergeometric functions.

5.2. Good basis for the quartic. In this section we would like to give the good basis v for the case when $d_1=d_2=d_3=d_4=1/4$. This is the case when our orbifold Y is a quotient of the quartic $K3$-surface by a finite group. The computations are very long, so here we only give the final answer. The details will appear elsewhere.

Let us fix a set of monomial holomorphic forms $x^{\beta_i}dx/Q$ $(1\leq i\leq N=81)$. We split the indices β_i into blocks which fall into one of the following four types:

(0) $\{(0,0,0,0),(1,1,1,1),(2,2,2,2)\}$.
(n) All permutations of $\{(n,0,0,0),(n+1,1,1,1)\}$, where $1\leq n\leq 3$.
(n,n) All permutations of $\{(n,n,0,0),(0,0,4-n,4-n)\}$, where $1\leq n\leq 2$.
(m,n) All permutations of $\{(m,n,0,0)\}$, where $1\leq m\neq n\leq 3$.

Put $D=Q\partial_Q Q=Q^2\partial_Q+Q$, $h_k^{(a)}=1+\frac{1}{2^a}+\cdots+\frac{1}{k^a}$ $(a=1,2)$. The formulas for $I^{\beta_i}(Q,z)$ are the following:

Type (0). We have

$$\begin{aligned}
I^{(0,0,0,0)}(Q,z)&=f_{0;1}(Q)+f_{0;2}(Q)pz^{-1}+f_{0;3}(Q)pz^{-2}/2,\\
I^{(1,1,1,1)}(Q,z)&=zDI(Q,z),\\
I^{(2,2,2,2)}(Q,z)&=(zD)^2I(Q,z),
\end{aligned}$$

where

$$f_{0;1}(Q)=1+\sum_{k=1}^{\infty}\frac{(4k)!}{(k!)^4}Q^{4k},$$

$$f_{0;2}(Q)=\log Q+\sum_{k=1}^{\infty}\frac{(4k)!}{(k!)^4}Q^{4k}(\log Q+h_{4k}^{(1)}-h_k^{(1)}),$$

$$f_{0;3}(Q)=(\log Q)^2+\sum_{k=1}^{\infty}\frac{(4k)!}{(k!)^4}Q^{4k}\Big((\log Q+h_{4k}^{(1)}-h_k^{(1)})^2-h_{4k}^{(2)}+\frac{1}{4}h_k^{(2)}\Big).$$

The functions $f_{0;1},f_{0;2}$, and $f_{0;3}$ are solutions to the 3rd order Picard–Fuchs equation

$$(Q\partial_Q)^3f=4^4Q^4(Q\partial_Q+1)(Q\partial_Q+2)(Q\partial_Q+3)f,$$

satisfying the quadratic relation $f_{0;2}^2=f_{0;1}f_{0;3}$. The equation $\mathrm{Im}(4^4Q^4)=0$ defines 4 lines that split $\Sigma=\mathbb{C}\setminus\{0,\pm\frac{1}{4},\pm\frac{\mathrm{i}}{4}\}$ into 8 sectors. One can check that the above generalized hypergeometric equation is a symmetric square of a Gauss hypergeometric equation and that the function $\tau_0(Q) := f_{0;2}(Q)/f_{0;1}(Q)$ coincides with the corresponding Schwartz triangular map. Furthermore, $\tau_0(Q)$ is a multi-valued analytic function on Σ and its restriction to each of the 8 sectors defined above consists of a set of holomorphic branches. The various holomorphic branches map the 8 sectors of Σ into hyperbolic triangles that provide a triangulation of the half-plane

$$\mathbb{H}_0 := \{\tau_0 \in \mathbb{C} \mid \operatorname{Re}(\tau_0) < 0\}. \tag{5.2}$$

We refer to [**21**] for some background on the Schwartz triangular map.

The part of the good basis corresponding to the Type (0) block takes the form

$$\begin{aligned}
v_{0,0,0,0} &= dx/(Qf_{0;1}),\\
v_{1,1,1,1} &= \Big(x_1\cdots x_4 - z\phi(Q)\Big)(1-4^4Q^4)^{1/2}dx/Q^2,\\
v_{2,2,2,2} &= \Big((x_1\cdots x_4)^2 - z\Big(\frac{2Q}{1-4^4Q^4}+\phi(Q)\Big)\,x_1\cdots x_4 + \frac{z^2}{2}\Big(\frac{Q^2}{1-4^4Q^4}+\phi(Q)^2\Big)\Big)\times\\
&\quad\times(1-4^4Q^4)f_{0;1}dx/Q^3,
\end{aligned}$$

where $\phi(Q) = Df_{0;1}/f_{0;1}$.

Type (n): $1 \le n \le 3$. We have

$$\begin{aligned}
I^{(n,0,0,0)}(Q,z) &= f_{n;1}(Q)\mathbf{1}_{\frac{n}{4},0,0,0} + f_{n;2}(Q)z^{-1}\mathbf{p}\mathbf{1}_{\frac{n}{4},0,0,0}\\
I^{(n+1,1,1,1)}(Q,z) &= zQ\partial_Q Q\, I^{(n,0,0,0)}(Q,z),
\end{aligned}$$

where

$$\begin{aligned}
f_{n;1} &= 1 + \sum_{k=1}^{\infty} \frac{(4k)!}{(k!)^3\prod_{i=1}^{k}(i-n/4)}Q^{4k}\\
f_{n;2} &= \log Q + \sum_{k=1}^{\infty} \frac{(4k)!}{(k!)^3\prod_{i=1}^{k}(i-n/4)}Q^{4k}\Big(\log Q + h_{4k} - \frac{3}{4}h_k - \Big(\frac{1}{4-n}+\cdots+\frac{1}{4k-n}\Big)\Big).
\end{aligned}$$

After the substitution $x = 4^4Q^4$, the above two functions become solutions to the hypergeometric equation

$$x(1-x)\partial_x^2 f + (c-(a+b+1)x)\partial_x f - abf = 0,$$

with parameters $a = 1 - i/4$, $b = 1 - j/4$, and $c = 1$, where $\{i,j\} = \{1,2,3\} - \{n\}$. Similarly to the previous case, the holomorphic branches of the Schwartz map $\tau_n(Q) := f_{n;2}(Q)/f_{n;1}(Q)$ will map the 8 sectors of Σ into hyperbolic triangles that provide a triangulation of the half-plane

$$\mathbb{H}_n := \{\tau_n \in \mathbb{C} \mid \operatorname{Re}(\tau_n) < -a_n\}, \tag{5.3}$$

where $a_n = \frac{3}{4}\ln 2$ for $n = 1, 3$ and $a_2 = \frac{1}{2}\ln 2$.

For the good basis we get

$$\begin{aligned}
v_{n,0,0,0} &= dx/(Qf_{n;1})\\
v_{n+1,1,1,1} &= (x_1^{n+1}x_2x_3x_4 - z\phi_n(Q)x_1^n)\,\frac{f_{n;1}dx}{Q\det(I_n)},
\end{aligned}$$

where $\phi_n(Q) = Df_{n;1}/f_{n;1}$ and $\det(I_n) = f_{n;1}(Df_{n;2}) - f_{n;2}(Df_{n;1})$.

Type (n,n): $1 \le n \le 2$. We have

$$I^{(n,n,0,0)} = f_{n,n;1}(Q)\mathbf{1}_{\frac{n}{4},\frac{n}{4},0,0} + z^{n-2} f_{n,n;2}(Q)\mathbf{1}_{0,0,1-\frac{n}{4},1-\frac{n}{4}}$$
$$I^{(0,0,4-n,4-n)} = f_{4-n,4-n;1}(Q)\mathbf{1}_{0,0,1-\frac{n}{4},1-\frac{n}{4}} + z^{2-n} f_{4-n,4-n;2}(Q)\mathbf{1}_{\frac{n}{4},\frac{n}{4},0,0},$$

where

$$f_{n,n;1}(Q) = 1 + \sum_{k=1}^{\infty} \frac{(4k)!}{(k!)^2 \prod_{i=1}^{k}(i-n/4)^2} Q^{4k}$$
$$f_{n,n;2}(Q) = \sum_{k=0}^{\infty} \frac{(4k+n)!}{(k!)^2 \prod_{i=0}^{k}(i+n/4)^2} Q^{4k+n}.$$

If $n = 2$, then the hypergeometric function can be expressed in terms of radicals

$$f_{2,2;1}(Q) = \frac{1}{\sqrt{2}}\Big(\frac{1}{1-4^4Q^4} + \frac{1}{(1-4^4Q^4)^{1/2}}\Big)^{1/2}$$
$$f_{2,2;2}(Q) = \frac{1}{\sqrt{2}}\Big(\frac{1}{1-4^4Q^4} - \frac{1}{(1-4^4Q^4)^{1/2}}\Big)^{1/2}.$$

There is no need of Birkhoff factorization, and for the good basis we get

$$v_{2,2,0,0} = \Big(x_1^2x_2^2 - \frac{f_{2,2;2}(Q)}{f_{2,2;1}(Q)} x_3^2x_4^2\Big) \frac{f_{2,2;1}(Q)dx}{Q\det(I_{2,2})},$$

where $\det(I_{2,2}) = (f_{2,2;1})^2 - (f_{2,2;2})^2$. Suppose that $n = 1$. Then we get that $f_{1,1;i}$ are solutions to the hypergeometric equation with parameters $a = 1/2, b = 1/4, c = 3/4$. More precisely,

$$f_{1,1;1}(Q) = F(1/2, 1/4; 3/4, x),$$
$$f_{1,1;2}(Q) = 4x^{1/4} F(3/4, 1/2; 5/4; x),$$

where $x = 4^4Q^4$. The various holomorphic branches of the Schwartz map $\tau_{1,1}(Q) := f_{1,1;2}(Q)/f_{1,1;1}(Q)$ will map the 8 sectors of Σ into hyperbolic triangles that provide a triangulation of the disk

$$\mathbb{H}_{1,1} = \{\tau_{1,1} \in \mathbb{C} \mid |\tau_{1,1}| < a_{1,1}\}, \tag{5.4}$$

where $a_{1,1} = \Big(\frac{\Gamma(1/4)}{\Gamma(3/4)}\Big)^2$.

For the good basis we get

$$v_{1,1,0,0} = x_1x_2dx/(Qf_{1,1;1})$$
$$v_{0,0,3,3} = \Big(x_3^3x_4^3 - z\frac{f_{3,3;2}(Q)}{f_{1,1;1}(Q)} x_1x_2\Big) \frac{f_{1,1,1}dx}{Q\det(I_{1,1})},$$

where $\det(I_{1,1}) = f_{1,1;1}f_{3,3;1} - f_{1,1;2}f_{3,3;2}$.

Type (m,n): $1 \le m \ne n \le 3$. This is the simplest type. We have

$$I^{(m,n,0,0)}(Q,z) = (1-4^4Q^4)^{1/2-(m+n)/4}\mathbf{1}_{\frac{m}{4},\frac{n}{4},0,0}.$$

The corresponding part of the good basis is

$$v_{m,n,0,0} = x_1^m x_2^n (1-4^4Q^4)^{-1/2+(m+n)/4} dx/Q.$$

5.3. Analytic continuation. Finally, we would like to discuss the analytic continuation of the total ancestor potential $\mathcal{A}_{\omega,v}(\hbar,\mathbf{q})$ under analytic continuation along a closed loop $\gamma\in\pi_1(\Sigma)$. We will assume that γ is such that the part of the good basis of type $(2,2)$ or (m,n) is invariant. Geometrically this corresponds to pulling back the total ancestor potential, viewed as a function on $Q\in\Sigma$, to some finite covering Σ' of Σ and that we are interested in the analytic continuation along a closed loop on Σ'.

For each fixed type $(\alpha)\in\{(0),(1),(2),(3),(1,1)\}$, we have constructed a half-plane or a disk $\mathbb{H}_\alpha$ equipped with a triangulation via an appropriate Schwartz triangular map $Q\mapsto\tau_\alpha(Q)$. Let us denote by $V_\alpha\subset\mathbb{H}_\alpha$ the subset of vertices in the corresponding triangulation. Note that the remaining vertices are on the boundary of $\mathbb{H}_\alpha$. Then we have the following diagram

$$\begin{array}{ccc} \widetilde{\Sigma} & \xrightarrow{\ \tau_\alpha\ } & \mathbb{H}_\alpha\setminus V_\alpha \\ \big\downarrow & \overset{p_\alpha}{\swarrow} & \big\downarrow \\ \Sigma & \longrightarrow & \mathbb{C}\setminus\{0,1\}, \end{array}$$

where $\widetilde{\Sigma}$ is the universal cover of Σ, the bottom arrow is the map $Q\mapsto 4^4Q^4$, and p_α is uniquely defined so that the diagram is commutative. The analytic continuation of $\tau_\alpha(Q)$ defines a monodromy representation

$$\chi_\alpha:\pi_1(\Sigma)\to\operatorname{Aut}(\mathbb{H}_\alpha),\quad \chi_\alpha(\gamma,\tau_\alpha)=\frac{a\tau_\alpha+b}{c\tau_\alpha+d},$$

and we have $\mathbb{H}_\alpha\setminus V_\alpha\cong\widetilde{\Sigma}/\operatorname{Ker}(\chi_\alpha)$. In other words, p_α is the monodromy covering space of the monodromy representation χ_α. Therefore, the pullbacks of the multi-valued analytic functions $f_{\alpha;1}$ and $f_{\alpha;2}$ on $\mathbb{H}_\alpha\setminus V_\alpha$ are analytic functions. Let us introduce the non-compact Riemann surface

$$\mathcal{M}=\Big\{\tau\in\prod_\alpha(\mathbb{H}_\alpha\setminus V_\alpha)\mid p_0(\tau_0)=p_\alpha(\tau_\alpha)\ \forall\ \alpha\Big\}.$$

Note that $\mathcal{M}$ is non-singular because p_α are covering maps. In particular, the induced map $p:\mathcal{M}\to\Sigma$ is also a covering, and we can view $\mathcal{M}$ as the monodromy covering space for the representation $\chi:\pi_1(\Sigma)\to\prod_\alpha\operatorname{Aut}(\mathbb{H}_\alpha)$ defined by

$$\chi(\gamma,\tau)_\alpha:=\chi_\alpha(\gamma,\tau_\alpha).$$

We have the following commutative diagram

$$\begin{array}{ccc} \widetilde{\Sigma} & \xrightarrow{\ \tau\ } & \mathcal{M} \\ \big\downarrow & \overset{p}{\swarrow} & \big\downarrow \\ \Sigma & \longrightarrow & \mathbb{C}\setminus\{0,1\}. \end{array}$$

Suppose that $\gamma\in\pi_1(\Sigma)$ and that $\chi_\alpha(\gamma,\tau_\alpha)=\frac{a\tau_\alpha+b}{c\tau_\alpha+d}$ is the corresponding monodromy transformation of τ_α along γ. Put $j_\alpha(\gamma,\tau_\alpha)=c\tau_\alpha+d$ for the standard automorphy factor. Then the analytic continuation transforms the good basis v and the volume form ω into respectively $v\,R(\gamma,\tau,z)\,B(\tau)$ and $\omega\, j_0(\gamma,\tau_0)^{-1}$, where R is a symplectic transformation of the type $1+O(z)$ and $B(\tau)$ is a diagonal matrix whose diagonal entries have the form $j_\alpha(\gamma,\tau_\alpha)^m$ with $m\in\{0,\pm1\}$. Recalling formulas (3.1) and (3.2), we get that the total ancestor potential $\mathcal{A}_{\omega,v}(\hbar,\mathbf{q})$ transforms into

$$((R(\gamma,\tau,-z)^{-1})\widehat{\ }\mathcal{A}_{\omega,v})(\hbar j_0(\gamma,\tau_0)^2, j_0(\gamma,\tau_0)B\mathbf{q}).$$

Suppose for a moment that $R=1$. Then the above transformation rule can be interpreted as follows: The total ancestor potential is the exponential of the infinite sum of all possible terms of the type

$$\hbar^{g-1}\langle\phi_{i_1}\psi^{k_1},\dots,\phi_{i_n}\psi^{k_n}\rangle_{g,l}(Q)q_{k_1,i_1}\cdots q_{k_l,i_l}.$$

The correlator is a multivalued analytic function in Q which pullbacks via $p:\mathcal{M}\to\Sigma$ to an analytic function on $\mathcal{M}$. Suppose that $Q=p(\tau)$ where $\tau\in\mathcal{M}$ is arbitrary. If we analytically continue along γ, then by definition, we will obtain the value of the correlator at $\gamma\tau:=\chi(\gamma,\tau)$. Our transformation formula implies that

$$\langle\phi_{i_1}\psi^{k_1},\dots,\phi_{i_n}\psi^{k_n}\rangle_{g,l}(\gamma\tau)=\prod_\alpha j_\alpha^{m_\alpha}\ \langle\phi_{i_1}\psi^{k_1},\dots,\phi_{i_n}\psi^{k_n}\rangle_{g,l}(\tau),$$

where m_α are some integers. We get that the Gromov–Witten invariants transform as automorphic forms on $\mathcal{M}$.

In general, $R\neq1$, and in order to arrange for the above transformation rules, we have to modify the definition of the total ancestor potential. It turns out that if we choose as an opposite subspace the complex conjugate of $\widehat{\mathbb{H}}_+(f_Q,\mathbb{C})$ (see [**9**], Proposition 4.12), we get a total ancestor potential $\widetilde{\mathcal{A}}_{\omega,v}(\hbar,\mathbf{q})_{\tau,\overline{\tau}}$ with the following properties:

1) $\lim_{\overline{\tau}\to0}\widetilde{\mathcal{A}}_{\omega,v}(\hbar,\mathbf{q})_{\tau,\overline{\tau}}=\mathcal{A}_{\omega,v}(\hbar,\mathbf{q})$.
2) $\widetilde{\mathcal{A}}_{\omega,v}(\hbar,\mathbf{q})_{\tau,\overline{\tau}}$ transforms as an automorphic form on $\mathcal{M}$.
3) $\widetilde{\mathcal{A}}_{\omega,v}(\hbar,\mathbf{q})_{\tau,\overline{\tau}}$ satisfies the Holomorphic anomaly equations (see [**9**], Proposition 5.1).

Acknowledgments

I would like to thank the organizers of the workshops "Crossing the walls in enumerative geometry 2018," Snowbird, UT, USA and "Current developments in geometry 2018," Novosibirsk, Russia, where most of the material in these lectures were presented. I would like also to thank the referee for pointing out many typos and making useful suggestions.

References

[1] Enrico Arbarello, Maurizio Cornalba, and Phillip A. Griffiths, *Geometry of algebraic curves. Volume II*, Grundlehren der Mathematischen Wissenschaften [Fundamental Principles of Mathematical Sciences], vol. 268, Springer, Heidelberg, 2011. With a contribution by Joseph Daniel Harris. MR2807457

[2] Alejandro Adem, Johann Leida, and Yongbin Ruan, *Orbifolds and stringy topology*, Cambridge Tracts in Mathematics, vol. 171, Cambridge University Press, Cambridge, 2007. MR2359514

[3] V. I. Arnol′d, S. M. Guseĭn-Zade, and A. N. Varchenko, *Singularities of differentiable maps. Vol. II*, Monographs in Mathematics, vol. 83, Birkhäuser Boston, Inc., Boston, MA, 1988. Monodromy and asymptotics of integrals; Translated from the Russian by Hugh Porteous; Translation revised by the authors and James Montaldi. MR966191

[4] Tom Coates, Alessio Corti, Hiroshi Iritani, and Hsian-Hua Tseng, *Some applications of the mirror theorem for toric stacks*, Adv. Theor. Math. Phys. **23** (2019), no. 3, 767–802, DOI 10.4310/ATMP.2019.v23.n3.a4. MR4049075

[5] Boris Dubrovin, *Geometry of 2D topological field theories*, Integrable systems and quantum groups (Montecatini Terme, 1993), Lecture Notes in Math., vol. 1620, Springer, Berlin, 1996, pp. 120–348, DOI 10.1007/BFb0094793. MR1397274

[6] Alexander B. Givental, *Semisimple Frobenius structures at higher genus*, Internat. Math. Res. Notices **23** (2001), 1265–1286, DOI 10.1155/S1073792801000605. MR1866444

[7] Alexander B. Givental, *Gromov-Witten invariants and quantization of quadratic Hamiltonians* (English, with English and Russian summaries), Mosc. Math. J. **1** (2001), no. 4, 551–568, 645, DOI 10.17323/1609-4514-2001-1-4-551-568. Dedicated to the memory of I. G. Petrovskii on the occasion of his 100th anniversary. MR1901075

[8] Claus Hertling, *Frobenius manifolds and moduli spaces for singularities*, Cambridge Tracts in Mathematics, vol. 151, Cambridge University Press, Cambridge, 2002. MR1924259

[9] H. Iritani, T. Milanov, Y. Ruan, and Y. Shen, *Gromov–Witten theory of quotient of Fermat Calabi–Yau varieties*, arXiv:1605.088852.

[10] Maxim Kontsevich, *Intersection theory on the moduli space of curves and the matrix Airy function*, Comm. Math. Phys. **147** (1992), no. 1, 1–23. MR1171758

[11] C. Li, Si Li, and K. Saito, *Primitive forms via polyvector fields*, Preprint arXiv:1311.1659

[12] B. Malgrange, *Déformations de systèmes différentiels et microdifférentiels* (French), Mathematics and physics (Paris, 1979/1982), Progr. Math., vol. 37, Birkhäuser Boston, Boston, MA, 1983, pp. 353–379. MR728429

[13] Bernard Malgrange, *Deformations of differential systems. II*, J. Ramanujan Math. Soc. **1** (1986), no. 1-2, 3–15. MR945599

[14] Todor Milanov, *Analyticity of the total ancestor potential in singularity theory*, Adv. Math. **255** (2014), 217–241, DOI 10.1016/j.aim.2014.01.009. MR3167482

[15] Morihiko Saito, *On the structure of Brieskorn lattice* (English, with French summary), Ann. Inst. Fourier (Grenoble) **39** (1989), no. 1, 27–72. MR1011977

[16] Kyoji Saito and Atsushi Takahashi, *From primitive forms to Frobenius manifolds*, From Hodge theory to integrability and TQFT tt*-geometry, Proc. Sympos. Pure Math., vol. 78, Amer. Math. Soc., Providence, RI, 2008, pp. 31–48, DOI 10.1090/pspum/078/2483747. MR2483747

[17] Constantin Teleman, *The structure of 2D semi-simple field theories*, Invent. Math. **188** (2012), no. 3, 525–588, DOI 10.1007/s00222-011-0352-5. MR2917177

[18] K. Saito, *On periods of primitive integrals, I*, Preprint RIMS (1982).

[19] Kyoji Saito, *The higher residue pairings $K_F^{(k)}$ for a family of hypersurface singular points*, Singularities, Part 2 (Arcata, Calif., 1981), Proc. Sympos. Pure Math., vol. 40, Amer. Math. Soc., Providence, RI, 1983, pp. 441–463. MR713270

[20] Edward Witten, *Two-dimensional gravity and intersection theory on moduli space*, Surveys in differential geometry (Cambridge, MA, 1990), Lehigh Univ., Bethlehem, PA, 1991, pp. 243–310. MR1144529

[21] Masaaki Yoshida, *Hypergeometric functions, my love*, Aspects of Mathematics, E32, Friedr. Vieweg & Sohn, Braunschweig, 1997. Modular interpretations of configuration spaces. MR1453580

Kavli IPMU (WPI), UTIAS, The University of Tokyo, Kashiwa, Chiba 277-8583, Japan

Contemporary Mathematics
Volume **763**, 2021
https://doi.org/10.1090/conm/763/15328

Counting points with Berglund–Hübsch–Krawitz mirror symmetry

Ursula Whitcher

ABSTRACT. We give an expository discussion of recent work using Berglund–Hübsch–Krawitz mirror symmetry to describe the structure of point counts on algebraic varieties over finite fields.

1. Arithmetic mirror symmetry

Can one exploit intuition and constructions from mirror symmetry to prove theorems about the arithmetic of varieties? In this expository paper, we describe work building on observations made by Candelas, de la Ossa, and Rodriguez Villegas around the turn of the millennium. We focus on a particular mirror symmetry construction known as Berglund–Hübsch–Krawitz mirror symmetry. Our primary aim is to give members of the mirror symmetry research community a taste of the interactions of this subject with number theory. Our approach is generally low-tech and focused on specific examples. We hope that more arithmetically minded readers will appreciate the concrete nature of the constructions, some of which involve varieties of arbitrarily high dimension, and will be inspired to seek unifying principles.

Let us begin by considering a classical example, the Legendre family of elliptic curves, given by the equation:

$$X_\psi : y^2 = x(x-1)(x-\psi).$$

If we view X_ψ as a family of curves defined over the complex numbers, we see that each smooth member of the family admits a holomorphic form which is unique up to scaling. By choosing a connection on the family, we may fix the choice of scaling and thus the holomorphic form ω_ψ in a consistent way. The holomorphic form ω_ψ satisfies a differential equation depending on the parameter ψ, the *Picard–Fuchs differential equation*, which characterizes the changes in the complex structure of the elliptic curves as we move through the family.

Alternatively, we may view X_ψ as a family of elliptic curves defined over a finite field $\mathbb{F}_p$. In this case, we may consider a different invariant, the *trace of Frobenius* a_p, which controls the number of points on X_ψ:

2020 *Mathematics Subject Classification.* Primary 11G42; Secondary 14J28, 14J32, 14J33, 14G17.

Key words and phrases. mirror symmetry, finite field hypergeometric functions.

$$a_p = 1 + p - \#X_\psi(\mathbb{F}_p).$$

In 1958, Igusa showed that the complex and finite-field aspects of the Legendre family are intimately related. The Picard–Fuchs equation for the holomorphic form ω_ψ is *hypergeometric*, with solution

$$_2F_1\left(\frac{1}{2}, \frac{1}{2}; 1 \mid \psi\right) = \sum_{n=1}^{\infty} \frac{(\frac{1}{2})_n^2}{(n!)^2}\psi^n.$$

Meanwhile, the trace of Frobenius satisfies a truncated hypergeometric formula, with the same coefficients:

$$a_p \equiv (-1)^{\frac{p-1}{2}} \sum_{n=1}^{(p-1)/2} \frac{(\frac{1}{2})_n^2}{(n!)^2}\psi^n \pmod{p}.$$

Here the generalized hypergeometric equation is given by

$$F(\boldsymbol{\alpha}; \boldsymbol{\beta} \mid z) = \sum_{k=0}^{\infty} \frac{(\alpha_1)_k \cdots (\alpha_n)_k}{(\beta_1)_k \cdots (\beta_m)_k} z^k \in \mathbb{Q}[[z]], \tag{1.1}$$

where $(x)_k$ is the rising factorial

$$(x)_k = x(x+1)\cdots(x+k-1) = \frac{\Gamma(x+k)}{\Gamma(x)}.$$

Igusa's result inspired extensive work in number theory highlighting relationships between cohomology, point counting, and hypergeometric structures. We shall focus on the relationship between Igusa's observation and mirror symmetry.

Viewed as a complex manifold, an elliptic curve has a nonvanishing holomorphic form that is unique up to scaling. Simply connected manifolds with this property are known as *Calabi-Yau manifolds*. Due to their geometric structure, they play a key role in formulations of string theory. One may ask whether arithmetic properties of elliptic curves like the ones Igusa observed extend to Calabi-Yau manifolds in higher dimensions.

Let us consider the diagonal pencil of Calabi-Yau $n-1$-folds in $\mathbb{P}^n$ given by

$$X_\psi \colon x_0^{n+1} + \cdots + x_n^{n+1} - (n+1)\psi x_0 \cdots x_n = 0.$$

Dwork studied this pencil in the case $n = 3$ in [**Dwo69**]; he described a relationship between the Picard-Fuchs equation and the family's arithmetic properties, which he used to illuminate the structure of his theory of p-adic cohomology. Thus, the diagonal pencil in $\mathbb{P}^n$ is usually called the Dwork pencil in the number theory literature, while in the physics literature it is generally referred to as the Fermat pencil. In an ecumenical spirit, and following chronological order, we will refer to it as the Fermat-Dwork pencil.

When $n = 4$, the Picard–Fuchs equation for the holomorphic form on X_ψ has hypergeometric solution $_4F_3\left(\frac{1}{5}, \frac{1}{5}, \frac{1}{5}, \frac{1}{5}; \frac{4}{5}, \frac{3}{5}, \frac{2}{5} \mid \psi^5\right)$. The hypergeometric structure of this Picard-Fuchs equation is used in the proof of the quintic mirror theorem (see [**CK99**, Chapter 11] for an expository treatment). Candelas–de la Ossa–Rodriguez Villegas showed in [**CDRV00**] that over a field of prime order, $\#X_\psi(\mathbb{F}_p) \pmod{p}$ is given by a truncation of the hypergeometric series.

Furthermore, Candelas–de la Ossa–Rodriguez Villegas made a direct link between the arithmetic structure of X_ψ and the structure of the Greene-Plesser mirror Y_ψ. Recall that we can organize information about the number of points on a variety over a finite field in a generating function.

DEFINITION 1.1. Let $X/\mathbb{F}_q$ be an algebraic variety over the finite field of $q = p^s$ elements. The *zeta function* of X is

$$Z(X/\mathbb{F}_q, T) := \exp\left(\sum_{s=1}^{\infty} \#X(\mathbb{F}_{q^s})\frac{T^s}{s}\right) \in \mathbb{Q}[[T]].$$

For a smooth projective hypersurface X in $\mathbb{P}^n$, we have

$$Z(X,T) = \frac{P_X(T)^{(-1)^n}}{(1-T)(1-qT)\cdots(1-q^{n-1}T)}, \tag{1.2}$$

with $P_X(T) \in \mathbb{Q}[T]$. The degree of P_X is determined by the Betti numbers of X. If X is a Calabi-Yau hypersurface in $\mathbb{P}^n$, P_X has at most one root that is a p-adic unit, termed the *unit root*. The value of this root determines $\#X(\mathbb{F}_q) \pmod q$.

If X and Y are mirror Calabi-Yau threefolds, we can expect a relationship between the degrees of certain factors of $Z(X/\mathbb{F}_q, T)$ and $Z(Y/\mathbb{F}_q, T)$ due to the interchange of Hodge numbers. In some cases one can be more specific. Candelas, de la Ossa, and Rodriguez Villegas showed in [**CDRV01**] that for the Fermat-Dwork quintic pencil X_ψ and the Greene–Plesser mirror Y_ψ, P_{X_ψ} and P_{Y_ψ} share a common factor of degree 4.

We will be particularly interested in detailed analyses of K3 surfaces, so let us move one dimension down and note corresponding results for the K3 surface case. Let X_ψ be the Fermat–Dwork quartic pencil. In [**Kad04**], Shabnam Kadir reports a computation by Xenia de la Ossa, building on Dwork's results, that gives the number of points over a field of prime order p:

$$P_{X_\psi}(T) = R_\psi(T)Q_\psi^3(T)S_\psi^6(T). \tag{1.3}$$

Here (with choices of $\pm$ depending on p and ψ), $R_\psi(T) = (1 \pm pT)(1 - a_\psi T + p^2T)$, $Q_\psi(T) = (1 \pm pT)(1 \pm pT)$, and $S_\psi(T) = (1 - pT)(1 + pT)$ when $p \equiv 3 \bmod 4$ and $(1 \pm pT)^2$ otherwise. We'll discuss point counts for this family over general $\mathbb{F}_q$ in Sections 4 and 5.

Let Y_ψ be the mirror family to quartics in $\mathbb{P}^3$ (constructed using Greene–Plesser and the Fermat pencil). Then de la Ossa and Kadir showed:

$$Z(Y_\psi/\mathbb{F}_p, T) = \frac{1}{(1-T)(1-pT)^{19}(1-p^2T)R_\psi(T)}. \tag{1.4}$$

The factor $R_\psi(T)$ corresponds to periods of the holomorphic form and its derivatives, and is invariant under mirror symmetry.

One can relate $R_\psi(T)$ to classical arithmetic objects using a natural geometric transformation called a Shioda–Inose structure (see [**Dol96, ES08, Nas17**]). Each of the mirror quartics Y_ψ admits an involution ι which acts symplectically, that is, preserves the holomorphic form. The quotient Y_ψ/ι is singular. Resolving the singularities yields a new K3 surface. For a general Shioda–Inose structure, the new K3 surface is the Kummer surface associated to an abelian surface. In the case of mirror quartics, the abelian surface is a product of elliptic curves E_ψ and E'_ψ related by a 2-isogeny:

$$Y_\psi \to Y_\psi/\iota \leftarrow \mathrm{Km}(E_\psi \times E'_\psi).$$

Such pairs of elliptic curves are parametrized by the modular curve $X_0(2)/w_2$, where w_2 is an Atkin-Lehner map. Thus, one can study $R_\psi(T)$ using properties of the modular curve.

One immediately asks how far the point-counting results for the Fermat-Dwork quartic and quintic pencils and their mirrors can be generalized. Daqing Wan considered the Fermat-Dwork pencils X_ψ and their mirrors Y_ψ in any dimension. In [**Wan06**], he showed that for any ψ where X_ψ is smooth, the unit roots of X_ψ and Y_ψ coincide. Thus, we obtain a congruence of point counts:

$$\#X_\psi(\mathbb{F}_q) \equiv \#Y_\psi(\mathbb{F}_q) \pmod{q}. \tag{1.5}$$

Kadir studied a two-parameter family of octic Calabi-Yau threefolds and its generalized Greene–Plesser mirror in [**Kad04**] and [**Kad06**]. She showed that over fields of prime order, the zeta functions of the octic threefolds and their mirrors have a common factor, using an explicit point count and techniques from toric varieties. In [**Kad06**], Kadir argues that a similar computation will show in general that if $X_{\psi_1,\ldots,\psi_r}$ is a generalized Fermat pencil of Calabi-Yau varieties in a Gorenstein Fano weighted projective space and $Y_{\psi_1,\ldots,\psi_r}$ is the Greene-Plesser mirror, $Z(X_{\psi_1,\ldots,\psi_r}/\mathbb{F}_p, T)$ and $Z(Y_{\psi_1,\ldots,\psi_r}/\mathbb{F}_p, T)$ share a common factor.

Building on such analysis, Kloosterman studied deformations of generalized Fermat hypersurfaces in weighted projective spaces by monomials other than $x_0 \cdots x_n$ in [**Klo07**]. He showed that the corresponding Picard–Fuchs equations are of generalized hypergeometric form, and used this result to analyze the factorization structure of the zeta function. More general deformations of diagonal hypersurfaces are also important for purely arithmetic reasons; for example, in [**PT15**] Pancratz and Tuitman describe an algorithm to compute the zeta function of a projective hypersurface using a p-adic version of the Picard–Fuchs differential equation for a one-parameter deformation of a diagonal hypersurface.

The evidence for arithmetic mirror symmetry phenomena for Greene–Plesser mirror symmetry raises the question of whether other mirror constructions also have arithmetic implications. In [**MW16**], Magyar and the present author described experimental point-counting results for Calabi-Yau hypersurfaces in Fano toric varieties and gave a conjectural characterization of pencils satisfying a congruence like Equation 1.5. The authors of [**HLYY18**] use Hasse–Witt matrices to extract information about the unit roots of toric Calabi-Yau hypersurfaces.

In the following sections, we discuss arithmetic implications of Berglund–Hübsch–Krawitz (BHK) mirror symmetry. We begin by reviewing the BHK mirror construction in Section 2. In Section 3, we describe a combinatorial trick from [**Per13**] for counting points on the affine variety determined by an invertible polynomial. Though elementary, the method draws on aspects of the BHK construction in a way that hints at a deeper cohomological interpretation. In Section 4, we use intuition from mirror symmetry to identify common factors in zeta functions of different Calabi-Yau pencils. Our exposition follows [**DKSSVW18**], though we also discuss work of Kloosterman in [**Klo18**]. We focus on a specific set of K3 surface examples in Section 5; we exploit the hypergeometric structure of Picard–Fuchs equations and point counts to describe an explicit motivic deconstruction for these families, following [**DKSSVW20**].

2. Berglund-Hübsch-Krawitz mirrors

Let us recall the procedure for the Berglund-Hübsch-Krawitz (BHK) mirror symmetry construction. Consider a polynomial F_A that is the sum of $n+1$ monomials in $n+1$ variables

$$F_A = \sum_{i=0}^{n} \prod_{j=0}^{n} x_j^{a_{ij}}. \tag{2.1}$$

We view F_A as determined by an integer matrix $A = (a_{ij})$, so each row of the matrix corresponds to a monomial.

DEFINITION 2.1. We say F_A is *invertible* if the matrix A is invertible, there exist positive integers called *weights* r_j so that $d := \sum_{j=0}^{n} r_j a_{ij}$ is the same constant for all i, and the polynomial F_A has exactly one critical point, namely at the origin.

DEFINITION 2.2. We say an invertible polynomial F_A satisfies the *Calabi-Yau condition* if $d = \sum_{j=0}^{n} r_j$.

If a polynomial is invertible and the Calabi-Yau condition is satisfied, the weights determine a weighted projective space $\mathbb{WP}^n(r_0, \ldots, r_n)$ and F_A determines a Calabi-Yau hypersurface X_A in this weighted projective space. Alternatively, one may view F_A as a function from $\mathbb{C}^{n+1}$ to $\mathbb{C}$; this point of view yields a *Landau–Ginzburg model.*

Kreuzer and Skarke proved in [**KS92**] that any invertible polynomial F_A can be written as a sum of invertible potentials, each of which must be of one of the three *atomic types*:

$$\begin{aligned} W_{\text{Fermat}} &:= x^a, \\ W_{\text{loop}} &:= x_1^{a_1}x_2 + x_2^{a_2}x_3 + \ldots + x_{m-1}^{a_{m-1}}x_m + x_m^{a_m}x_1, \text{ and} \\ W_{\text{chain}} &:= x_1^{a_1}x_2 + x_2^{a_2}x_3 + \ldots x_{m-1}^{a_{m-1}}x_m + x_m^{a_m}. \end{aligned} \tag{2.2}$$

We will be interested in three subgroups of the torus $(\mathbb{C}^*)^{n+1}$ related to F_A. Let $\mathrm{Aut}(A)$ be the diagonal symmetries of F_A, that is, those $(\lambda_0, \ldots, \lambda_n) \in (\mathbb{C}^*)^{n+1}$ such that $F_A(\lambda_0 x_0, \ldots, \lambda_n x_n) = F_A(x_0, \ldots, x_n)$ for all $(x_0, \ldots, x_n) \in \mathbb{C}^{n+1}$. Note that $\mathrm{Aut}(F_A)$ is a finite abelian group, and the coordinates of each element of $\mathrm{Aut}(F_A)$ are roots of unity. One may compute $\mathrm{Aut}(F_A)$ by using the fact that it is generated by the columns of the matrix A^{-1}: if r_{ij} are the coordinates of column ρ_j, then $(e^{2\pi i r_{0j}}, \ldots, e^{2\pi i r_{nj}})$ is the corresponding element of $\mathrm{Aut}(F_A)$. Let $SL(F_A) \subset (\mathbb{C}^*)^{n+1}$ be the diagonal symmetries of F_A with the property that the product of coordinates $\lambda_0 \cdots \lambda_n$ is 1. Let $J(F_A)$ be the trivial diagonal symmetries, that is, the subgroup of $SL(F_A)$ that acts trivially on $\mathbb{WP}^n(r_0, \ldots, r_n)$. Then $SL(F_A)/J(F_A)$ acts nontrivially and *symplectically* on X_A; that is, it fixes the holomorphic $n-1$-form.

Now, suppose we start with a Calabi-Yau manifold X_A corresponding to an invertible matrix A. To construct the Berglund–Hübsch–Krawitz (BHK) mirror of X_A, we take the transpose matrix A^T. This matrix is also invertible; we refer to its weights as *dual weights.* Consider the polynomial F_{A^T}. Let $\widetilde{G^T} =$

$SL(F_{A^T})/J(F_{A^T})$. We obtain a dual orbifold $X_{A^T}/\widetilde{G^T}$ as the mirror of X_A. In general, for any group H such that $J(F_A) \subset H \subset SL(F_A)$, one may define the Berglund-Hübsch-Krawitz mirror of the orbifold $X_A/\widetilde{H}$. Under this more general construction, BHK duality is a true duality: the mirror of the mirror yields the original orbifold. Furthermore, BHK duality for the polynomial F_A extends naturally to the pencil of hypersurfaces described by

$$F_A - (d^T)\psi x_0 \cdots x_n,$$

where $d^T = \sum q_i$ is the sum of the dual weights.

Berglund–Hübsch–Krawitz mirror symmetry involves correspondences of algebras, as well as spaces. Fan, Jarvis, and Ruan constructed a Gromov–Witten style theory, FJRW theory, for Landau–Ginzburg models; see [**FJR13**] for a detailed discussion. This theory involves Frobenius algebras constructed from the elements of G. Chiodo and Ruan proved in [**CR11**] that there is an isomorphism between the FJRW theory of X/G and the Chen-Ruan orbifold cohomology of $X_A/\widetilde{G}$. When $X_A/\widetilde{G}$ and $X_{A^T}/\widetilde{G^T}$ admit crepant resolutions, this isomorphism can be used to recover the classical mirror relationship between Hodge diamonds.

3. An affine point count

In his dissertation, [**Per13**], Peruničić showed that if $\det A \mid (p-1)$, then there is a formula for the number of points on the affine variety cut out by F_A in $\mathbb{F}_p^{n+1}$ (mod p) that depends only on the matrix A^T. Though the argument reduces to elementary combinatorics, it highlights objects of interest in BHK mirror symmetry and hints at deeper connections between the arithmetic of X_A and the structure of its mirror. Let us discuss the details, and correct a small error in [**Per13**], as a warmup.

PROPOSITION 3.1 (Theorem 4.3.2, [**Per13**]). *Let F_A be an invertible polynomial, and suppose* $\det A \mid (p-1)$. *For any rational vector $\vec{\xi} = (\xi_0, \dots, \xi_n)$, set* $\operatorname{age}(\vec{\xi}) = \sum \xi_i$ *and*

$$\nu(\xi) = \frac{(p-1)!}{\prod_{i=0}^{n}((p-1)\xi_i)!}.$$

Let Ξ be the set of positive integer linear combinations of the columns of $(A^T)^{-1}$:

$$\Xi = \{(A^T)^{-1}\vec{v} \mid v_0, \dots, v_n \in \mathbb{Z}^+\}.$$

Then the number of points $\nu(A)$ on the affine hypersurface $\{\vec{x} \in \mathbb{F}_p^n \mid F_a(\vec{x}) = 0\}$ satisfies the equivalence

$$\nu(A) \equiv (-1)^n \sum_{\substack{\xi \in \Xi \\ \operatorname{age}(\xi)=1}} \nu(\xi) \pmod{p}.$$

The proof of Proposition 3.1 uses the fact that $\nu \equiv \sum_{\vec{x} \in \mathbb{F}_p^n} (1 - F_A(\vec{x})^{p-1})$ (mod p), together with the multinomial theorem. Note that the Calabi-Yau condition is not required: though we will focus on that case here, Proposition 3.1 applies to more general invertible polynomials.

EXAMPLE 3.2. Let $F_A = x_0^2x_1 + x_1^2x_2 + x_2^3$. The weights of A are $(1,1,1)$, and the weights of A^T are $(2,1,1)$. We have $\det A = 12$; let us take $p = 73$. The set Ξ contains one element, the vector $(\frac{1}{2}, \frac{1}{4}, \frac{1}{4})$. We compute:

$$\begin{aligned}\nu(A) &\equiv (-1)^2 \frac{72!}{(72/2)!(72/4)!(72/4)!} \pmod{73} \\ &\equiv 67 \pmod{73}.\end{aligned}$$

One can check that $\nu(A) = 5761$, which is indeed equivalent to 67 (mod 73).

EXAMPLE 3.3. Let $F_A = x_0^2x_1 + x_1^5 + x_2^5 + x_3^5$. The weights of A are $(2,1,1,1)$, and the weights of A^T are $(5,1,2,2)$. We have $\det A = 250$; let us take $p = 251$. The set Ξ contains one element, the vector $(\frac{1}{2}, \frac{1}{10}, \frac{1}{5}, \frac{1}{5})$. We compute:

$$\begin{aligned}\nu(A) &\equiv (-1)^3 \frac{250!}{(250/2)!(250/10)!(250/5)!(250/5)!} \pmod{251} \\ &\equiv 6 \pmod{251}.\end{aligned}$$

Since the columns of $(A^T)^{-1}$ generate the group $\mathrm{Aut}(A^T)$, it is natural to try to rephrase Proposition 3.1 in terms of elements of $\mathrm{Aut}(A^T)$.

DEFINITION 3.4. Let $g = (e^{2\pi i \xi_0}, \ldots, e^{2\pi i \xi_n})$ be a diagonal symmetry written in canonical form, that is, assume $0 \le \xi_i < 1$ for each ξ_i. Then the *age* of g is given by

$$\mathrm{age}(g) = \sum_{i=0}^{n} \xi_i.$$

We shall identify g with the corresponding ξ vector $(\xi_0, \ldots, \xi_n)$, when confusion will not arise.

REMARK 3.5. The version of Proposition 3.1 given in Peruničić's dissertation ([**Per13**, Theorem 4.3.2]) takes a sum over all group elements $g \in \mathrm{Aut}(A^T)$ such that $\mathrm{age}(g) = 1$. However, this statement is in error: such a sum can lead to extraneous terms. For example, if we take $F_A = x_0^2x_1 + x_1^2x_2 + x_2^3$ as in Example 3.2, then $\mathrm{Aut}(A^T)$ contains the group element given by $(0, \frac{1}{2}, \frac{1}{2})$, which contributes a gratuitous $\frac{(p-1)!}{(p-1)!^2/4}$ to the sum.

DEFINITION 3.6. If $g \in \mathrm{Aut}(A)$ acts nontrivially on each of the variables $x_0, \ldots, x_n$, that is, the corresponding ξ vector contains no zeros, then we say g is *narrow*. Otherwise, we say g is *broad*.

If $\xi \in \Xi$, the element g of $\mathrm{Aut}(A)$ determined by ξ must be narrow, because the columns of $(A^T)^{-1}$ are linearly independent. When A (and thus A^T) satisfies the Calabi-Yau condition, there is precisely one element of Ξ, the vector given by $(\frac{q_0}{d^T}, \ldots, \frac{q_n}{d^T})$, where $q_0, \ldots, q_n$ are the dual weights. The corresponding group element generates $J(F_{A^T})$, the trivial symmetries of the associated weighted projective space.

Chiodo and Ruan observe in [**CR11**] that, under their Landau–Ginzburg/Calabi–Yau correspondence, each narrow element of G^T yields a cohomology class generated by hyperplanes in $H_{CR}^{p,q}(X_{A^T}/\widetilde{G^T})$. In particular, for any G satisfying $J(F_A) \subset G \subset SL(F_A)$, we have $J(F_{A^T}) \subset G^T$. Because it has age 1, the generator of $J(F_A)$ given by $(\frac{q_0}{d^T}, \ldots, \frac{q_n}{d^T})$ yields an element of $H_{CR}^{n-1,n-1}(X_{A^T}/\widetilde{G^T})$.

Under mirror symmetry, we obtain a corresponding element of $H^{n-1,1}_{CR}(X_A/\widetilde{G})$. In particular, when $G = J(F_A)$, we may work in ordinary rather than Chen–Ruan cohomology. In this case, Proposition 3.1 tells us that for certain primes p, we may compute the number of points on the affine cone over $X_A \pmod p$, and thus the number of points on $X_A \pmod p$, in terms of information associated to a particular class in $H^{n-1,1}(X_A)$. In the following sections, we shall observe such an association in a different way, by noting the relationship between a particular factor of the zeta function whose root is the unit root and an element of $H^{n-1,1}(X_A)$ corresponding to a deformation of the holomorphic form. We will also give more general, hypergeometric formulas for many point counts.

In [**AP15**], Aldi and Peruničić pursue another strategy for studying the arithmetic and cohomological structure of the mirror correspondence. Borisov gave a vertex algebra formulation of BHK mirror symmetry in [**Bor13**] that unifies rings associated to both the A- and B-models in a single algebraic structure. Aldi and Peruničić realize this vertex algebra construction in the setting of p-adic D-modules, and show that it can be made compatible with the Frobenius action. The challenge is then to link this structure to specific arithmetic or geometric predictions.

4. Common factors

In the early 1990s, Greene, Plesser, and Roan showed in [**GPR92**] that one may construct a mirror family to smooth quintics in $\mathbb{P}^4$ using discrete group quotients of pencils other than the Fermat pencil. Similarly, the Fermat pencil is not the only highly symmetric pencil one can use to construct the mirror to smooth quartics in $\mathbb{P}^3$. The alternatives are certain invertible pencils; we list them in Table 1.

Family	Equation	$SL(F_A)/J(F_A)$
F_4	$x_0^4 + x_1^4 + x_2^4 + x_3^4 - 4\psi x_0x_1x_2x_3$	$(\mathbb{Z}/4\mathbb{Z})^2$
$\mathsf{F}_2\mathsf{L}_2$	$x_0^4 + x_1^4 + x_2^3x_3 + x_3^3x_2 - 4\psi x_0x_1x_2x_3$	$\mathbb{Z}/8\mathbb{Z}$
$\mathsf{F}_1\mathsf{L}_3$	$x_0^4 + x_1^3x_2 + x_2^3x_3 + x_3^3x_1 - 4\psi x_0x_1x_2x_3$	$\mathbb{Z}/7\mathbb{Z}$
$\mathsf{L}_2\mathsf{L}_2$	$x_0^3x_1 + x_1^3x_0 + x_2^3x_3 + x_3^3x_2 - 4\psi x_0x_1x_2x_3$	$\mathbb{Z}/4\mathbb{Z} \times \mathbb{Z}/2\mathbb{Z}$
L_4	$x_0^3x_1 + x_1^3x_2 + x_2^3x_3 + x_3^3x_0 - 4\psi x_0x_1x_2x_3$	$\mathbb{Z}/5\mathbb{Z}$

TABLE 1. Symmetric quartic pencils

The zeta functions of these pencils were studied in [**DKSSVW18**]. For each ψ such that the corresponding K3 surfaces $X_{\diamond,\psi}$ are smooth and nondegenerate, the zeta functions $Z(X_{\diamond,\psi}/\mathbb{F}_q, T)$ share a common factor $R_\psi(T)$ of degree 3 (see [**DKSSVW18**, Theorem 5.1.3], and note that we have suppressed the dependence on q).

For $\mathbb{F}_q$ containing sufficiently many roots of unity, it holds that

$$Z(X_{\diamond,\psi}/\mathbb{F}_q, T) = \frac{1}{(1-T)(1-qT)^{19}(1-q^2T)R_\psi(T)}. \tag{4.1}$$

We say the zeta functions $Z(X_{\diamond,\psi})$ are *potentially equal*. Let Y_ψ be the family of mirror quartics constructed by the Greene–Plesser quotient of the Fermat pencil. Then $Z(X_{\diamond,\psi})$ and $Z(Y_\psi)$ are potentially equal for any $\diamond$.

In this case, we have used the physical intuition that the pencils $X_{\diamond,\psi}$ have the same mirror, in the sense of Greene–Plesser–Roan, to extract arithmetic consequences. Since all of the pencils listed in Table 1 are invertible, one may ask whether this phenomenon holds for other cases of the BHK mirror construction: if X_{A^T} and X_{B^T} have common properties, do X_A and X_B share arithmetic properties? The results of [**DKSSVW18**] show that for projective invertible pencils, one can detect common properties using only the matrices A^T and B^T:

THEOREM 4.1. [**DKSSVW18**] *Let $X_{A,\psi}$ and $X_{B,\psi}$ be invertible pencils of Calabi-Yau $(n-1)$-folds in $\mathbb{P}^n$. Suppose A and B have the same dual weights $(q_0,\dots,q_n)$, and let $d^T = q_0+\cdots+q_n$. Then for each $\psi \in \mathbb{F}_q$ such that $\gcd(q,(n+1)d^T) = 1$ and the fibers $X_{A,\psi}$ and $X_{B,\psi}$ are nondegenerate and smooth, the polynomials $P_{X_{A,\psi}}(T)$ and $P_{X_{B,\psi}}(T)$ have a common factor $R_\psi(T) \in \mathbb{Q}[T]$ with*

$$\deg R_\psi(T) \geq D(q_0,\dots,q_n),$$

where $D(q_0,\dots,q_n)$ is the degree of the Picard–Fuchs equation for the holomorphic form.

By results of Gährs in [**Gäh13**], the Picard–Fuchs equation for the holomorphic form depends only on the dual weights, and is hypergeometric. The proof of Theorem 4.1 uses this fact together with Dwork's p-adic cohomology theory. Though the details of the proof depend on properties of exponential sums, the intuition is that the zeta function can be calculated as the characteristic polynomial of the Frobenius action on p-adic cohomology, and that one can identify a subspace of this cohomology space corresponding to the holomorphic form and its derivatives.

One may also use A^T and B^T to detect coincidences of unit roots. This gives less information about the structure of the zeta function, but requires fewer conditions on the field $\mathbb{F}_q$.

PROPOSITION 4.2. [**DKSSVW18**] *Let $F_A(x)$ and $F_B(x)$ be invertible polynomials in $n+1$ variables satisfying the Calabi–Yau condition. Suppose A^T and B^T have the same weights. Then for all $\psi \in \mathbb{F}_q$ and in all characteristics (including when $p \mid d^T$), either the unit root of $X_{A,\psi}$ is the same as the unit root of $X_{B,\psi}$, or neither variety has a nontrivial unit root.*

COROLLARY 4.3. [**DKSSVW18**] *Let $F_A(x)$ and $F_B(x)$ be invertible polynomials in $n+1$ variables satisfying the Calabi–Yau condition. Suppose A^T and B^T have the same weights. Then for any fixed $\psi \in \mathbb{F}_q$ and in all characteristics (including $p \mid d^T$) the $\mathbb{F}_q$-rational point counts for fibers $X_{A,\psi}$ and $X_{B,\psi}$ are congruent as follows:*

$$\#X_{A,\psi} \equiv \#X_{B,\psi} \pmod{q}.$$

In these cases, the unit root is determined by a formal power series depending on

$$_DF_{D-1}\left(\alpha_i;\beta_j \mid (\textstyle\prod_i q_i^{-q_i})\psi^{-d^T}\right)$$

where the parameters α_i and β_j depend only on the dual weights q_i. This follows from results of Miyatani in [**Miy15**] when $X_{A,\psi}$ is smooth and $\psi \neq 0$, or Adolphson and Sperber in [**AS16**], in general. Thus, there is a truncated hypergeometric formula for $\#X_\psi(\mathbb{F}_q) \pmod{q}$, generalizing the results of Igusa for the Legendre family.

Kloosterman showed in [**Klo18**] that one can extend Theorem 4.1 to a broader class of hypersurfaces: he allows for invertible polynomials A and B that do not necessarily satisfy the Calabi-Yau condition, and permits more general one-parameter monomial deformations. The resulting common factor of $P_{X_{A,\psi}}(T)$ and $P_{X_{B,\psi}}(T)$ may be of larger degree than the common factor identified in Theorem 4.1.

The idea of Kloosterman's proof is to use a Fermat space Y_ψ covering both $X_{A,\psi}$ and $X_{B,\psi}$, rather than relying on the hypergeometric structure. This construction, which originated in [**Shi86**], is called a Shioda map. In mirror symmetry, it has been used to study "multiple mirror" phenomena, as in [**Kel13**]. Kloosterman studies Shioda maps in more generality, for deformations of invertible polynomials corresponding to hypersurfaces in weighted projective spaces, and shows that when Y_ψ is smooth, the characteristic polynomial of Frobenius acting on a certain subspace of $H^{n-1}(Y_\psi)$ divides the characteristic polynomials of Frobenius on both $H^{n-1}(X_{A,\psi})$ and $H^{n-1}(X_{B,\psi})$. In the projective case, no cancellation occurs, so the result for characteristic polynomials yields a factor of the zeta function.

5. Point counting and hypergeometric formulas

Theorem 4.1 shows that, for invertible pencils describing hypersurfaces X in $\mathbb{P}^n$, the piece of middle cohomology corresponding to the holomorphic form and its derivatives corresponds to a factor of P_X. This raises the question of whether one can describe other factors of P_X in a similar fashion. For the families of K3 surfaces $X_{\diamond,\psi}$ with $\diamond \in \{\mathsf{F}_4, \mathsf{F}_2\mathsf{L}_2, \mathsf{F}_1\mathsf{L}_3, \mathsf{L}_2\mathsf{L}_2, \mathsf{L}_4\}$, this correspondence is worked out completely in [**DKSSVW20**].

Let us take the Dwork-Fermat quartic pencil F_4 and the $\mathsf{L}_2\mathsf{L}_2$ family as examples. The polynomials P_X for several different values of ψ when $p = q = 281$ are given in Table 2 and Table 3. We computed this data using code written by Edgar Costa and described in [**CT14**].

ψ	F_4
0	$(1-281T)^{19}(1+462T+281^2T^2)$
1, 53, 228, 280	not smooth
2, 106, 175, 279	$(1-281T)^3(1+281T)^{16}(1+238T+281^2T^2)$
3, 122, 159, 278	$(1-281T)^{19}(1+78T+281^2T^2)$
4, 69, 212, 277	$(1-281T)^3(1+281T)^{16}(1-434T+281^2T^2)$
5, 16, 265, 276	$(1-281T)^{13}(1+281T)^6(1+418T+281^2T^2)$
6, 37, 244, 275	$(1-281T)^3(1+281T)^{16}(1-50T+281^2T^2)$
7, 90, 191, 274	$(1-281T)^3(1+281T)^{16}(1+238T+281^2T^2)$
8, 138, 143, 273	$(1-281T)^3(1+281T)^{16}(1-50T+281^2T^2)$
9, 85, 196, 272	$(1-281T)^3(1+281T)^{16}(1-50T+281^2T^2)$
10, 32, 249, 271	$(1-281T)^5(1+281T)^{16}$

TABLE 2. Examples of P_X for F_4 when $q = 281$

One can compute a Picard–Fuchs equation for any primitive differential form, not just the holomorphic form. We obtain hypergeometric (or trivial) differential equations organized by the group of symmetries.

The Picard–Fuchs equations for the Dwork-Fermat quartic are given by the following hypergeometric equations, as described for example in [**Dwo69**] and [**Klo07**], and reviewed in [**DKSSVW20**].

ψ	P_X
0	$(1-281T)^{19}(1+462T+281^2T^2)$
1, 53, 228, 280	not smooth
2, 106, 175, 279	$(1-281T)^{11}(1+281T)^8(1+238T+281^2T^2)$
3, 122, 159, 278	$(1-281T)^{19}(1+78T+281^2T^2)$
4, 69, 212, 277	$(1-281T)^{15}(1+281T)^4(1-434T+281^2T^2)$
5, 16, 265, 276	$(1-281T)^{13}(1+281T)^6(1+418T+281^2T^2)$
6, 37, 244, 275	$(1-281T)^{15}(1+281T)^4(1-50T+281^2T^2)$
7, 90, 191, 274	$(1-281T)^{11}(1+281T)^8(1+238T+281^2T^2)$
8, 138, 143, 273	$(1-281T)^{15}(1+281T)^4(1-50T+281^2T^2)$
9, 85, 196, 272	$(1-281T)^{15}(1+281T)^4(1-50T+281^2T^2)$
10, 32, 249, 271	$(1-281T)^{17}(1+281T)^4$

TABLE 3. Examples of P_X for $\mathsf{L}_2\mathsf{L}_2$ when $q=281$

PROPOSITION 5.1. *The primitive middle-dimensional cohomology group $H^2_{\text{prim}}(X_{\mathsf{F}_4,\psi},\mathbb{C})$ has 21 periods whose Picard–Fuchs equations are hypergeometric differential equations as follows:*

- *3 periods are annihilated by $D(\frac{1}{4},\frac{1}{2},\frac{3}{4};1,1,1\mid\psi^{-4})$*
- *6 periods are annihilated by $D(\frac{1}{4},\frac{3}{4};1,\frac{1}{2}\mid\psi^{-4})$*
- *12 periods are annihilated by $D(\frac{1}{2};1\mid\psi^{-4})$.*

Several of the periods for the $\mathsf{L}_2\mathsf{L}_2$ family satisfy trivial Picard–Fuchs equations. The rest are hypergeometric:

PROPOSITION 5.2 ([**DKSSVW20**]). *The group $H^2_{\text{prim}}(X_{\mathsf{L}_2\mathsf{L}_2,\psi},\mathbb{C})$ has periods whose Picard–Fuchs equations are hypergeometric differential equations as follows:*

- *3 periods are annihilated by $D(\frac{1}{4},\frac{1}{2},\frac{3}{4};1,1,1\mid\psi^{-4})$*
- *2 periods are annihilated by $D(\frac{1}{4},\frac{3}{4};1,\frac{1}{2}\mid\psi^{4})$*
- *8 periods are annihilated by $D(\frac{1}{8},\frac{3}{8},\frac{5}{8},\frac{7}{8};0,\frac{1}{4},\frac{1}{2},\frac{3}{4}\mid\psi^{4})$*

To organize the corresponding point-counting information, we use an incomplete L-series. Let S be the set of bad primes for X_ψ, and define

$$L_S(X_{\diamond,\psi},s)=\prod_{p\notin S}P_{X_\psi,p}(p^{-s})^{-1} \tag{5.1}$$

which is convergent for $s\in\mathbb{C}$ in a right half-plane.

The main theorem of [**DKSSVW20**] writes $L_S(X_{\diamond,\psi},s)$ explicitly in terms of *finite field hypergeometric functions* whose parameters are consistent with the Picard–Fuchs hypergeometric parameters. Such functions have been studied by many authors, under different hypotheses on the parameters; [**DKSSVW20**] allows for a further weakening of the hypotheses.

Recall that one may define the rising factorials used in hypergeometric functions as ratios of gamma functions. The notion analogous to the gamma function over a finite field is a *Gauss sum.* Let p be prime, let $q=p^r$, and abbreviate $q-1$ as $q^\times$.

DEFINITION 5.3. Let $\omega\colon\mathbb{F}_q^\times\to\mathbb{C}^\times$ be a generator of the character group on $\mathbb{F}_q^\times$, and let $\Theta\colon\mathbb{F}_q\to\mathbb{C}^\times$ be a nontrivial additive character. For $m\in\mathbb{Z}$, we define the Gauss sum $g(m)$ as

$$g(m) = \sum_{x \in \mathbb{F}_q^\times} \omega(x)^m \Theta(x). \tag{5.2}$$

Here, the combination of the multiplicative and additive character is reminiscent of the definition of $\Gamma(z)$ using the integral $\int_0^\infty x^{z-1} e^{-x}\, dx$.

Let $\boldsymbol{\alpha} = \{\alpha_1, \ldots, \alpha_d\}$ and $\boldsymbol{\beta} = \{\beta_1, \ldots, \beta_d\}$ be multisets of d rational numbers. Suppose that $\boldsymbol{\alpha}$ and $\boldsymbol{\beta}$ are *disjoint modulo* $\mathbb{Z}$, that is, $\alpha_i - \beta_j \notin \mathbb{Z}$ for all $i, j \in \{1, \ldots, d\}$.

We now define a finite field hypergeometric sum. We follow work of Greene [**Gre87**] and Katz [**Kat90**, p. 258] but normalize using the convention of [**McC13**, Definition 3.2] and Beukers–Cohen–Mellit [**BCM15**, Definition 1.1].

DEFINITION 5.4. Suppose that

$$q^\times \alpha_i, q^\times \beta_i \in \mathbb{Z} \tag{5.3}$$

for all $i = 1, \ldots, d$. For $t \in \mathbb{F}_q^\times$, we define a finite field hypergeometric sum by

$$H_q(\boldsymbol{\alpha}, \boldsymbol{\beta} \mid t) = -\frac{1}{q^\times} \sum_{m=0}^{q-2} \omega((-1)^d t)^m G(m + \boldsymbol{\alpha} q^\times, -m - \boldsymbol{\beta} q^\times) \tag{5.4}$$

where

$$G(m + \boldsymbol{\alpha} q^\times, -m - \boldsymbol{\beta} q^\times) = \prod_{i=1}^{d} \frac{g(m + \alpha_i q^\times) g(-m - \beta_i q^\times)}{g(\alpha_i q^\times) g(-\beta_i q^\times)} \tag{5.5}$$

for $m \in \mathbb{Z}$.

The divisibility condition given in Equation 5.3 is restrictive. In [**BCM15**], Beukers, Cohen, and Mellit gave an alternative definition of a finite field hypergeometric sum. We first define the *field of definition* associated to hypergeometric parameters.

DEFINITION 5.5 ([**BCM15**]). The field of definition $K_{\boldsymbol{\alpha}, \boldsymbol{\beta}} \subset \mathbb{C}$ associated to $\boldsymbol{\alpha}, \boldsymbol{\beta}$ is the field generated by the coefficients of the polynomials

$$\prod_{j=1}^{d} (x - e^{2\pi i \alpha_j}) \text{ and } \prod_{j=1}^{d} (x - e^{2\pi i \beta_j}). \tag{5.6}$$

In particular, if $\prod_{j=1}^d (x - e^{2\pi i \alpha_j})$ and $\prod_{j=1}^d (x - e^{2\pi i \beta_j})$ are products of cyclotomic polynomials, and thus have coefficients in $\mathbb{Z}$, we say $\boldsymbol{\alpha}, \boldsymbol{\beta}$ is *defined over* $\mathbb{Q}$.

We say that q is *good* for $\boldsymbol{\alpha}, \boldsymbol{\beta}$ if q is coprime to the least common denominator of $\boldsymbol{\alpha} \cup \boldsymbol{\beta}$.

DEFINITION 5.6 ([**BCM15**]). Suppose that $\boldsymbol{\alpha}, \boldsymbol{\beta}$ are defined over $\mathbb{Q}$ and q is good for $\boldsymbol{\alpha}, \boldsymbol{\beta}$. Choose $p_1, \ldots, p_r, q_1, \ldots, q_s \in \mathbb{Z}_{\geq 1}$ such that

$$\prod_{j=1}^{d} \frac{(x - e^{2\pi i \alpha_j})}{(x - e^{2\pi i \beta_j})} = \frac{\prod_{j=1}^r x^{p_j} - 1}{\prod_{j=1}^s x^{q_j} - 1}. \tag{5.7}$$

Let $D(x) = \gcd(\prod_{j=1}^r (x^{p_j} - 1), \prod_{j=1}^s (x^{q_j} - 1))$, let $M = \left(\prod_{j=1}^r p_j^{p_j}\right)\left(\prod_{j=1}^s q_j^{-q_j}\right)$, set $\epsilon = (-1)^{\sum_{j=1}^s q_j}$, and let $s(m) \in \mathbb{Z}_{\geq 0}$ be the multiplicity of the root $e^{2\pi i m/q^\times}$ in $D(x)$. Finally, abbreviate

$$g(\mathbf{p}m, -\mathbf{q}m) = g(p_1 m) \cdots g(p_r m) g(-q_1 m) \cdots g(-q_s m). \tag{5.8}$$

For $t \in \mathbb{F}_q^\times$, define the finite field hypergeometric sum associated to $\boldsymbol{\alpha}$ and $\boldsymbol{\beta}$ by

$$H_q(\boldsymbol{\alpha}, \boldsymbol{\beta} \mid t) = \frac{(-1)^{r+s}}{1-q} \sum_{m=0}^{q-2} q^{-s(0)+s(m)} g(\mathbf{p}m, -\mathbf{q}m) \omega(\epsilon M^{-1} t)^m. \tag{5.9}$$

By [**BCM15**, Theorem 1.3], Definitions 5.4 and 5.6 yield the same result in the cases where they both apply. However, situations may arise where neither definition applies directly. For example, analysis of the $\mathsf{F}_1\mathsf{L}_3$ family involves the hypergeometric parameters $\boldsymbol{\alpha} = \{\frac{1}{14}, \frac{9}{14}, \frac{11}{14}\}$ and $\boldsymbol{\beta} = \{0, \frac{1}{4}, \frac{3}{4}\}$. We cannot use Definition 5.6 since $(x - e^{2\pi i/14})(x - e^{18\pi i/14})(x - e^{22\pi i/14}) \notin \mathbb{Q}[x]$. When $q \equiv 1 \pmod{28}$, Definition 5.6 applies; otherwise, it does not. However, one may decompose these parameters into multisets where each of these definitions applies.

Definition 5.7 ([**DKSSVW20**]). We say that q is *splittable* for $\boldsymbol{\alpha}, \boldsymbol{\beta}$ if there exist partitions

$$\boldsymbol{\alpha} = \boldsymbol{\alpha}_0 \sqcup \boldsymbol{\alpha}' \text{ and } \boldsymbol{\beta} = \boldsymbol{\beta}_0 \sqcup \boldsymbol{\beta}' \tag{5.10}$$

where $\boldsymbol{\alpha}_0, \boldsymbol{\beta}_0$ are defined over $\mathbb{Q}$ and

$$q^\times \alpha_i', q^\times \beta_j' \in \mathbf{Z}$$

for all $\alpha_i' \in \boldsymbol{\alpha}'$ and all $\beta_j' \in \boldsymbol{\beta}'$.

Example 5.8. Let $\boldsymbol{\alpha} = \{\frac{1}{14}, \frac{9}{14}, \frac{11}{14}\}$ and $\boldsymbol{\beta} = \{0, \frac{1}{4}, \frac{3}{4}\}$, and let q be odd with $q \equiv 1 \pmod 7$. Then q is splittable for $\boldsymbol{\alpha}, \boldsymbol{\beta}$: we may take $\boldsymbol{\alpha}_0 = \emptyset$, $\boldsymbol{\alpha}' = \boldsymbol{\alpha}$ and $\boldsymbol{\beta}_0 = \boldsymbol{\beta}$, $\boldsymbol{\beta}' = \emptyset$.

A hybrid definition of a finite field hypergeometric sum that applies in the splittable case is given in [**DKSSVW20**].

With these definitions in hand, we may define an exponential generating series associated to a finite field hypergeometric series.

Definition 5.9. Let $t \in \mathbb{F}_q$. The finite field hypergeometric L-function associated to hypergeometric parameters $\boldsymbol{\alpha}$ and $\boldsymbol{\beta}$ is as follows.

$$L_\mathfrak{p}(H_\mathfrak{p}(\boldsymbol{\alpha}; \boldsymbol{\beta} \mid t), T) = \exp\left(-\sum_{r=1}^\infty H_{\mathfrak{p}^r}(\boldsymbol{\alpha}; \boldsymbol{\beta} \mid t) \frac{T^r}{r}\right) \in K[[T]]. \tag{5.11}$$

One may show that $L_\mathfrak{p}(H_\mathfrak{p}(\boldsymbol{\alpha}; \boldsymbol{\beta} \mid t), T)$ is a polynomial. Furthermore, the degree of this polynomial matches the order of the hypergeometric differential equation with the same parameters.

The main theorem of [**DKSSVW20**] describes the L-functions for each of the five families in terms of finite field hypergeometric L-functions. We state the results for F_4 and $\mathsf{L}_2\mathsf{L}_2$.

Theorem 5.10 ([**DKSSVW20**]). *Let* $t = \psi^{-4}$.

- *For the Fermat pencil* F_4*,*

$$\begin{aligned} L_S(X_{\mathsf{F}_4,\psi}, s) = L_S(H(\tfrac{1}{4}, \tfrac{1}{2}, \tfrac{3}{4}; 0,0,0 \mid t), s) \\ \cdot L_S(H(\tfrac{1}{4}, \tfrac{3}{4}; 0, \tfrac{1}{2} \mid t), s-1, \phi_{-1})^3 \\ \cdot L_S(H(\tfrac{1}{2}; 0 \mid t), \mathbb{Q}(\sqrt{-1}), s-1, \phi_{\sqrt{-1}})^6 \end{aligned}$$

where

$$\phi_{-1}(p) = \left(\frac{-1}{p}\right) = (-1)^{(p-1)/2} \quad \textit{is associated to } \mathbb{Q}(\sqrt{-1}) \mid \mathbb{Q}, \textit{ and}$$

$$\phi_{\sqrt{-1}}(\mathfrak{p}) = \left(\frac{i}{\mathfrak{p}}\right) = (-1)^{(\mathrm{Nm}(\mathfrak{p})-1)/4} \quad \textit{is associated to } \mathbb{Q}(\zeta_8) \mid \mathbb{Q}(\sqrt{-1}).$$

- *For the pencil* $\mathsf{L}_2\mathsf{L}_2$*,*

$$\begin{aligned} L_S(X_{\mathsf{L}_2\mathsf{L}_2,\psi}, s) = L_S(H(\tfrac{1}{4}, \tfrac{1}{2}, \tfrac{3}{4}; 0,0,0 \mid t), s) \\ \cdot \zeta_{\mathbb{Q}(\sqrt{-1})}(s-1)^4 L_S(H(\tfrac{1}{4}, \tfrac{3}{4}; 0, \tfrac{1}{2} \mid t), s-1, \phi_{-1}) \\ \cdot L_S(H(\tfrac{1}{8}, \tfrac{3}{8}, \tfrac{5}{8}, \tfrac{7}{8}; 0, \tfrac{1}{4}, \tfrac{1}{2}, \tfrac{3}{4} \mid t), \mathbb{Q}(\sqrt{-1}), s-1, \phi_{\sqrt{-1}}\phi_\psi) \end{aligned}$$

where

$$\phi_\psi(p) = \left(\frac{\psi}{p}\right) \quad \textit{is associated to } \mathbb{Q}(\sqrt{\psi}) \mid \mathbb{Q}.$$

Note that the finite field hypergeometric parameters in Theorem 5.10 match the Picard–Fuchs hypergeometric parameters in Propositions 5.1 and 5.2. The trivial Picard–Fuchs equations for $\mathsf{L}_2\mathsf{L}_2$ correspond to the zeta function $\zeta_{\mathbb{Q}(\sqrt{-1})}(s-1)^4$. The proof of Theorem 5.10 proceeds by explicit computation with finite field hypergeometric sums, using the corresponding Picard–Fuchs parameters as a guide.

Intuitively, one expects that as long as the discrete group of symmetries $SL(F_A)/J(F_A)$ commutes with the action of Frobenius, we will obtain subspaces of cohomology, with each block corresponding to both a Picard–Fuchs differential equation and a factor of P_X. However, such a factorization is only guaranteed over $\overline{\mathbb{Q}}$. To predict factors of P_X over $\mathbb{Z}$, one must study the structure of the associated L-functions more carefully.

EXAMPLE 5.11 ([**DKSSVW20**]). Let $Q_{\diamond,\psi,q}(T) = P_{X_{\diamond,\psi}}/R_\psi$. For smooth members of the F_4 and $\mathsf{L}_2\mathsf{L}_2$ families, the polynomials $Q_{\diamond,\psi,q}(T)$ factor over $\mathbb{Q}[T]$ as follows:

(5.12)

Family	Factorization	Hypothesis
F_4	$(\deg 2)^3(\deg 1)^{12}$	$q \equiv 1 \pmod 4$
	$(\deg 2)^3(\deg 2)^6$	$q \equiv 3 \pmod 4$
$\mathsf{L}_2\mathsf{L}_2$	$(1-qT)^8(\deg 2)(\deg 4)^2$	$q \equiv 1 \pmod 4$
	$(1-q^2T^2)^4(\deg 2)(\deg 8)$	$q \equiv 3 \pmod 4$

The polynomials $Q_{\diamond,\psi,q}(T)$ may factor further, depending on specific values of ψ.

Acknowledgments

The author thanks the Crossing the Walls in Enumerative Geometry conference and the Fields Institute for their hospitality, and the anonymous referee for providing detailed and helpful comments. Edgar Costa and Amanda Francis shared useful and illuminating code. Conversations with Tyler Kelly, Adriana Salerno, and John Voight were, as ever, both delightful and enlightening.

References

[AS16] Alan Adolphson and Steven Sperber, *Distinguished-root formulas for generalized Calabi-Yau hypersurfaces*, Algebra Number Theory **11** (2017), no. 6, 1317–1356, DOI 10.2140/ant.2017.11.1317. MR3687099

[AP15] Marco Aldi and Andrija Peruničić, *p-adic Berglund-Hübsch duality*, Adv. Theor. Math. Phys. **19** (2015), no. 5, 1115–1139, DOI 10.4310/ATMP.2015.v19.n5.a5. MR3487652

[BCM15] Frits Beukers, Henri Cohen, and Anton Mellit, *Finite hypergeometric functions*, Pure Appl. Math. Q. **11** (2015), no. 4, 559–589, DOI 10.4310/PAMQ.2015.v11.n4.a2. MR3613122

[Bor13] Lev A. Borisov, *Berglund-Hübsch mirror symmetry via vertex algebras*, Comm. Math. Phys. **320** (2013), no. 1, 73–99, DOI 10.1007/s00220-013-1717-y. MR3046990

[CR11] Alessandro Chiodo and Yongbin Ruan, *LG/CY correspondence: the state space isomorphism*, Adv. Math. **227** (2011), no. 6, 2157–2188, DOI 10.1016/j.aim.2011.04.011. MR2807086

[CT14] Edgar Costa and Yuri Tschinkel, *Variation of Néron-Severi ranks of reductions of K3 surfaces*, Exp. Math. **23** (2014), no. 4, 475–481, DOI 10.1080/10586458.2014.947054. MR3277943

[CDRV00] Philip Candelas, Xenia de la Ossa, and Fernando Rodriguez-Villegas, *Calabi-Yau manifolds over finite fields. II*, Calabi-Yau varieties and mirror symmetry (Toronto, ON, 2001), Fields Inst. Commun., vol. 38, Amer. Math. Soc., Providence, RI, 2003, pp. 121–157. MR2019149

[CDRV01] Philip Candelas, Xenia de la Ossa, and Fernando Rodriguez-Villegas, *Calabi-Yau manifolds over finite fields. II*, Calabi-Yau varieties and mirror symmetry (Toronto, ON, 2001), Fields Inst. Commun., vol. 38, Amer. Math. Soc., Providence, RI, 2003, pp. 121–157. MR2019149

[CK99] David A. Cox and Sheldon Katz, *Mirror symmetry and algebraic geometry*, Mathematical Surveys and Monographs, vol. 68, American Mathematical Society, Providence, RI, 1999. MR1677117

[Dol96] I. V. Dolgachev, *Mirror symmetry for lattice polarized K3 surfaces*, J. Math. Sci. **81** (1996), no. 3, 2599–2630, DOI 10.1007/BF02362332. Algebraic geometry, 4. MR1420220

[DKSSVW18] Charles F. Doran, Tyler L. Kelly, Adriana Salerno, Steven Sperber, John Voight, and Ursula Whitcher, *Zeta functions of alternate mirror Calabi-Yau families*, Israel J. Math. **228** (2018), no. 2, 665–705, DOI 10.1007/s11856-018-1783-0. MR3874856

[DKSSVW20] Charles F. Doran, Tyler L. Kelly, Adriana Salerno, Steven Sperber, John Voight, and Ursula Whitcher, *Hypergeometric decomposition of symmetric K3 quartic pencils*, Res. Math. Sci. **7** (2020), no. 2, Paper No. 7, 81, DOI 10.1007/s40687-020-0203-3. MR4078177

[Dwo69] B. Dwork, *p-adic cycles*, Inst. Hautes Études Sci. Publ. Math. **37** (1969), 27–115. MR294346

[ES08] N. D. Elkies and M. Schütt, K3 *families of high Picard rank*, unpublished notes, 2008. `http://www2.iag.uni-hannover.de/~schuett/K3-fam.pdf`

[FJR13] Huijun Fan, Tyler Jarvis, and Yongbin Ruan, *The Witten equation, mirror symmetry, and quantum singularity theory*, Ann. of Math. (2) **178** (2013), no. 1, 1–106, DOI 10.4007/annals.2013.178.1.1. MR3043578

[Gäh13] Swantje Gährs, *Picard-Fuchs equations of special one-parameter families of invertible polynomials*, Arithmetic and geometry of K3 surfaces and Calabi-Yau threefolds, Fields Inst. Commun., vol. 67, Springer, New York, 2013, pp. 285–310, DOI 10.1007/978-1-4614-6403-7_9. MR3156419

[GPR92] B. R. Greene, M. R. Plesser, and S.-S. Roan, *New constructions of mirror manifolds: probing moduli space far from Fermat points*, Essays on mirror manifolds, Int. Press, Hong Kong, 1992, pp. 408–448. MR1191435

[Gre87] John Greene, *Hypergeometric functions over finite fields*, Trans. Amer. Math. Soc. **301** (1987), no. 1, 77–101, DOI 10.2307/2000329. MR879564

[HLYY18] A. Huang, B. Lian, S.-T. Yau, and C. Yu, *Hasse-Witt matrices, unit roots and period integrals*, arXiv:1801.01189 (2018).

[Igu58] Jun-ichi Igusa, *Class number of a definite quaternion with prime discriminant*, Proc. Nat. Acad. Sci. U.S.A. **44** (1958), 312–314, DOI 10.1073/pnas.44.4.312. MR98728

[Kad04] S. Kadir, *The arithmetic of Calabi–Yau manifolds and mirror symmetry*, D.Phil. thesis, Univ. of Oxford, 2004. arXiv: hep-th/0409202

[Kad06] Shabnam Kadir, *Arithmetic mirror symmetry for a two-parameter family of Calabi-Yau manifolds*, Mirror symmetry. V, AMS/IP Stud. Adv. Math., vol. 38, Amer. Math. Soc., Providence, RI, 2006, pp. 35–86. MR2282954

[Kat90] Nicholas M. Katz, *Exponential sums and differential equations*, Annals of Mathematics Studies, vol. 124, Princeton University Press, Princeton, NJ, 1990. MR1081536

[Kel13] Tyler L. Kelly, *Berglund-Hübsch-Krawitz mirrors via Shioda maps*, Adv. Theor. Math. Phys. **17** (2013), no. 6, 1425–1449. MR3262528

[Klo07] Remke Kloosterman, *The zeta function of monomial deformations of Fermat hypersurfaces*, Algebra Number Theory **1** (2007), no. 4, 421–450, DOI 10.2140/ant.2007.1.421. MR2368956

[Klo18] Remke Kloosterman, *Zeta functions of monomial deformations of Delsarte hypersurfaces*, SIGMA Symmetry Integrability Geom. Methods Appl. **13** (2017), Paper No. 087, 22, DOI 10.3842/SIGMA.2017.087. MR3720376

[KS92] Maximilian Kreuzer and Harald Skarke, *On the classification of quasihomogeneous functions*, Comm. Math. Phys. **150** (1992), no. 1, 137–147. MR1188500

[MW16] C. Magyar and U. Whitcher, *Strong arithmetic mirror symmetry and toric isogenies*, Proceedings of the AMS Special Session on Higher Genus Curves and Fibrations of Higher Genus Curves in Mathematical Physics and Arithmetic Geometry, 117–129, Contemp. Math. **703**, Amer. Math. Soc., Providence, RI, 2018. MR3782462

[McC13] Dermot McCarthy, *The trace of Frobenius of elliptic curves and the p-adic gamma function*, Pacific J. Math. **261** (2013), no. 1, 219–236, DOI 10.2140/pjm.2013.261.219. MR3037565

[Miy15] Kazuaki Miyatani, *Monomial deformations of certain hypersurfaces and two hypergeometric functions*, Int. J. Number Theory **11** (2015), no. 8, 2405–2430, DOI 10.1142/S1793042115501122. MR3420753

[Nas17] B. Naskręcki, *On a certain hypergeometric motive of weight 2 and rank 3*, arXiv:1702.07738 (2017).

[PT15] Sebastian Pancratz and Jan Tuitman, *Improvements to the deformation method for counting points on smooth projective hypersurfaces*, Found. Comput. Math. **15** (2015), no. 6, 1413–1464, DOI 10.1007/s10208-014-9242-8. MR3413626

[Per13] Andrija Peruničić, *Arithmetic Aspects of Berglund-Hubsch Duality*, ProQuest LLC, Ann Arbor, MI, 2013. Thesis (Ph.D.)–Brandeis University. MR3192946

[Shi86] Tetsuji Shioda, *An explicit algorithm for computing the Picard number of certain algebraic surfaces*, Amer. J. Math. **108** (1986), no. 2, 415–432, DOI 10.2307/2374678. MR833362

[Wan06] Daqing Wan, *Mirror symmetry for zeta functions*, Mirror symmetry. V, AMS/IP Stud. Adv. Math., vol. 38, Amer. Math. Soc., Providence, RI, 2006, pp. 159–184. With an appendix by C. Douglas Haessig. MR2282960

MATHEMATICAL REVIEWS, 416 FOURTH ST, ANN ARBOR, MICHIGAN 48103

Contemporary Mathematics
Volume **763**, 2021
https://doi.org/10.1090/conm/763/15322

Variations on the theme of quantum Lefschetz

Honglu Fan and Yuan-Pin Lee

Abstract. In this companion piece to our previous work in Geom. Topol. 23 (2019), we sketch some variations on the main results. In particular,

- the recursions in the previous work, which we interpreted as the quantum Lefschetz, are reformulated in terms of Givental's quantization formalism, or equivalently, a summation of *finitely many* graphs;
- varieties of modification of the auxilliary spaces (masterspaces) for the fixed point localization are given, leading to different (looking) recursions;
- applications of this circle of ideas are used to derive (apparently) new relations of Gromov–Witten invariants.

0. Introduction

Establishing a *quantum Lefschetz hyperplane theorem* (QLHT) is an important problem in the Gromov–Witten theory (GW theory for short). One often encounters enumerative problems on hypersurfaces or complete interesections whose ambient spaces have much easier corresponding enumerative problems. The celebrated quintic threefold in P^4 is one such example, where the GW invariants of P^4 have been completely determined by A. Givental in [**11**]. From the functoriality point of view, the QLHT and the quantum Leray–Hirsch are the two pillars in the study of the functoriality of GW theory [**15**, **17**]. The quantum Leray-Hirsch for toric fibration was established in genus zero [**2**, **16**], and in higher genus [**8**]. However, the quantum Lefschetz for higher genus was only available in a very limited form. For Calabi–Yau threefolds, there have been various approaches, most of which put a special focus on the quintic hypersurface of P^4. In genus 1, there are works in [**19**, **21**] among others. In higher genus there is an ongoing *mixed spin P-field* approach pioneered by H.-L. Chang, S. Guo, J. Li, W.-P. Li and C.-C. Liu in [**3**–**6**]. S. Guo, F. Janda and Y. Ruan in [**12**, **13**] proposed a new method which determines higher genus invariants of quintic 3-folds via twisted invariants of the ambient space P^4 plus some "effective invariants" which are finite for a fixed genus. In the same paper, the authors used this method in genus 2 to prove the BCOV holomorphic anomaly conjecture [**1**], and we expect this method to be generalized to higher genera. However, beyond the Calabi–Yau threefolds, there was no proposal to a QLHT until our previous work.

The research of the first author was supported by SwissMAP and the grant ERC-2012-AdG-320368-MCSK in the group of Rahul Pandharipande at ETH Zürich.

The research of the second author was partially supported by the NSF.

The goal of this paper is to amplify a few points made in [**9**]. Therefore, this is a companion piece and will tacitly assume that the readers are familiar with the previous paper. In the previous paper, we derived a recursion relation which computes the GW invariants of an ample hypersurface in terms of (twisted) GW invariants of the ambient space and a finite number of "initial data" for each genus. The initial data are finite numbers of GW invariants on the hypersurface of low degrees. In this paper, we will reinterpret the recursion relation in the frame work of Givental's quantization formalism [**10, 11**]. By way of the quantization formalism, the recursion can be written as a summation of finite hypergraphs for each fixed genus. This interpretation simplifies when the hypersurface in question is a Calabi–Yau threefold, but it applies in general cases. Another feature of this approach is its flexibility of allowing primitive class insertions in computing the GW invariants of the hypersurfaces (see Remark 1.2). This is sketched in Section 1.

The second point is to explore the varieties of related but different recursions using the same technique: localization on the masterspace. In utilizing different masterspaces, we reached different recursions. In Section 2.1, the GW theory of the hypersurfaces was related to different twisted theories of the ambient spaces. In Section 2.2, the original masterspace for the quintic is, by way of K-equivalence (crepant transformation), modified to an orbifold masterspace, where one of the fixed points becomes the orbifold point $B\mu_5$. The GW theory of quintic is thus related to the orbifold theory of $B\mu_5$ recursively.

In Section 3, we flip the coin and consider the case of when the hypersurface has trivial GW theory. The aformentioned relation thus becomes a relation of GW invariants of the ambient space. This works, for example, if the ambient space is a Fano threefold and the hypersurface is a K3 suface, representing the canonical divisor. We wrote down some sample equations in this case and hope to come back to applications of these types of equations in the future.

1. Quantum Lefschetz in generating functions and finite graphs

The goal of this section is to reformulate some results from [**9**] in terms of generating functions. We assume that the readers are familiar with [**9**]. Whenever possible, we follow the notations in [**9**] and [**8, 18**].

1.1. Recursions in Givental's quantization formalism. Recall the setup in [**9**]. Let $D \subset X$ be a smooth hypersurface in a smooth projective variety X and let

$$\mathfrak{X} = Bl_{D\times\{0\}}X \times P^1$$

be the "masterspace" associated to (X, D). There is a birational morphism $\mathfrak{X} \to X \times P^1$, which can be composed with projections to X and P^1. We denote the first composition by $p : \mathfrak{X} \to X$ and the second by $\pi : \mathfrak{X} \to P^1$. The fiber

$$\mathfrak{X}_0 := \pi^{-1}(\{0\}) \cong X \cup_D P_D(\mathcal{O} \oplus \mathcal{O}(D))$$

is the union of X and $P_D(\mathcal{O}\oplus\mathcal{O}(D))$. These two pieces glue transversally along the hypersurface $D \subset X$ and the section

$$D \cong P_D(\mathcal{O}(D)) \subset P_D(\mathcal{O} \oplus \mathcal{O}(D)).$$

As in [**9**], we use the following notations

- $X_\infty := \pi^{-1}(\{\infty\}) \cong X$;
- X_0 is the irreducible component of $\mathfrak{X}_0$, which is isomorphic to X;

- $D_0 := P_D(\mathcal{O}) \subset P_D(\mathcal{O} \oplus \mathcal{O}(D)) \subset \mathfrak{X}_0$.

Endow P^1 with the $\mathbb{C}^*$ action such that it has weight -1 on T_0P^1. Since $\mathfrak{X}$ is the blow-up of a fixed locus, *there is an induced* $\mathbb{C}^*$ *action on* $\mathfrak{X}$. This action acts trivially on X but scales the fibers of the P^1 fibration $P_D(\mathcal{O} \oplus \mathcal{O}(D))$. Under this $\mathbb{C}^*$ action, the fixed loci are

$$\text{(a)}\, D_0, \quad \text{(b)}\, X_0, \quad \text{(c)}\, X_\infty.$$

Also, as in [**9**], when the $\mathbb{C}^*$ acts on the fibers on E by "positive" scaling, sending a vector v to λv, we write E^+ for the equivariant bundle and use the notation $\langle \cdots \rangle_{g,n,\beta}^{X,E^+}$ for the corresponding equivariant twisted invariant. Similarly, $\langle \cdots \rangle_{g,n,\beta}^{X,E^-}$ stands for invariants with inverse scaling action $v \mapsto \lambda^{-1} v$. In general, if these distinctions are not called for, $\langle \cdots \rangle_{g,n,\beta}^{X,tw}$ will stand for general twisted invariants.

Let $\{e_\mu\}$ be a basis of $H(X)$. The equivariant total descendant potential of X is denoted by

$$\mathcal{D}_{eq}^X(t) = \exp\left(\sum_{g=0}^{\infty} \hbar^{g-1} F_g^X(t)\right),$$

where

$$F_g^X(t) = \sum_\beta \sum_{n=0}^{\infty} \frac{q^\beta \langle t, \ldots, t\rangle_{g,n,\beta}^{\mathfrak{X}}}{n!}, \quad t := \sum_{k,\mu} t_k^\mu e_\mu \psi^k. \tag{1.1}$$

We note that the summation over n includes $n = 0$. By the results of [**8,10,11**] we have

$$\mathcal{D}_{eq}^{\mathfrak{X}} = \widehat{M}^{\mathfrak{X}} \left(\mathcal{D}^D \mathcal{D}^{X_0} \mathcal{D}^{X_\infty}\right), \tag{1.2}$$

where $\widehat{M}$ is an operator coming from the quantization of quadratic hamiltonians [**10**, **14**] and similar to the one in [**8**, Theorem 1.4]. Note that here our one-dimensional orbits of the $\mathbb{C}^*$ action are not isolated, but they do not depend on any parameter, much like the situation in [**8**]. Therefore, the formula there holds without any modification. We don't really need to know the exact formula for $\widehat{M}^{\mathfrak{X}}$. What we need to know is that $\widehat{M}^{\mathfrak{X}}$ is a product of a few operators, the most important of which is $\widehat{R}$

$$\widehat{R} = \exp\left(\frac{\hbar}{2} \sum_{\mu_1,\mu_2} \sum_{k_1,k_2} \Delta_{k_1 k_2}^{\mu_1 \mu_2} \frac{\partial}{\partial t_{k_1}^{\mu_1}} \frac{\partial}{\partial t_{k_2}^{\mu_2}}\right), \tag{1.3}$$

where Δ can be obtained from certain generating functions of genus zero invariants, or equivalently, the fundamental solution of the Dubrovin connection. When $\widehat{R}$ acts on the product of $\mathcal{D} = \exp F$, it can be organized by *hypergraphs* where Δ accounts for the edge factor and $\partial_{k_1}^{\mu_1} \partial_{k_2}^{\mu_2}$ connects two (not necessarily distinct) *hypervertices* on which two differentiations act. See Equation (17) in [**8**] and Part II of [**18**] for more details.

Implicit in the Equation (1.2) is the dependence on $\mathrm{NE}(\mathfrak{X})$ and $H(\mathfrak{X})$. There is a natural map

$$(\iota_{D_0})_* : \mathrm{NE}(D) \to \mathrm{NE}(\mathfrak{X}),$$

where $\iota_{D_0} : D_0 \to \mathfrak{X}$ is the embedding of the fixed component. Let $H(D)_X$ be the subgroup of $H(D)$, pulled back from $H(X)$. There is a natural lifting

$$H(D)_X \to H(\mathfrak{X}),$$

by pulling back the class in X to $\mathfrak{X}$. Denote their images by $\mathrm{NE}(\mathfrak{X})_D$ and $H(\mathfrak{X})_D$ respectively. Let $\{T_\mu\}$ now stand for a basis of $H(D)_X$. By abusing the notation, the same symbols β_D and $t_D := \sum_{\mu,k} t_k^\mu T_\mu \psi^k$ will be used for the classes and their images under the above maps. We now *restrict the variables of GW theory on* $\mathfrak{X}$ *from* $\mathrm{NE}(\mathfrak{X})$ *to* $\mathrm{NE}(D)$ *(for Novikov variables* q^{β_D}*) and* $H(\mathfrak{X})$ *to* $H(D)$ *(for* t_D*).* Let $\mathcal{D}^{X_\infty,\mathcal{O}^+}$ be the total descendant potential of X_∞ twisted by $\mathcal{O}$ with fiberwise $\mathbb{C}^*$ action of weight 1.

PROPOSITION 1.1. *With the restriction of* $t = t_D$ *and* $q^\beta = q^{\beta_D}$*, we have*

$$\mathcal{D}^{\mathfrak{X}}_{eq} = \widehat{M}\left(\mathcal{D}^D \mathcal{D}^{X_0}\right) \mathcal{D}^{X_\infty,\mathcal{O}^+}, \tag{1.4}$$

where $\widehat{M}$ *acts only on the first two* $\mathcal{D}$*'s, and the last factor is the descendant potential twisted by the Euler class of the trivial bundle.*

PROOF. This follows from (1.2). The only difference is the single factor of $\mathcal{D}^{X_\infty,\mathcal{O}^+}$, on which the operator $\widehat{M}$ does not act. As in [**9**, § 2.3], the restriction of the curve classes implies that there will be no curve connecting X_∞ with $\mathfrak{X}_0$. The contribution from X_∞, therefore, is a single factor by itself as twisted invariants. We also note that, due to the quantum Riemann–Roch theorem in [**7**], there is an (explicit) operator $\widehat{\Gamma}^{X_\infty}$ such that

$$\mathcal{D}^{X_\infty,\mathcal{O}^+} = \left(\widehat{\Gamma}^{X_\infty}\right)^{-1} \mathcal{D}^{X_\infty}, \tag{1.5}$$

where Γ depends on the twisting data. Similarly, (1.5) applies to D and X_0. $\left(\widehat{\Gamma}^{X_\infty}\right)^{-1}$ was part of the ingredient in $\widehat{M}^{\mathfrak{X}}$ and the only one which concerns X_∞. Removing $\left(\widehat{\Gamma}^{X_\infty}\right)^{-1}$ from $\widehat{M}^{\mathfrak{X}}$, we get $\widehat{M}$ acting only on the first two factors. The process also takes $\mathcal{D}^{X_\infty}$ back to $\mathcal{D}^{X_\infty,\mathcal{O}^+}$, as the localization formula dictates. □

The recursion in [**9**] can be phrased in this context as follows: Let

$$I := \langle \Omega \rangle^D_{g,n,\beta_D}, \quad \Omega = \prod_i T_{\mu_i} \psi_i^{k_i}$$

be the GW invariant in question on D. That is, I is the coefficient of $q^{\beta_D} \prod_i t_{k_i}^{\mu_i}$ in $F_g^D(t_D)$, or equialently, the coefficient of $\hbar^{g-1} q^{\beta_D} \prod_i t_{k_i}^{\mu_i}$ in $\log \mathcal{D}^D$. As explained in [**9**, Lemma 2.14], due to the difference of the virtual dimensions

$$\mathrm{vdim}^{\mathfrak{X}}_{g,n,\beta_D} - \mathrm{vdim}^D_{g,n,\beta_D} = 2 - 2g + (\beta_D, D)^X, \tag{1.6}$$

the corresponding GW invariant $\langle \Omega \rangle^{\mathfrak{X}}_{g,n,\beta_D} = 0$ whenever

$$2 - 2g + (\beta_D, D)^X > 0. \tag{1.7}$$

Therefore, the coefficient of $\hbar^{g-1} q^{\beta_D} \prod_i t_{k_i}^{\mu_i}$ of $\log \mathcal{D}^{\mathfrak{X}}_{eq}$ vanishes.

The vanishing gives an equation on the right hand side of (1.4). Treating all (twisted) invariants on X as known, this equation's leading term is exactly $(-1)^{1-g} \lambda^{-2+2g-(\beta_D,D)} I$. That is, I can be written in terms of invaraints on D and $\mathfrak{X}$ of "lower orders", with invariants on X as coefficients. Since invariants on $\mathfrak{X}$ can be expressed in terms of invaraints on D of equal or lower orders, this gives a recursion. The initial conditions are decided by (1.7) when the vanishing, or equivalently the inequality in (1.7), no longer holds.

REMARK 1.2 (On insertions of primitive classes). In Proposition 1.1, we restrict the parametrized insertion t to t_D. When t is a general class in $H(\mathfrak{X})$, Equation (1.4) still holds, although in this case the degree bound (1.7) is more complicated.

Since our primary objective is to obtain the GW invariants for D, the insertion classes missing from the restriction $t = t_D$ are the primitive classes $H_{prim}(D)$ in $H^d(D)$, where $d = \dim_{\mathbb{C}}(D)$. Recall that $\mathfrak{X}$ is a blow-up of $X \times P^1$. Denote the exceptional divisor by E and the inclusion $\iota_E : E \hookrightarrow \mathfrak{X}$. E is a projective bundle over D and admits a projection $\pi : E \to D$. By the blowup formula of cohomology, $H^{d+2}(\mathfrak{X})$ is generated by the images of pushforwards from $H^d(E)$ and of pullbacks from $H^{d+2}(X \times P^1)$. A convenient choice for $H_{prim}(D)$ in (1.4) can be obtained as follows:

Given a collection of $\omega_i \in H_{prim}(D), 1 \leq i \leq m$, consider

$$\tilde{\omega}_i = (\iota_E)_* \pi^* \omega_i.$$

It is easy to see that $\tilde{\omega}_i|_{D_0} = -\lambda\omega_i$. Similar to the above analysis,

$$\langle \tilde{\omega}_1, \ldots, \tilde{\omega}_m, t_D, \ldots, t_D \rangle^{\mathfrak{X}}_{g,n+m,\beta_D} = (-\lambda)^m \langle \omega_1, \ldots, \omega_m, t_D, \ldots, t_D \rangle^D_{g,n+m,\beta_D} + \cdots. \tag{1.8}$$

This leads to a modified degree condition: The left hand side of (1.8) vanishes if

$$2 - 2g + (\beta_D, D)^X > m. \tag{1.9}$$

As before, the vanishing of the LHS produces a relation. Despite the worse degree condition, it produces a way to compute the GW invariants of D with primitive insertions, whereas the methods in [**4, 13**] so far have not. These relations can be combined with the monodromy invariance requirement of GW invariants. More detailed study is in progress in a separate project.

1.2. Calabi–Yau threefolds as the divisor D. Simplifications occur in case D is a Calabi–Yau threefold, when the n-pointed GW invariants can be reconstructed from $n = 0$ GW invariants (with no insertion). The generating function for $n = 0$ invariants in genus g, called genus g partition function, is a function of the Novikov variable q and is independent of t, the insertion variables. See [**9**, § 1.3] for two closed formulae relating n-pointed generating functions to the partition function. Consequently, $\mathcal{D}^D(t_D)$ is completely determined by $\mathcal{D}^D(t_D = 0)$, for which we will determine a simplified form of recursion relation.

For simplicity, we spell out the case when $D \subset X$ is the quintic threefold in P^4, although all discussions extend to general CY3 without difficulty, albeit with more complicated notations.

Since $t_D = 0$, F_g^D depends only on q^{β_D}. Due to (1.6) and the fact that $\mathrm{vdim}^D = 0$, $F_g^{\mathfrak{X}}(t_D = 0, q)$ is a polynomial in q of degree $\leq \frac{2g-2}{5}$, denoted by $P_g(q)$. Since the information of $P_g(q)$ is equivalent to our initial conditions in [**9**, Theorem 1.1], in the case of CY3, we can assume that $\mathcal{D}^{\mathfrak{X}}_{eq}$ is the initial data. Hence, one can rephrase Proposition 1.1 as follows:

COROLLARY 1.3.

$$\left[\widehat{M(q)}\left(\mathcal{D}^D \mathcal{D}^{X_0}\right)\right]_{t_D=0} = e^{\sum_{g=0}^{\infty} \hbar^{g-1} P_g(q)} \left(\mathcal{D}^{X^\infty, \mathcal{O}^+}\right)^{-1}.$$

Note that in the above equation, P_g and $\mathcal{D}^X$ are the initial data, and the equation completely determines $\mathcal{D}^D$ from the initial data.

PROOF. By (1.5), the generating functions of twisted GW invariants are determined by those of the untwisted GW invariants. The coefficient of q^d of the logarithm of the LHS of the equation in Corollary 1.3 has the single highest order term for D as $\langle\cdot\rangle_{g,0,d}$ with a nonzero constant coefficient. □

REMARK 1.4. We remark that the operator $\widehat{M(q)}$ involves differentiation with respect to t, denoted by $\widehat{R}$ in (1.3). One can only set $t_D = 0$ after the differentiation. Hence, one has to use $n \neq 0$ pointed GW invariants for D. However, as already explained, they are all explicitly determined by 0-points ones.

Secondly, one can formally rewrite (1.4) as

$$\mathcal{D}^D = \left(\mathcal{D}^{X_0}\right)^{-1} \widehat{M}^{-1} \left(\mathcal{D}^{\mathfrak{X}}_{eq} \left(\mathcal{D}^{X^\infty, \mathcal{O}^+}\right)^{-1}\right).$$

When one restricts to the CY3 case with $t_D = 0$, $\mathcal{D}^{\mathfrak{X}}_{eq}$ will apparently need to use more than $e^{\sum_{g=0}^\infty \hbar^{g-1} P_g(q)}$ for the input, due to the extra insertions created by the differentiation in $\widehat{M}$. However, the differentiation in $\widehat{M}$ carries a degree in q, and the apparently extra information is actually in the lower order and, by induction, was already determined by (1.4).

1.3. Remarks on hypergraphs. In a series of papers [**10**, **11**], A. Givental devised a summation over *hypergraphs* via the localization techniques. Initially spelled out for the case when the torus fixed points are isolated, this has been extended to more general cases, e.g., [**8**], which we will follow. See also [**14**, **18**].

According to Givental's organizations of the localization in terms of the contributions from hypergraphs, there are only a *finite* number of hypergraphs for each genus. The above relation can be written in terms of finite hypergraphs. These hypergraphs are the decorated graphs of the dual graphs of strata in $\overline{M}_{g,0}$. The decoration assigns to each hypervertex a connected component of fixed loci, to which the torus fixed curve maps. The hyperedges are created by $\hat{R}$ acting on a product of $\mathcal{D}^D$ and $\mathcal{D}^X$. The facts that $\hat{R}$ is an exponential function of quadratic derivatives in (1.3) and $\mathcal{D}^Y = \exp F^Y$ creates hypergraphs via Wick's formula. Here, we use Y to refer to a general variety independent of the context (X, D). The finiteness of the hypergraph in each power of $\hbar$, accounting for genus, has to do with the fact that the domain curves are connected. We very briefly sketch the graphs. The interested readers can consult [**8**, **10**, **11**, **18**].

In the equivariant setting, as we are now, one can interpret the hypergraphs in terms of the fixed point localization. For simplicity, assume $g \geq 2$. The fixed point loci can be labeled by graphs, with vertices as either nodes or a connected union of irreducible components mapping to a connected component of the fixed loci Y^T. An edge connects two (not necessarily distinct) vertices if they intersect. (Contributions of) fixed loci of $\overline{M}_{g,n}(Y, \beta)$ can be indexed by their images under $\mathrm{st} : \overline{M}_{g,n}(Y, \beta) \to \overline{M}_{g,0}$. We call a vertex *a hypervertex* if it stablizes to an irreducible component of the stable $(g, 0)$ curve. The (necessarily genus zero) trees of vertices and edges which map to the intersecting points of hypervertices are called *hyperedges.* [1] A hypergraph is a connected graph consisting of hypervertices and hyperedges, with

[1]Here our terminology is different from that of [**8**, **10**, **11**] and is consistent with [**18**]. What we call vertices and edges here are called joints and legs, while hypervertices and hyperedges are called vertices and edges there.

decoration (g_i, f_i) on the the i-th hypervertice, where f_i is the connected component of fixed loci Y^T associated to the sub-curve, and g_i is the genus of its normalization.

It is easy to see that there are only finitely many hypergraphs for each fixed genus g. There are only finitely many dual stable graphs for $\overline{M}_{g,0}$, and the possibility of decoration is finite. For example, in our case, there are three (single-vertex) hypergraphs with no hyperedges, corresponding to three connected components of fixed loci D_0, X_0, X_∞. When the number of hyperedges is nonzero, the hypervertices must be associated with either D_0 or X_0, due to the constraint on the curve class.

Hence, the recursion relation in the previous subsections can also be phrased as one associated to only finitely many graphs for each genus, as in [**12, 13**], or in the work of Chang, Guo, Li in [**3, 4**].

REMARK 1.5. We note that Givental's quantization can be generalized to semisimple case without equivariant theory. This is the Givental–Teleman classification of semisimple theories. See [**11, 20**].

2. Modifications of the masterspace

In this section, we describe a few modifications to our original masterspace $\mathfrak{X}$ and, consequently, different recursion relations. Even though it is unclear whether these new recursions are easier to "solve", they bring other types of (twisted, orbifold) invariants into play and may be of interest on their own.

2.1. Hypersurface and other twisted theories. We relate hypersurface invariants with invariants of X twisted by other line bundles.

Choose a line bundle L over X. Instead of blowing up $X \times P^1$, we consider a P^1-bundle $\mathcal{P}_L = P_X(\mathcal{O} \oplus L))$. There are two sections, $(\mathcal{P}_L)_0$ and $(\mathcal{P}_L)_\infty$, both of which are isomorphic to X, with normal bundles L and $L^\vee$, respectively. Denote the hypersurface in $(\mathcal{P}_L)_0$ by D_0. Define

$$\mathfrak{X}(L) = \mathrm{Bl}_{D_0} P_L$$

There is a $\mathbb{C}^*$ action on P_L, and it induces an action on $\mathfrak{X}(L)$. As before, denote the strict transforms of $(P_L)_0$ and $(P_L)_\infty$ by X_0 and X_∞, respectively. There is another connected component of fixed locus in the exceptional locus, isomorphic to D. By a slight abuse of notation, denote it as D_0. To sum it up, the fixed loci under the induced $\mathbb{C}^*$ action are again,

$$\text{(a)}\, D_0, \quad \text{(b)}\, X_0, \quad \text{(c)}\, X_\infty.$$

The difference comes from their normal bundles. One identifies that

(1) $N_{D_0/\mathfrak{X}(L)} \cong L \oplus \mathcal{O}_D(D) \otimes L^\vee$,
(2) $N_{X_0/\mathfrak{X}(L)} \cong \mathcal{O}_D(-D) \otimes L$,
(3) $N_{X_\infty/\mathfrak{X}(L)} \cong L^\vee$.

When L is a trivial line bundle, we recover our previous construction.

Consider a curve class $\beta \in \mathrm{NE}(\mathfrak{X}(L)))$ lying in the subgroup $\mathrm{NE}(D_0) \hookrightarrow \mathrm{NE}(\mathfrak{X}(L))$. The GW invariants of genus g with curve class β can be recursively determined by lower terms if the following is satisfied:

$$\mathrm{vdim}(\overline{M}_{g,n}(\mathfrak{X}(L), \beta)) > \mathrm{vdim}(\overline{M}_{g,n}(D_0, \beta)). \tag{2.1}$$

This is explained in [**9**, Section 2.3]. Briefly, for a set of insertions (restrictions from X) whose degrees add up to $\mathrm{vdim}(\overline{M}_{g,n}(D_0, \beta))$, we can always find a lift in

$\mathfrak{X}$. If we integrate the lift against $[\overline{M}_{g,n}(\mathfrak{X}(L),\beta)]^{\mathrm{vir}}$, the integral vanishes due to dimensional reason. Equation (2.1) happens when

$$\int_\beta (c_1(L) + c_1(\mathcal{O}_D(D) \otimes L^\vee)) = \int_\beta c_1(D) > 2(g-1), \tag{2.2}$$

where β is regarded as a curve class on D_0. Thus, we conclude that the degree condition does not change under the modification. However, the recursion relations now use different twisted invariants of X.

EXAMPLE 2.1. When $X = P^4$, D is a smooth quintic hypersurface and when $L = \mathcal{O}(1)$, we conclude that quintic invariants with degree $d > (2g-2)/5$ can be recursively determined by invariants of P^4 twisted by $\mathcal{O}(-4)$ and by $\mathcal{O}(-1)$ plus some lower degree quintic invariants, whereas the "original" master space has the quintic invariants related to invariants of P^4 twisted $\mathcal{O}$ and $\mathcal{O}(-5)$.

2.2. Quintic 3-fold and Hurwitz–Hodge integrals. The construction in this subsection, as it stands now, only applies to the case when there exists a crepant contraction $\pi : \mathfrak{X} \to \overline{\mathfrak{X}}$ such that $X_0 \subset \mathfrak{X}$ is contracted to a point with quotient singularity. That is, there is a K-equivalence, or crepant transformation, between $\mathfrak{X}$ and an orbifold crepant resolution $\tilde{\overline{\mathfrak{X}}} \to \overline{\mathfrak{X}}$. Abusing the notation, as is standard in the literature, we identify $\tilde{\overline{\mathfrak{X}}}$ with $\overline{\mathfrak{X}}$. This is the case for the quintic 3-folds inside P^4, as we shall explain. An ongoing project of the first author with Longting Wu continues further developments.

The construction may start with the deformation to the normal cone $\mathfrak{X}$ when X is P^4 and D is the quintic hypersurface. In this case, $X_0 \cong P^4$ and $N_{X_0/\mathfrak{X}} = \mathcal{O}_{P^4}(-5)$. Therefore, locally (in fact, Zariski-locally), X_0 is a local P^4. Let U be the Zariski neighborhood of X_0 isomorphic to the total space of $\mathcal{O}(-5)$. It is well-known that $\mathcal{O}(-5)$ is a crepant resolution of the smooth Deligne–Mumford stack $[\mathbb{C}^5/\mu_5]$. In particular,

$$U \setminus X_0 \cong [\mathbb{C}^5/\mu_5] \setminus [\mathrm{pt}/\mu_5].$$

As a result, there is a contraction

$$\pi : \mathfrak{X} \to \overline{\mathfrak{X}},$$

where X_0 is contracted to $[\mathrm{pt}/\mu_5]$. More concretely, what we do here is simply replacing U by $[\mathbb{C}^5/\mu_5]$.

Alternatively, one can describe $\overline{\mathfrak{X}}$ as follows. Consider the weighted projective space $P(5,1,1,1,1,1)$. P^4 embeds into it in the following way:

$$P^4 \cong P(1,1,1,1,1) \subset P(5,1,1,1,1,1).$$

One can check that the normal bundle of this P^4 is $\mathcal{O}(5)$. Consider the quintic hypersurface $Q \subset P^4$. The normal bundle of Q inside $P(5,1,1,1,1,1)$ is now $\mathcal{O}(5) \oplus \mathcal{O}(5)$. We claim that

$$\overline{\mathfrak{X}} \cong \mathrm{Bl}_Q\, P(5,1,1,1,1,1).$$

The only $[\mathrm{pt}/\mu_5]$ corresponds to the stacky point in $\overline{\mathfrak{X}}$, and the strict transform of P^4 corresponds to X_∞. The exceptional divisor is isomorphic to $Q \times P^1$, and intersects with X_∞ at a quintic threefold. Such $Q \times P^1$ also appears in $\mathfrak{X}$ as the strip connecting Q_0 and X_∞. Thus, the other end that does not intersect with X_∞ corresponds to Q_0 of $\overline{\mathfrak{X}}$. A rigorous identification is left to the readers.

REMARK 2.2. In fact, it is possible to realize both $\mathfrak{X}$ and $\overline{\mathfrak{X}}$ as invariant hypersurfaces of GIT quotients under a certain action of $(\mathbb{C}^*)^3$ on $\mathbb{C}^9$. The crepant contraction $\mathfrak{X} \to \overline{\mathfrak{X}}$ is a consequence of varying stability conditions. The GIT phase of $\overline{\mathfrak{X}}$ occurred in a stimulating discussion in our other project with M. Shoemaker.

On $\overline{\mathfrak{X}}$, there is a $\mathbb{C}^*$ action inherited from $\mathfrak{X}$. The fixed loci are Q_0, $[\mathrm{pt}/\mu_5]$, and X_∞. Note that under the birational contraction of $\mathfrak{X}$, a neighborhood of Q_0 is left untouched. Therefore, we can similarly choose a curve class $\beta_Q \in \mathrm{NE}(Q_0) \subset \mathrm{NE}(\overline{\mathfrak{X}})$ and apply localization. Let d be the degree of β_Q as a curve class in P^4. Under the same bound $d > (2g-2)/5$, invariants of Q can be recursively determined by GW invariants of $[\mathbb{C}^5/\mu_5]$ ($\mathbb{C}^*$ acting with weights $(1,1,1,1,1)$). The whole process is completely parallel to [**9**], and the invariants of X_∞ will vanish by virtual-dimensional reason.

3. New relations for the ambient space X

So far, the technique has been used to find relations for the hypersurfaces, assuming that the data from the ambient space is known. Now we turn the idea on its head and show that our construction can also be used to find relations for the ambient spaces under certain favorable conditions. These favorable conditions are designed so that the GW invariants of D will not enter the recursion relations. We show two examples in this section.

Both examples use the same set-up as before. Consider that the deformation to the normal cone of a smooth pair (X, D) and that the same $\mathbb{C}^*$-action has fixed loci D_0, X_0, X_∞. However, the curve classes β no longer come from D_0. Our strategy is to choose the conditions, which might involve the geometry of (X, D) or the numerical condition on the curve class β such that the GW invariants on D_0 will not appear in recursion relation.

3.1. Adding fiber classes. Let $f \in \mathrm{NE}(\mathfrak{X})$ be the curve class corresponding to a fiber on the exceptional divisor between D_0 and X_0. Choose a curve class $\beta_X \in \mathrm{NE}(X_0) \subset \mathrm{NE}(\mathfrak{X})$. Suppose $\alpha_1, \ldots, \alpha_n \in H^*(\mathfrak{X})$ are insertions. Since we only care about relations on X_0, localization on $\overline{M}_{g,n}(\mathfrak{X}, \beta_X + kf)$ yields a relation when

$$\sum_{i=1}^{n} \deg(\alpha_i) < (1-g)(\dim(X) - 2) + \int_{\beta_X} c_1(\mathcal{O}(-D)) + k + n.$$

Since we do not wish to involve invariants on D_0, certain conditions have to be satisfied. For example, a relation can be obtained if $\beta_X + kf$ is not numerically equivalent to a connected curve with an irreducible component in D_0. (If $(X, D) = (P^4, \text{quintic})$, this is the case when $k \le 4$.) We note that this is a very strong condition which limits applications of our method.

EXAMPLE 3.1. We try to give a complete and self-contained statement in the following special case. Let X be a smooth variety and D be a divisor such that $(\gamma, D) > 1$ for any nonzero $\gamma \in NE(X)$. Fix a nonzero $\beta_X \in NE(X)$ such that

$$(\beta_X, c_1(TX) - D) + 1 > 0.$$

Let $N = dim(X)$. In twisted GW theory, we give a weight-1 fiberwise $\mathbb{C}^*$-action on the twisting bundles. Let λ be the equivariant parameter.

PROPOSITION 3.2. (1) *(D is not necessarily effective) Assuming the above conditions, we have*

$$\langle\frac{1}{\lambda-D-\psi}\rangle_{1,1,\beta_X}^{X,\mathcal{O}(-D)}=-\frac{1}{24\lambda^2}\langle\frac{D\cdot c_{N-1}(TX)}{\lambda-D-\psi}\rangle_{0,1,\beta_X}^{X,\mathcal{O}(-D)}-\frac{1}{24\lambda}\langle\frac{c_{N-1}(TX)}{\lambda-D-\psi}\rangle_{0,1,\beta_X}^{X,\mathcal{O}(-D)}.$$

(2) *Let D be a smooth effective divisor; assuming the above conditions, we have*

$$\langle\frac{D}{\lambda-D-\psi}\rangle_{1,1,\beta_X}^{X,\mathcal{O}(-D)}=-\frac{1}{24\lambda}\langle\frac{D\cdot c_{N-1}(TX)}{\lambda-D-\psi}\rangle_{0,1,\beta_X}^{X,\mathcal{O}(-D)}-\frac{1}{24}\langle\frac{D\cdot c_{N-2}(TX)}{\lambda-D-\psi}\rangle_{0,1,\beta_X}^{X,\mathcal{O}(-D)}.$$

Relation (2) is derived in the above setup using $k=1$. We note that relation (1) is, however, derived by directly localizing on $P_X(\mathcal{O}\oplus\mathcal{O}(D))$ and, hence, not directly related to our discussion. We include it as a comparison to (2).

3.2. Hypersurfaces with trivial GW theory. Let X be a Fano 3-fold whose $-K_X$ is base-point free. If we choose D to be a smooth member of $|-K_X|$, it is a K3-surface by the adjunction formula, and hence has vanishing GW invariants except for certain degree 0 invariants. In principle, relations on X can be produced if the choice $\beta=\beta_X+kf\in\mathrm{NE}(\mathfrak{X})$ (β_X, f as in Section 3.1) and insertions $\alpha_1,\ldots,\alpha_n\in H^*(\mathfrak{X})$ satisfy

$$\sum_{i=1}^{n}\deg(\alpha_i)<(1-g)+k+n.$$

Since it is independent on γ and k can be arbitrarily large, this is not a very serious constraint. For example, the following sample relation holds for all GW invariants of Fano threefolds whose aniticanonical divisors are base-point free.

PROPOSITION 3.3. *If X is a Fano 3-fold with base-point free anticanonical divisor and $\beta_X\in\mathrm{NE}(X)$ is a nonzero curve class, the following relation holds without any extra conditions.*

$$\langle\frac{-K_X}{\lambda+K_X-\psi}\rangle_{1,1,\beta_X}^{X,\mathcal{O}(K_X)}=-\frac{1}{24\lambda}\langle\frac{-K_X\cdot c_2(TX)}{\lambda+K_X-\psi}\rangle_{0,1,\beta_X}^{X,\mathcal{O}(K_X)}-\frac{1}{24}\langle\frac{K_X^2}{\lambda+K_X-\psi}\rangle_{0,1,\beta_X}^{X,\mathcal{O}(K_X)}.$$

There are likely more equations of this type for any Fano threefolds. We wish to be able to report some interesting applications of these equations in the future.

Acknowledgments

We wish to thank T. Jarvis, N. Priddis and Y. Ruan, the organizers of the Snowbird Conference and School, for the invitations to attend the conference and to submit the article for its proceedings. Part of the work was done while both authors were in Snowbird. We are also grateful to M. Shoemaker for stimulating discussions.

References

[1] M. Bershadsky, S. Cecotti, H. Ooguri, and C. Vafa, *Kodaira-Spencer theory of gravity and exact results for quantum string amplitudes*, Comm. Math. Phys. **165** (1994), no. 2, 311–427. MR1301851

[2] Jeffrey Steven Brown, *Gromov-Witten invariants of toric fibrations*, ProQuest LLC, Ann Arbor, MI, 2010. Thesis (Ph.D.)–University of California, Berkeley. MR2941530

[3] H.-L. Chang, S. Guo, and J. Li, *BCOV's Feynman rule of quintic \$3\$-folds* (2018). arXiv:1810.00394.

[4] H.-L. Chang, S. Guo, and J. Li, *Polynomial structure of Gromov-Witten potential of quintic \$3\$-folds via NMSP* (2018). arXiv:1809.11058.

[5] H.-L. Chang, S. Guo, W.-P. Li, and J. Zhou, *Genus one GW invariants of quintic threefolds via MSP localization* (2017). 1711.10118.

[6] H.-L. Chang, J. Li, W.-P. Li, and C.-C. M. Liu, *An effective theory of GW and FJRW invariants of quintics Calabi-Yau manifolds* (2016). arXiv:1603.06184.

[7] Tom Coates and Alexander Givental, *Quantum Riemann-Roch, Lefschetz and Serre*, Ann. of Math. (2) **165** (2007), no. 1, 15–53, DOI 10.4007/annals.2007.165.15. MR2276766

[8] T. Coates, A. Givental, and H.-H. Tseng, *Virasoro constraints for toric bundles* (2015). arXiv:1508.06282.

[9] Honglu Fan and Yuan-Pin Lee, *Towards a quantum Lefschetz hyperplane theorem in all genera*, Geom. Topol. **23** (2019), no. 1, 493–512, DOI 10.2140/gt.2019.23.493. MR3921324

[10] Alexander B. Givental, *Gromov-Witten invariants and quantization of quadratic Hamiltonians* (English, with English and Russian summaries), Mosc. Math. J. **1** (2001), no. 4, 551–568, 645, DOI 10.17323/1609-4514-2001-1-4-551-568. Dedicated to the memory of I. G. Petrovskii on the occasion of his 100th anniversary. MR1901075

[11] Alexander B. Givental, *Semisimple Frobenius structures at higher genus*, Internat. Math. Res. Notices **23** (2001), 1265–1286, DOI 10.1155/S1073792801000605. MR1866444

[12] S. Guo, F. Janda, and Y. Ruan, *A mirror theorem for genus two Gromov–Witten invariants of quintic threefolds* (2017). arXiv:1709.07392.

[13] S. Guo, F. Janda, and Y. Ruan, *Structure of Higher Genus Gromov-Witten Invariants of Quintic 3-folds* (2018). arXiv:1812.11908.

[14] Y.-P. Lee, *Notes on axiomatic Gromov-Witten theory and applications*, Algebraic geometry—Seattle 2005. Part 1, Proc. Sympos. Pure Math., vol. 80, Amer. Math. Soc., Providence, RI, 2009, pp. 309–323, DOI 10.1090/pspum/080.1/2483940. MR2483940

[15] Y.-P. Lee, H.-W. Lin, and C.-L. Wang, *Birational Transformation and degeneration in Gromov–Witten theory, reported by the first author in the AMS summer institute* (2015Jul.)

[16] Yuan-Pin Lee, Hui-Wen Lin, and Chin-Lung Wang, *Invariance of quantum rings under ordinary flops II: A quantum Leray-Hirsch theorem*, Algebr. Geom. **3** (2016), no. 5, 615–653, DOI 10.14231/AG-2016-027. MR3568339

[17] Yuan-Pin Lee, Hui-Wen Lin, and Chin-Lung Wang, *Quantum cohomology under birational maps and transitions*, String-Math 2015, Proc. Sympos. Pure Math., vol. 96, Amer. Math. Soc., Providence, RI, 2017, pp. 149–168. MR3751817

[18] Y.-P. Lee and R. Pandharipande, *Frobenius manifolds, Gromov–Witten theory, and Virasoro constraints* (2004). unfinished book available at https://www.math.utah.edu/~yplee/research/.

[19] Bumsig Kim and Hyenho Lho, *Mirror theorem for elliptic quasimap invariants*, Geom. Topol. **22** (2018), no. 3, 1459–1481, DOI 10.2140/gt.2018.22.1459. MR3780438

[20] C. Teleman, *The structure of 2D semi-simple field theories*, Invent. Math. **188** (2012), no. 3, 525–588. MR2917177

[21] Aleksey Zinger, *The reduced genus 1 Gromov-Witten invariants of Calabi-Yau hypersurfaces*, J. Amer. Math. Soc. **22** (2009), no. 3, 691–737, DOI 10.1090/S0894-0347-08-00625-5. MR2505298

D-Math, ETH Zürich, Rämistrasse 101, 8092, Zürich, Switzerland
Email address: honglu.fan@math.ethz.ch

Department of Mathematics, University of Utah, Salt Lake City, Utah 84112-0090
Email address: yplee@math.utah.edu

Contemporary Mathematics
Volume **763**, 2021
https://doi.org/10.1090/conm/763/15341

Type II extremal transitions in Gromov-Witten theory

Rongxiao Mi

Abstract. This is an expository paper on the recent progress in understanding the change of Gromov-Witten theory under Type II extremal transitions. We propose a conjectural framework using the language of quantum D-modules, which is verified in several particular examples of Type II extremal transitions.

1. Introduction

Extremal transitions are a type of topological surgery, serving as the building blocks of the topological changes of Calabi-Yau 3-folds. The goal of this paper is to explore very leisurely the recent study of Type II extremal transitions in Gromov-Witten theory.

Gromov-Witten theory originates from mirror symmetry, a physical duality coming from two different types of string theory. Mirror symmetry has received a lot of interest in the mathematical community over the past few decades. Probably the most intriguing reason to study mirror symmetry is that it yields a highly non-trivial correspondence between two rather different mathematical theories, which are called A-model theory and B-model theory, respectively. The A-model theory involves the so-called Gromov-Witten invariants, which (virtually) count the holomorphic curve on a smooth projective variety. Computing Gromov-Witten invariants presents an important yet challenging problem in geometry and physics. However, mirror symmetry allows one to compute these invariants by appealing to the B-model theory, i.e. studying certain period integrals in complex geometry but on a different manifold (called mirror manifold). The existence of a Calabi-Yau 3-fold X and its mirror manifold $\hat{X}$ is predicted by mirror symmetry and proved for many examples, but is it possible to prove that mirror symmetry holds for all Calabi-Yau 3-folds?

To answer this question, we need to understand a bit about the classification of Calabi-Yau 3-folds. This is a fundamental problem in algebraic geometry. However, a construction by Batyrev reveals 473,800,776 families of Calabi-Yau 3-folds, which makes solving this problem nearly impossible. A very popular classifying scheme in the 1990s was to classify the Calabi-Yau 3-folds up to surgeries. There are two kinds of surgeries involved: flops and extremal transitions. A flop is a birational surgery, which usually arises as different resolutions of a singular variety. It is known that any birationally equivalent Calabi-Yau 3-folds are connected through

2010 *Mathematics Subject Classification.* Primary 14N35, 14J32; Secondary 81T30.
Key words and phrases. Calabi-Yau 3-folds, extremal transitions, Gromov-Witten theory.

a sequence of flops. On the other hand, extremal transitions are a non-birational surgery. An influential conjecture (known as Reid's Fantasy [**15**]) asserts that any two different birational classes of Calabi-Yau 3-folds are connected through extremal transitions. One strong support for this conjecture may be found in [**1**] where examples of Calabi-Yau 3-folds in weighted $\mathbb{P}^4$ are indeed connected through extremal transitions.

In order to extend mirror symmetry to a larger class of Calabi-Yau 3-folds, Li-Ruan [**11**] initiated a program to study how mirror symmetry behaves under surgeries on Calabi-Yau 3-folds, and they also proposed the mirror surgery conjecture, which roughly says the following: starting from one mirror pair of Calabi-Yau 3-folds $(X, \hat{X})$, if another Calabi-Yau 3-fold Y is obtained from X through a flop/extremal transition, then there is conjecturally a reverse flop/extremal transition that allows one to obtain the mirror manifold $\hat{Y}$ from $\hat{X}$. To implement this program, the first step would be to study the change of Gromov-Witten theory under flops and under extremal transitions.

In the case of flops, Li-Ruan [**11**] showed that a flop between two Calabi-Yau 3-folds induces an equivalence of quantum cohomology rings up to an appropriate analytic continuation of certain quantum variables in the extended Kähler moduli space. The quantum cohomology rings can be viewed as encoding the genus zero Gromov-Witten theory. Thus, Li-Ruan's result shows that the Gromov-Witten theory is invariant under flops in some sense. Their result has also been generalized to higher dimensions by Lee-Lin-Wang [**7–9**].

In sharp contrasts to flops, the change of the Gromov-Witten theory under extremal transitions is very complicated. This is because extremal transitions involve two different types of operations: a birational contraction followed by a complex smoothing or the reverse. One cannot expect the Gromov-Witten theory to be invariant under extremal transitions due to the asymmetry alluded to in this picture. It is not very clear what kind of statement one can prove for the change of Gromov-Witten theory under extremal transitions. In this paper, we will explore a conjectural framework proposed in [**13, 14**] that relates two quantum D-modules (See Section 3) under extremal transitions, and then we will give several examples involving Calabi-Yau 3-folds.

Plan of this paper. In Section 2, we will introduce the notion of extremal transitions and Li-Ruan's result on conifold transitions [**11**]. In Section 3, we will give a short background of the quantum D-modules and propose a conjectural framework that relates two quantum D-modules under an extremal transition. In Section 4, we will give some examples involving Calabi-Yau 3-folds, which verify the conjectural framework in Section 3. In Section 5, we will make some concluding remarks.

2. Extremal transitions and a theorem of Li-Ruan

DEFINITION 2.1. Let X and Y be smooth algebraic varieties over $\mathbb{C}$. The process of going from X to Y (or from Y to X) is called an extremal transition if Y is obtained by applying a birational contraction $\pi : X \to \overline{Y}$ followed by a smoothing $\overline{Y} \rightsquigarrow Y$.

$$\begin{array}{ccc} X & & \\ \downarrow{\scriptstyle \pi} & \searrow & \\ \overline{Y} & \rightsquigarrow & Y \end{array}$$

EXAMPLE 2.2. The following diagram gives an example of conifold transitions,

$$\begin{array}{ccc} \mathcal{O}_{\mathbb{P}^1}(-1) \oplus \mathcal{O}_{\mathbb{P}^1}(-1) & & \\ \downarrow{\scriptstyle \pi} & \searrow & \\ \overline{Y} & \rightsquigarrow & SL_2(\mathbb{C}) \end{array}$$

where $\overline{Y} := \{xw - yz = 0\} \subseteq \mathbb{C}^4$ is called a conifold. The morphism π is a birational contraction of $\mathbb{P}^1$. Indeed, we have

$$\mathcal{O}_{\mathbb{P}^1}(-1) \oplus \mathcal{O}_{\mathbb{P}^1}(-1) = \left\{ (x, y, z, w), [p_0 : p_1]; \begin{vmatrix} x & y \\ p_0 & p_1 \end{vmatrix} = \begin{vmatrix} z & w \\ p_0 & p_1 \end{vmatrix} = 0 \right\} \subseteq \mathbb{C}^4 \times \mathbb{P}^1,$$

and π is induced by the projection to the first factor $\mathbb{C}^4 \times \mathbb{P}^1 \to \mathbb{C}^4$. On the other hand, one may view $\overline{Y}$ as the set of 2 by 2 singular complex matrices. So $\overline{Y} \rightsquigarrow SL_2(\mathbb{C})$ is a smoothing.

REMARK. In Example 2.2, $\mathcal{O}_{\mathbb{P}^1}(-1) \oplus \mathcal{O}_{\mathbb{P}^1}(-1)$ is topologically an $\mathbb{R}^4$-bundle over S^2, while $SL_2(\mathbb{C})$ is topologically homeomorphic to T^*S^3. Thus, the conifold transition is a topological surgery from $\mathbb{R}^4 \times S^2$ to $S^3 \times \mathbb{R}^3$.

DEFINITION 2.3. An extremal transition $X \to \overline{Y} \rightsquigarrow Y$ is called primitive if the birational contraction $\pi : X \to \overline{Y}$ cannot be factored in the algebraic category.

According to P.M.H. Wilson's classification [**16**] of birational contractions, there are three types of primitive extremal transitions:

- Type I (small transition): π contracts a union of curves
- Type II: π contracts a divisor to a point
- Type III: π contracts a divisor to a curve

In their pioneering work, Li-Ruan [**11**] studied the change of Gromov-Witten invariants between two Calabi-Yau 3-folds that are related by a conifold transition, where the normal bundle to the curve that gets contracted is $\mathcal{O}_{\mathbb{P}^1}(-1) \oplus \mathcal{O}_{\mathbb{P}^1}(-1)$. Their result can be directly generalized to all type I extremal transitions between Calabi-Yau 3-folds because a small transition can be deformed into a collection of conifold transitions, and Gromov-Witten invariants are deformation invariants.

To describe Li-Ruan's result, we need to fix our notations first. Recall that for any smooth projective variety X and a fixed curve class $\beta \in H_2(X, \mathbb{Q})$, there is a moduli space $\overline{\mathcal{M}}_{g,n}(X, \beta)$ parametrizing the isomorphic class of stable maps $f : C \to X$, where C is a genus g nodal curve with n-markings and $f_*[C] = \beta$. Let ev_i be the i-th evaluation maps, then for $\gamma_i \in H^*(X, \mathbb{Q})$, the Gromov-Witten invariants are defined as the integration

$$\langle \gamma_1, \cdots, \gamma_n \rangle_{g,\beta}^X := \int_{[\overline{\mathcal{M}}_{g,n}(X,\beta)]^{vir}} \mathrm{ev}_1^* \gamma_1 \cup \cdots \cup \mathrm{ev}_n^* \gamma_n,$$

where $[\overline{\mathcal{M}}_{g,n}(X, \beta)]^{vir}$ is the virtual fundamental class supported in $H_{2d}(X, \mathbb{Q})$ and $d = \int_\beta c_1(X) + (\dim X - 3)(1 - g) + n$ is the virtual dimension. Intuitively, the above Gromov-Witten invariant can be intepreted as the counting of the genus g

nodal curves on X that pass through the submanifolds determined by γ_i. However, the Gromov-Witten invariants are rational numbers in general.

When the target is a Calabi-Yau 3-fold X, i.e. $c_1(X) = 0$ and $H^1(X, \mathcal{O}_X) = 0$, the virtual dimension of $\overline{\mathcal{M}}_{g,0}(X, \beta)$ is zero. The Gromov-Witten invariants of X can be reduced to the following numbers:

$$N^X_{g,\beta} := \int_{[\overline{\mathcal{M}}_{g,0}(X,\beta)]^{vir}} 1.$$

If two Calabi-Yau 3-folds X and Y are related by a conifold transition, Li-Ruan [**11**] proved the following theorem:

THEOREM 2.4. *The conifold transition induces a surjective map* $\pi : H_2(X, \mathbb{Q}) \to H_2(Y, \mathbb{Q})$ *such that for all nonzero* $\beta \in H_2(Y, \mathbb{Q})$*, we have*

$$N^Y_{g,\beta} = \sum_{\pi(\beta')=\beta} N^X_{g,\beta'}.$$

REMARK. Li-Ruan's theorem shows that the Gromov-Witten invariant of Y can be computed using the Gromov-Witten invariants of X, but not the other way around. This manifests the asymmetric nature of extremal transitions. While it is very difficult to compute higher genus Gromov-Witten invariants in general, this theorem tells us that all higher genus Gromov-Witten invariants between X and Y are related by this simple formula.

3. A conjectural framework for extremal transitions

While Type I extremal transitions are studied in many works [**5, 6, 10**], no such work had been done for Type II and Type III extremal transitions in the last twenty years due to the lack of good formulations. New progress was made in 2017, when the author studied the change of Gromov-Witten theory under cubic extremal transitions [**13**], which is the first example of Type II extremal transitions. A conjectural framework has been proposed for the change of the Gromov-Witten theory under an extremal transition, using the language of quantum D-modules. This framework is then verified for several examples involving Calabi-Yau 3-folds (see Section 4).

For a mirror pair $(X, \hat{X})$, mirror symmetry yields a correspondence between the quantum connection of X and the Gauss-Manin connection of $\hat{X}$. Recall that, given a smooth projective variety X, and letting $T_1, \cdots, T_r$ be a basis of $H^2(X, \mathbb{C}^*)$ with coordinates $q_1, \cdots, q_r$, one can define an integrable connection on the following trivial bundle,

$$H^2(X, \mathbb{C}^*) \times H^{2*}(X, \mathbb{C}) \to H^2(X, \mathbb{C}^*),$$

$$\nabla_i := q_i \frac{\partial}{\partial q_i} - T_i \star (...),$$

where $\star$ is the small quantum product, a deformation of ordinary cup product. The structure constants of the small quantum product are generating series of genus-zero Gromov-Witten invariants of X. The variables $q_1, \cdots, q_r$ are called quantum variables, which can be viewed as the local coordinates of the extended Kähler moduli space of X. This integrable connection gives rise to a holonomic D-module, which is called the quantum D-module of X. One may view the quantum D-module as encoding the genus zero Gromov-Witten theory of X.

When X is a hypersurface or a complete intersection inside a GIT quotient (which is the case for most known examples of Calabi-Yau 3-folds), it is more natural to study the twisted Gromov-Witten invariants, which may be viewed as the Gromov-Witten invariants of X whose cohomology inputs come from the ambient variety. We have an analogous notion of the ambient part quantum D-module, whose definition is more involved[1]. We will not give a full definition here, but we only remark that when a mirror theorem is available, this ambient part quantum D-module may be identified as the cyclic D-module generated by the I-function, which is usually a cohomology-valued hypergeometric series arising from the solution to the Picard-Fuchs equations in the B-model theory. As a result, the ambient part quantum D-modules for most known examples of Calabi-Yau 3-folds can be computed explicitly.

Suppose two smooth projective varieties X and Y are related by a primitive extremal transition. Let $\mathcal{H}(X)$ and $\mathcal{H}(Y)$ denote the (ambient part) quantum D-modules of X and Y, respectively. In [**13**], the following conjectural framework is proposed:

CONJECTURE 3.1. *One may perform analytic continuation of $\mathcal{H}(X)$ over the extended Kähler moduli to obtain a D-module $\bar{\mathcal{H}}(X)$. There is a divisor E and a submodule $\bar{\mathcal{H}}^E(X) \subseteq \bar{\mathcal{H}}(X)$ with maximum trivial E-monodromy such that*

$$\bar{\mathcal{H}}^E(X)|_E \simeq \mathcal{H}(Y),$$

where $\bar{\mathcal{H}}^E(X)|_E$ is the restriction to E.

This conjecture is based on several observations. Due to the asymmetric nature of extremal transitions, we expect that $\mathcal{H}(Y)$ can be recovered from $\mathcal{H}(X)$, as suggested by Li-Ruan's result (Theorem 2.5). Normally the rank of $\mathcal{H}(X)$ is going to be higher than that of $\mathcal{H}(Y)$, and $\mathcal{H}(X)$ would also involve more quantum variables. To relate the quantum variables, a procedure of analytic continuation is needed, as is the case in the study of flops. So the quantum variables of $\mathcal{H}(Y)$ would be identified as a subset of those of $\overline{\mathcal{H}}(X)$. Since the rank of $\overline{\mathcal{H}}(X)$ is still higher than that of $\mathcal{H}(Y)$, one expects to find a restriction of $\overline{\mathcal{H}}(X)$ to a submodule so that the extra quantum variables can be eliminated. To make this restriction work, the submodule must have trivial monodromy around a certain divisor E in the extended Kähler moduli space associated to X. It turns out that in many particular examples, the submodule is exactly the monodromy invariant part around E.

4. Examples of Type II extremal transitions

In this section, we will give several particular examples of Type II extremal transitions, for which Conjecture 3.1 is verified.

EXAMPLE 4.1. The first example studied is the local model of cubic extremal transitions. Let X and Y be Calabi-Yau 3-folds and $X \to \overline{Y} \rightsquigarrow Y$ be a cubic extremal transition, i.e. $X \to \overline{Y}$ contracts a cubic surface S in X to a triple point in Y. Assuming the curves on S generate an extremal ray in the Mori cone of X, then according to a result of M. Gross [**3**], $\overline{Y}$ can always be smoothed into a Calabi-Yau 3-fold Y. The local model is obtained by first considering the local surgery: Take a tubular neighborhood of S in X, which may be identified with the

[1] One may refer to [**4**, **12**, **13**] for a definition.

normal bundle $N_{S/X}$. Since X is a 3-fold with $K_X = \mathcal{O}_X$, by adjunction formula we have

$$N_{S/X} = K_X.$$

Contracting the zero section in $N_{S/X}$, we obtained a cone over S, whose local equation is

$$x^3 + y^3 + z^3 + w^3 = 0.$$

Deforming this equation, we obtain a smoothing

$$x^3 + y^3 + z^3 + w^3 = t \quad (t \neq 0),$$

which is a cubic 3-fold in $\mathbb{C}^4$. Compactifying both sides, we take $X' := \mathbb{P}(K_S \oplus \mathcal{O})$, and Y' is a cubic 3-fold in $\mathbb{P}^4$. This is the local model of cubic extremal transitions.

REMARK. The reason to consider the local model is due to the fact that the local model is usually toric and hence computable using standard methods of localization. When the local model is understood, a natural next step is to use degeneration techniques to globalize the local result. This approach has been adopted in many works.

EXAMPLE 4.2. The global example of cubic extremal transitions appearing in [**2**] is obtained in the following way. Let Y_0 be a smooth quintic hypersurface in $\mathbb{P}^4$ defined by the following equation:

$$x_0^2(tx_0^3 + x_1^3 + x_2^3 + x_3^3 + x_4^3) + x_0 f(x_1, \cdots, x_4) + g(x_1, \cdots, x_4) = 0,$$

where $t \neq 0$ and f, g are generic homogenous polynomials in degree 4 and 5, respectively. Y_0 may be deformed into $\overline{Y}_0$, a singular variety defined by

$$x_0^2(x_1^3 + x_2^3 + x_3^3 + x_4^3) + x_0 f(x_1, \cdots, x_4) + g(x_1, \cdots, x_4) = 0.$$

Since f and g are generic, $\overline{Y}_0$ has a unique singularity at $P := [1 : 0 : 0 : 0 : 0]$ (a triple point). Let X_0 be the blow-up of $\overline{Y}_0$ at P, then $X_0 \to \overline{Y}_0$ is a crepant resolution, contracting a cubic surface. Therefore, the process $X_0 \to \overline{Y}_0 \rightsquigarrow Y_0$ is an example of cubic extremal transitions between Calabi-Yau 3-folds.

EXAMPLE 4.3. One may also consider the local model of Type II extremal transitions in degree 4, where the birational map contracts a del Pezzo surface of degree-4 to a point with (2,2) complete intersection singularity. Let E be the degree-4 del Pezzo surface inside a Calabi-Yau 3-fold X, which gets contracted to a point under π. Consider a tubular neighborhood around E, which can be identified with its normal bundle $N_{E/X}$. By adjunction formula, we have $N_{E/X} \simeq K_E$. Under the birational morphism π, the divisor is contracted to a point with the singularity given by a complete intersection of two polynomials with quadratic leading terms. Smoothing out the singularity locally yields a (2,2) complete intersection in $\mathbb{P}^5$. We compactify $N_{E/X} \simeq K_E$ and take $X' := \mathbb{P}(K_E \oplus \mathcal{O})$, while Y' is a (2,2) complete intersection in $\mathbb{P}^5$. The passage from X' to Y' is the local model of Type II extremal transitions in degree 4.

EXAMPLE 4.4. We may construct a Calabi-Yau example of Type II extremal transitions in degree 4. We start with two quadratic homogenous polynomials $f(x_1, x_2, x_3, x_4, x_5), g(x_1, x_2, x_3, x_4, x_5)$ that form a complete intersection. Let Y_1 be the projective variety in $\mathbb{P}^5$ defined by the following two equations:

$$f(x_1, x_2, x_3, x_4, x_5) + tx_0^2 = 0,$$

$$x_0^2(g(x_1, x_2, x_3, x_4, x_5) + tx_0^2) + x_0 g_1(x_1, x_2, x_3, x_4, x_5) + g_2(x_1, x_2, x_3, x_4, x_5) = 0,$$

where $t \neq 0$, g_1 and g_2 are generic homogenous polynomials in degree 3 and 4, respectively. Y_1 can be deformed into $\overline{Y}_1$, whose defining equations are:

$$f(x_1, x_2, x_3, x_4, x_5) = 0,$$
$$x_0^2 g(x_1, x_2, x_3, x_4, x_5) + x_0 g_1(x_1, x_2, x_3, x_4, x_5) + g_2(x_1, x_2, x_3, x_4, x_5) = 0.$$

Since g_1 and g_2 are generic, we may assume that $\overline{Y}_1$ has a unique singularity at $P := [1 : 0 : 0 : 0 : 0 : 0]$. Let X_1 be the blow-up of $\overline{Y}_1$ at P, then $X_1 \to \overline{Y}_1$ would be a crepant resolution, where the exceptional divisor is a del Pezzo surface of degree 4. Therefore, the process $X_1 \to \overline{Y}_1 \rightsquigarrow Y_1$ is an Calabi-Yau example of Type II extremal transitions in degree 4.

EXAMPLE 4.5. Another example of Type II extremal transitions in degree 4 is obtained in a similar fashion. Let f, g be as in Example 4.4. Take Y_2 to be the projective variety in $\mathbb{P}^5$ defined by

$$x_0(f(x_1, x_2, x_3, x_4, x_5) + tx_0^2) + f_1(x_1, x_2, x_3, x_4, x_5) = 0,$$
$$x_0(g(x_1, x_2, x_3, x_4, x_5) + tx_0^2) + g_1(x_1, x_2, x_3, x_4, x_5) = 0,$$

where $t \neq 0$ and f_1, g_1 are both homogenous polynomials in degree 3. Y_2 can be deformed into $\overline{Y}_2$ inside $\mathbb{P}^5$, whose defining equations are

$$x_0 f(x_1, x_2, x_3, x_4, x_5) + f_1(x_1, x_2, x_3, x_4, x_5) = 0,$$
$$x_0 g(x_1, x_2, x_3, x_4, x_5) + g_1(x_1, x_2, x_3, x_4, x_5) = 0.$$

By choosing f_1 and g_1 generically, we may assume that $\overline{Y}_2$ has a unique singularity at $P := [1 : 0 : 0 : 0 : 0 : 0]$. We then take X_2 to be the blow-up of $\overline{Y}_2$ at P, then $X_2 \to \overline{Y}_2$ is a crepant resolution, where the exceptional divisor is also a del Pezzo surface of degree 4. Thus the process $X_2 \to \overline{Y}_2 \rightsquigarrow Y_2$ is aother Calabi-Yau example of Type II extremal transitions in degree 4.

In [**13, 14**], the following theorem is proved:

THEOREM 4.6. *Conjecture 3.1 holds for Examples 4.1 to 4.5.*

REMARK. The common feature of Example 4.1 to 4.5 is that X and Y can be realized as complete intersections of some toric varieties $\widetilde{X}$ and $\widetilde{Y}$, respectively, and $\widetilde{X}$ can be obtained by blowing up $\widetilde{Y}$ at a point. In the proof of Theorem 4.6, we make use of the mirror theorem to turn the quantum D-modules into the Picard-Fuchs equations. There is a very close connection between the combinatoric data that defines X and Y, which allows us to find a relation between the I-functions for X and Y.

5. Conclusion

Understanding how Gromov-Witten theory changes under extremal transitions certainly presents an interesting and important field of study in mirror symmetry. With Conjecture 3.1 proposed, we expect that more examples will be verified, and new directions will be opened to understand how higher genus Gromov-Witten invariants are related under an extremal transition. Conjecture 3.1 also sheds some light on the mirror surgery conjecture [**11**], and we expect that there will be an analogous statement for mirror surgeries in the B-model theory. Eventually, this may help us better understand how generally mirror symmetry is preserved under extremal transitions among Calabi-Yau 3-folds.

Acknowledgments

The author thanks Tyler Jarvis, Nathan Priddis, and Yongbin Ruan for organizing the wonderful workshop in Snowbird. He thanks the editor for providing valuable comments and suggestions for improving this paper.

References

[1] Ti-Ming Chiang, Brian R. Greene, Mark Gross, and Yakov Kanter, *Black hole condensation and the web of Calabi-Yau manifolds*, Nuclear Phys. B Proc. Suppl. **46** (1996), 82–95, DOI 10.1016/0920-5632(96)00010-2. S-duality and mirror symmetry (Trieste, 1995). MR1411461

[2] M. Gross, D. Huybrechts, and D. Joyce, *Calabi-Yau manifolds and related geometries*, Universitext, Springer-Verlag, Berlin, 2003. Lectures from the Summer School held in Nordfjordeid, June 2001. MR1963559

[3] Mark Gross, *Deforming Calabi-Yau threefolds*, Math. Ann. **308** (1997), no. 2, 187–220, DOI 10.1007/s002080050072. MR1464900

[4] Hiroshi Iritani, *Quantum D-modules and equivariant Floer theory for free loop spaces*, Math. Z. **252** (2006), no. 3, 577–622, DOI 10.1007/s00209-005-0867-9. MR2207760

[5] Hiroshi Iritani and Jifu Xiao, *Extremal transition and quantum cohomology: examples of toric degeneration*, Kyoto J. Math. **56** (2016), no. 4, 873–905, DOI 10.1215/21562261-3664959. MR3568645

[6] Yuan-Pin Lee, Hui-Wen Lin, and Chin-Lung Wang, *Towards $A+B$ theory in conifold transitions for Calabi-Yau threefolds*, J. Differential Geom. **110** (2018), no. 3, 495–541, DOI 10.4310/jdg/1542423628. MR3880232

[7] Yuan-Pin Lee, Hui-Wen Lin, and Chin-Lung Wang, *Flops, motives, and invariance of quantum rings*, Ann. of Math. (2) **172** (2010), no. 1, 243–290, DOI 10.4007/annals.2010.172.243. MR2680420

[8] Yuan-Pin Lee, Hui-Wen Lin, and Chin-Lung Wang, *Invariance of quantum rings under ordinary flops I: Quantum corrections and reduction to local models*, Algebr. Geom. **3** (2016), no. 5, 578–614, DOI 10.14231/AG-2016-026. MR3568338

[9] Yuan-Pin Lee, Hui-Wen Lin, and Chin-Lung Wang, *Invariance of quantum rings under ordinary flops II: A quantum Leray-Hirsch theorem*, Algebr. Geom. **3** (2016), no. 5, 615–653, DOI 10.14231/AG-2016-027. MR3568339

[10] Yuan-Pin Lee, Hui-Wen Lin, and Chin-Lung Wang, *Quantum cohomology under birational maps and transitions*, String-Math 2015, Proc. Sympos. Pure Math., vol. 96, Amer. Math. Soc., Providence, RI, 2017, pp. 149–168. MR3751817

[11] An-Min Li and Yongbin Ruan, *Symplectic surgery and Gromov-Witten invariants of Calabi-Yau 3-folds*, Invent. Math. **145** (2001), no. 1, 151–218, DOI 10.1007/s002220100146. MR1839289

[12] Etienne Mann and Thierry Mignon, *Quantum $\mathcal{D}$-modules for toric nef complete intersections*, Internat. J. Math. **28** (2017), no. 6, 1750047, 49, DOI 10.1142/S0129167X17500471. MR3663796

[13] R. Mi, *Cubic Extremal Transition and Gromov-Witten Theory*, `arXiv:1711.11014` (2017).

[14] R. Mi, *Gromov-Witten theory under degree-4 Type II Extremal Transitions*, `arXiv:1810.05783` (2018).

[15] Miles Reid, *The moduli space of 3-folds with $K=0$ may nevertheless be irreducible*, Math. Ann. **278** (1987), no. 1-4, 329–334, DOI 10.1007/BF01458074. MR909231

[16] P. M. H. Wilson, *The Kähler cone on Calabi-Yau threefolds*, Invent. Math. **107** (1992), no. 3, 561–583, DOI 10.1007/BF01231902. MR1150602

Department of Mathematics, University of Michigan, Ann Arbor, Michigan, 48105
Email address: `rongxiao@umich.edu`

Contemporary Mathematics
Volume **763**, 2021
https://doi.org/10.1090/conm/763/15323

A lecture on holomorphic anomaly equations and extended holomorphic anomaly equations

Chiu-Chu Melissa Liu

1. Mirror symmetry for compact Calabi-Yau threefolds

Mirror symmetry relates the A-model on a compact Calabi-Yau threefold X, defined in terms of the symplectic structure on X, to the B-model on a mirror compact Calabi-Yau threefold Y, defined in terms of the complex structure on Y,

$$\text{A-model } (X,\omega) \overset{\text{Mirror Symmetry}}{\longleftrightarrow} \text{B-model } (Y,\Omega),$$

where ω is a Ricci flat Kähler form on X and Ω is a holomorphic volume form (i.e. a nowhere vanishing holomorphic 3-form) on Y.

The Hodge diamond of X is of the form

$$\begin{array}{ccccccc} & & & 1 & & & \\ & & 0 & & 0 & & \\ & 0 & & h^{2,2}(X) & & 0 & \\ 1 & & h^{2,1}(X) & & h^{1,2}(X) & & 1 \\ & 0 & & h^{1,1}(X) & & 0 & \\ & & 0 & & 0 & & \\ & & & 1. & & & \end{array}$$

The Hodge diamond of Y is of the same form. Let

$$\kappa := h^{1,1}(X) = h^{2,2}(X) = \dim_{\mathbb{C}} H^2(X;\mathbb{C}) = \dim_{\mathbb{C}} \mathcal{M}_A,$$

where $\mathcal{M}_A$ is the complexified Kähler moduli of X. Then

$$\kappa = h^{2,1}(Y) = h^{1,2}(Y) = \dim_{\mathbb{C}} H^1(Y,T_Y) = \dim_{\mathbb{C}} \mathcal{M}_B,$$

where $\mathcal{M}_B$ is the moduli of complex structures on Y.

2. A-model on compact Calabi-Yau threefolds

2.1. A-model topological closed strings: Gromov-Witten invariants.

Given a non-negative integer g and an effective curve class $\beta \in H_2(X;\mathbb{Z})$, let $\overline{\mathcal{M}}_g(X,\beta)$ be the moduli space of genus g degree β stable maps to X (with no marked points). Note that $\overline{\mathcal{M}}_g(X,\beta)$ is empty if $(g,\beta) = (0,0)$ or $(1,0)$. The

2010 *Mathematics Subject Classification.* Primary 53D37.

moduli $\overline{\mathcal{M}}_g(X,\beta)$ is a proper Deligne-Mumford stack equipped with a virtual fundamental class

$$[\overline{\mathcal{M}}_g(X,\beta)]^{\mathrm{vir}} \in H_0(\overline{\mathcal{M}}_g(X,\beta);\mathbb{Q}).$$

The genus g, degree β Gromov-Witten invariant of X is defined by

$$N_{g,\beta}^X := \int_{[\overline{\mathcal{M}}_g(X,\beta)]^{\mathrm{vir}}} 1 \in \mathbb{Q}$$

when $(g,\beta) \neq (0,0),(1,0)$. The integral sign in the above equation stands for the natural pairing between $[\overline{\mathcal{M}}_g(X,\beta)]^{\mathrm{vir}} \in H_0(\overline{\mathcal{M}}_g(X,\beta);\mathbb{Q})$ and $1 \in H^0(\overline{\mathcal{M}}_g(X,\beta);\mathbb{Q})$. For $g \geq 2$, we have

$$\overline{\mathcal{M}}_g(X,0) = \overline{\mathcal{M}}_g \times X,$$

and

$$N_{g,0}^X = \frac{(-1)^g}{2}\int_{\overline{\mathcal{M}}_g} \lambda_{g-1}^3 \int_X c_3(X) = \frac{(-1)^g |B_{2g}|\cdot|B_{2g-2}|}{4g(2g-2)\cdot(2g-2)!}\int_X c_3(X),$$

where B_{2m} are Bernoulli numbers.

We choose a basis $H_1,\ldots,H_k$ of $H^{1,1}(X) = H^2(X;\mathbb{C})$ such that each H_i is in $H^2(X;\mathbb{Z})$ and also in the nef cone (which is the closure of the Kähler cone of X). A complexified Kähler class is of the form

$$t = \sum_{i=1}^{\kappa} t^i H_i,$$

where $t^1,\ldots,t^\kappa \in \mathbb{C}$ are complexified Kähler parameters. The genus g Gromov-Witten potential of X, F_g^X, is a generating function of genus g Gromov-Witten invariants of X:

$$F_g^X(t) := \begin{cases} \frac{1}{6}\int_X t^3 + \sum_{\beta\neq 0} N_{0,\beta}^X \exp(\int_\beta t), & g=0;\\ -\frac{1}{24}\int_X c_2(X)t + \sum_{\beta\neq 0} N_{1,\beta}^X \exp(\int_\beta t), & g=1;\\ N_{g,0}^X + \sum_{\beta\neq 0} N_{g,\beta}^X \exp(\int_\beta t), & g\geq 2. \end{cases}$$

The sum $\sum_{\beta\neq 0}$ is over all non-zero effective classes. We may write

$$F_g^X = F_g^{X,\mathrm{classical}} + F_g^{X,\mathrm{quantum}},$$

where

$$F_g^{X,\mathrm{quantum}} = \sum_{\beta\neq 0} N_{g,\beta}^X \exp(\int_\beta t)$$

is the contribution from non-constant genus g stable maps to X. We have

$$\exp(\int_\beta t) = Q_1^{d_1}\cdots Q_\kappa^{d_\kappa},$$

where $Q_i = \exp(t^i)$ and $d_i = \int_\beta H_i \in \mathbb{Z}_{\geq 0}$. So $F_g^{X,\mathrm{quantum}}$ is a formal power series in $Q_1,\ldots,Q_\kappa$ with rational coefficients; it tends to zero at the large radius limit $Q_i \to 0$:

$$\lim_{Q_i\to 0} F_g^{X,\mathrm{quantum}}(t) = 0.$$

We have

$$\begin{aligned} \frac{\partial^3 F_0^X}{\partial t^i \partial t^j \partial t^k} &= (H_i \star H_j, H_k) \\ &= \int_X H_i \cup H_j \cup H_k + \sum_{\beta\neq 0}\Big(\int_\beta H_i \int_\beta H_j \int_\beta H_k\Big) N_{0,\beta}^X \exp(\int_\beta t), \end{aligned}$$

where $\star$ is the quantum product, where $\cup$ is the classical cup product, and where $(\ ,\ )$ is the Poincaré pairing on $H^*(X)$.

2.2. A-model topological open strings: Open Gromov-Witten invariants. Let $L \subset X$ be a closed oriented Lagrangian submanifold. Then the tangent bundle of L is trivial – recall that the tangent bundle of any orientable 3-manifold is trivial. In this paper, we further assume that L is a rational homology 3-sphere:

$$H_*(L;\mathbb{Q}) = H_*(S^3;\mathbb{Q}).$$

Then the map $H_2(X;\mathbb{Q}) \to H_2(X,L;\mathbb{Q})$ is an isomorphism, so $H_2(X;\mathbb{Z}) \to H_2(X,L;\mathbb{Z})$ has finite kernel and cokernel. Since $H_1(L;\mathbb{Z})$ is torsion and $H_0(L;\mathbb{Z}) = \mathbb{Z}$, we have $H^1(L;\mathbb{Z}) = 0$ by the universal coefficient theorem. In particular, the Maslov class $\mu(L) \in H^1(L;\mathbb{Z})$ (which is defined for any Lagrangian submanifold in a Calabi-Yau manifold) is zero. Given a pair (g,h), where g is a nonnegative integer and h is a positive integer, let $\overline{\mathcal{M}}_{(g,h)}(X,L,\beta)$ be the stable compactification of the moduli $\mathcal{M}_{g,h}(X,L,\beta)$, parametrizing holomorphic maps $u:(\Sigma,\partial\Sigma) \to (X,L)$, where Σ is a bordered Riemann surface with g handles and h holes, and $u_*[\Sigma] = \beta \in H_2(X,L;\mathbb{Z})$, where the domain Σ is oriented by its complex structure. Then $\overline{\mathcal{M}}_{g,h}(X,L,\beta)$ is a (usually singular) orbifold whose virtual dimension is zero. In some cases it is possible to define a virtual number $N^{X,L}_{(g,h),\beta}$ of points in $\overline{\mathcal{M}}_{g,h}(X,L,\beta)$; in general, $N^{X,L}_{(g,h),\beta}$ is a rational number (instead of an integer) due to the existence of orbifold points. We define generating functions of open Gromov-Witten invariants of the pair (X,L) by

$$F^{X,L}_{(g,h)}(t) = \sum_{\beta \neq 0} N^{X,L}_{(g,h),\beta} \exp\Big(\int_\beta t\Big),$$

where the sum is over nonzero relative homology classes $\beta \in H_2(X,L;\mathbb{Z})$. By assumption, for any $\beta \in H_2(X,L;\mathbb{Z})$ there exists a positive integer r such that $r\beta$ lies in the image of $H_2(X;\mathbb{Z}) \to H_2(X,L;\mathbb{Z})$, so $F^{X,L}_{(g,h)}(t)$ is a formal power series in $Q_1^{1/r_1},\dots,Q_\kappa^{1/r_\kappa}$ for some positive integers $r_1,\dots,r_\kappa$, and it tends to zero at the large radius limit:

$$\lim_{Q_i \to 0} F^{X,L}_{(g,h)}(t) = 0.$$

EXAMPLE 2.1. Let X be a quintic Calabi-Yau threefold with real coefficients, and let L be the real quintic. Then L is a Lagrangian submanifold of X, which is diffeomorphic to $\mathbb{RP}^3$, so it is orientable and is a rational homology sphere. The group homomorphism

$$H_2(X;\mathbb{Z}) = \mathbb{Z} \longrightarrow H_2(X,L;\mathbb{Z}) = \mathbb{Z}$$

is injective with cokernel $\mathbb{Z}/2\mathbb{Z}$. In this case $\kappa = 1$, and

$$F_g^{X,\text{quantum}} \in \mathbb{Q}[\![Q]\!], \quad F^{X,L}_{(g,h)} \in \mathbb{Q}[\![Q^{1/2}]\!],$$

where $Q = Q_1$.

3. Preliminaries on moduli of complex structures

3.1. The complex moduli and the vacuum line bundle. Recall that $\mathcal{M}_B$ is the moduli space of complex structures on Y, and that $\dim_{\mathbb{C}} \mathcal{M}_B = \kappa$. Let $q = (q_1,\dots,q_\kappa)$ be the local holomorphic coordinates on $\mathcal{M}_B$ such that $q = 0$ corresponds to a maximal unipotent monodromy point in the boundary of a (partial) compactification $\overline{\mathcal{M}}_B$ of $\mathcal{M}_B$. (In particular, $q = 0$ is in $\overline{\mathcal{M}}_B \setminus \mathcal{M}_B$.) Let $\mathcal{L}^\vee \to \mathcal{M}_B$ be the complex line bundle over $\mathcal{M}_B$ whose fiber over $q \in \mathcal{M}_B$ is $\mathcal{L}^\vee_q = H^0(Y_q, \Omega^3_{Y_q}) \cong \mathbb{C}$; then $\mathcal{L}^\vee$ is a holomorphic line bundle over $\mathcal{M}_B$, and its dual $\mathcal{L}$ is the vacuum line bundle in the physics literature such as [**13**] and [**4**]. The extended moduli space $\widetilde{\mathcal{M}}_B$ is the total space of the frame bundle of $\mathcal{L}^\vee$; it parametrizes pairs (Y_q, Ω), where Y_q corresponds to a point $q \in \mathcal{M}_B$ and Ω is a nonzero holomorphic 3-form on Y_q. So $p : \widetilde{\mathcal{M}}_B \to \mathcal{M}_B$ is a principal $\mathbb{C}^*$-bundle, and $\dim_{\mathbb{C}} \widetilde{\mathcal{M}}_B = \kappa + 1$.

3.2. The Torelli space. Let H denote the rank 2 lattice $\mathbb{Z} \oplus \mathbb{Z}$ equipped with the symplectic form $\begin{pmatrix} 0 & 1 \\ -1 & 0 \end{pmatrix}$. The Torelli space of Y is the moduli of the marked Calabi-Yau threefold (Y_q, γ), where Y_q corresponds to a point $q \in \mathcal{M}_B$ and the marking γ is an isometry from $H^{\oplus(\kappa+1)}$ to $H^3(X_q;\mathbb{Z})/\mathrm{Tor}$. Forgetting the marking γ defines a covering map $\pi : \mathcal{T} \to \mathcal{M}_B$, which is a principal $Sp(2\kappa+2;\mathbb{Z})$-bundle. Let $\widetilde{\mathcal{T}}$ be the fiber product:

$$\begin{array}{ccc} \widetilde{\mathcal{T}} & \longrightarrow & \widetilde{\mathcal{M}}_B \\ \big\downarrow & & \big\downarrow p \\ \mathcal{T} & \xrightarrow{\ \pi\ } & \mathcal{M}_B. \end{array}$$

Then $\widetilde{\mathcal{T}} \to \mathcal{T}$ is a principal $\mathbb{C}^*$-bundle that is the frame bundle of $\pi^*\mathcal{L}^\vee$, and $\widetilde{\mathcal{T}} \to \widetilde{\mathcal{M}}_B$ is a covering map that is a principal $Sp(2\kappa+2;\mathbb{Z})$ bundle.

3.3. The Hodge bundle $\mathbb{H}_{\mathbb{C}}$ and the Gauss-Manin connection. Let $\mathbb{H}_{\mathbb{Z}}$ be the local system of lattices on $\mathcal{M}_B$ whose fiber over $q \in \mathcal{M}_B$ is $H^3(Y_q;\mathbb{Z}) \cong \mathbb{Z}^{2k+2}$. Then $\mathbb{H}_{\mathbb{R}} = \mathbb{H}_{\mathbb{Z}} \otimes_{\mathbb{Z}} \mathbb{R}$ (resp. $\mathbb{H}_{\mathbb{C}} = \mathbb{H}_{\mathbb{Z}} \otimes_{\mathbb{Z}} \mathbb{C}$) is a flat real (resp. complex) vector bundle of rank $2\kappa + 2$ whose fiber at $q \in \mathcal{M}_B$ is $H^3(Y_q;\mathbb{R})$ (resp. $H^3(Y_q;\mathbb{C})$); the flat connection is known as the Gauss-Manin connection.

More explicitly, let

$$\nabla : \Omega^0(\mathcal{M}_B, \mathbb{H}_{\mathbb{C}}) = C^\infty(\mathcal{M}_B, \mathbb{H}_{\mathbb{C}}) \longrightarrow \Omega^1(\mathcal{M}_B, \mathbb{H}_{\mathbb{C}}) = C^\infty(\mathcal{M}_B, T^*M_B \otimes \mathbb{H}_{\mathbb{C}})$$

be the Gauss-Manin connection. Let U be an open subset on $\mathcal{M}_B$ such that $\mathbb{H}_{\mathbb{Z}}|_U$ is trivial; we choose a trivialization of $\mathbb{H}_{\mathbb{Z}}|_U$, or equivalently, a symplectic basis $\{\alpha_i, \beta^i : i = 0, 1, \ldots, \kappa\}$ of $H^3(Y_q;\mathbb{Z}) \cong H^{\oplus(\kappa+1)}$ for $q \in U$:

$$\int_{Y_q} \alpha_i \cup \alpha_j = \int_{Y_q} \beta^i \cup \beta^j = 0, \qquad \int_{Y_q} \alpha_i \cup \beta^j = -\int_{Y_q} \beta^j \cup \alpha_i = \delta_{ij}.$$

Then $\{\alpha_i, \beta^i\}$ is a frame of $\mathbb{H}_{\mathbb{C}}|_U$ and

$$\nabla \alpha_i = \nabla \beta^i = 0, \quad i = 0, 1, \ldots, \kappa.$$

Any C^∞ section $s \in C^\infty(U, \mathbb{H}_{\mathbb{C}})$ is of the form

$$s = \sum_{i=0}^{\kappa} (a^i \alpha_i + b_i \beta^i), \tag{3.1}$$

where a^i, b_i are complex-valued C^∞ functions on U. Then

$$\nabla s = \sum_{i=0}^{\kappa} (da^i \alpha_i + db_i \beta^i),$$

where $da^i, db_i \in \Omega^1(U, \mathbb{C})$ are C^∞ 1-forms on U.

We may write

$$\nabla = \nabla' + \nabla'',$$

where $\nabla' : \Omega^0(\mathcal{M}_B, \mathbb{H}_{\mathbb{C}}) \to \Omega^{1,0}(\mathcal{M}_B, \mathbb{H}_{\mathbb{C}})$ is a $(1,0)$-connection on the C^∞ complex vector bundle $\mathbb{H}_{\mathbb{C}}$ and where $\nabla'' : \Omega^0(\mathcal{M}_B, \mathbb{H}_{\mathbb{C}}) \to \Omega^{0,1}(\mathcal{M}_B, \mathbb{H}_{\mathbb{C}})$ is a $(0,1)$-connection on $\mathbb{H}_{\mathbb{C}}$. The $(0,1)$-connection ∇'' defines a holomorphic structure on $\mathbb{H}_{\mathbb{C}}$: a section $s \in C^\infty(U, \mathbb{H}_{\mathbb{C}})$ is holomorphic iff $\nabla'' s = 0$ iff

$$s = \sum_{i=0}^{\kappa} (a^i \alpha_i + b_i \beta^i),$$

where a^i, b_i are holomorphic functions on U.

3.4. Hodge filtration and the holomorphic polarization. We have the Hodge filtration

$$0 \subset \mathbb{F}^3 \subset \mathbb{F}^2 \subset \mathbb{F}^1 \subset \mathbb{F}^0 = \mathbb{H}_{\mathbb{C}},$$

where

$$\mathbb{F}^i_q = \bigoplus_{p \geq i} H^{p,3-p}(Y_q; \mathbb{C}).$$

The complex vector bundles $\mathbb{F}^1, \mathbb{F}^2, \mathbb{F}^3$ are (non-flat) holomorphic subbundles of $\mathbb{F}^0 = \mathbb{H}_{\mathbb{C}}$ of ranks $2\kappa + 1$, $\kappa + 1$, 1, respectively. In particular, $\mathbb{F}^3 = \mathcal{L}^\vee$ is the dual of the vacuum line bundle. We also have

$$\mathbb{H}_{\mathbb{C}} = \mathbb{F}^2 \oplus \overline{\mathbb{F}^2},$$

where

$$\mathbb{F}^2_q = H^{3,0}(Y_q; \mathbb{C}) \oplus H^{2,1}(Y_q; \mathbb{C}), \quad \overline{\mathbb{F}^2_q} = H^{1,2}(Y_q; \mathbb{C}) \oplus H^{0,3}(Y_q; \mathbb{C}).$$

For each $q \in U$,

$$H^3(Y_q; \mathbb{C}) = \mathbb{F}^2_q \oplus \overline{\mathbb{F}^2_q}$$

is the holomorphic polarization of the complex symplectic space $H^3(Y_q; \mathbb{C})$.

3.5. Real polarization. We choose a symplectic basis $\{A^i, B_i : i = 0, 1, \ldots, \kappa\}$ of $(H_3(Y; \mathbb{Z}), \cap)$, where $\cap$ is the intersection form, and let $\{\alpha_i, \beta^i : i = 0, 1, \ldots, \kappa\}$ be the dual symplectic basis of $(H^3(Y; \mathbb{Z}), Q)$, where

$$Q(\alpha, \beta) = \int_Y \alpha \cup \beta \in \mathbb{Z} \quad \text{for } \alpha, \beta \in H^3(Y; \mathbb{Z}).$$

We have

$$\delta^i_j = A^i \cap B_j = -B_j \cap A^i = \int_Y \alpha_i \cup \beta^j = -\int_Y \beta^j \cup \alpha_i = \int_{A^i} \alpha_j = \int_{B_j} \beta^i.$$

$$0 = A^i \cap A^j = B_i \cup B_j = \int_Y \alpha_i \cup \alpha_j = \int_Y \beta^i \cup \beta^j = \int_{A^i} \beta^j = \int_{B_j} \alpha_i.$$

For any $\phi \in H^3(X; \mathbb{R})$ we have

$$\phi = \sum_{i=0}^{\kappa} x^i \alpha_i + \sum_{i=0}^{\kappa} p_i \beta^i,$$

where

$$x^i = \int_{A^i} \phi, \quad p_i = \int_{B_i} \phi.$$

Then $\{x^i, p_i : i = 0, 1, \dots, k\}$ are Darboux coordinates of the real linear symplectic space $(H^3(Y;\mathbb{R}), Q)$. The linear symplectic form

$$\sum_{i=0}^{k} dx^i \wedge dp_i$$

on $H^3(Y;\mathbb{R})$ is independent of the choice of a symplectic basis. The integral symplectic basis $\{\alpha_i, \beta^i : i = 0, 1, \dots, \kappa\}$ extends to a flat frame of $\mathbb{H}_{\mathbb{R}}$ on an open neighborhood U of q_0 in $\mathcal{M}_B$, and $\{x^i, p_i : i = 0, 1, \dots, \kappa\}$ are flat fiber coordinates of $\mathbb{H}_{\mathbb{R}}|_U \cong U \times \mathbb{R}^{2\kappa+2}$.

The Gauss-Manin connection is compatible with the symplectic structure Q on $\mathbb{H}_{\mathbb{C}}$: given two C^∞ sections s_1, s_2 of $\mathbb{H}_{\mathbb{C}}$, and a C^∞ vector field ξ on $\mathcal{M}_B$, we have

$$L_\xi\left(Q(s_1, s_2)\right) = Q(\nabla_\xi s_1, s_2) + Q(s_1, \nabla_\xi s_2),$$

where

$$L_\xi : C^\infty(\mathcal{M}_B;\mathbb{C}) \to C^\infty(\mathcal{M}_B;\mathbb{C})$$

is the Lie derivative on C^∞ functions, and where

$$\nabla_\xi : C^\infty(\mathcal{M}_B, \mathbb{H}_{\mathbb{C}}) \to C^\infty(\mathcal{M}_B, \mathbb{H}_{\mathbb{C}})$$

is the covariant derivative defined by the Gauss-Manin connection.

3.6. Special homogeneous coordinates and the period map. Let $s : \widetilde{\mathcal{M}}_B \to p^*\mathcal{L}^\vee$ be the tautological section. Write

$$s(\xi) = \sum_{i=0}^{k} (x^i(\xi)\alpha_i + p_i(\xi)\beta^i).$$

Then $x^0(\xi), \dots, x^\kappa(\xi)$ are the local holomorphic coordinates on the extended complex moduli $\widetilde{\mathcal{M}}_B$ and the local homogeneous coordinates on the complex moduli $\mathcal{M}_B$; they are called "special homogeneous coordinates" in [**13**].

Let $V = \left(H^{\oplus(\kappa+1)}\right) \otimes_{\mathbb{Z}} \mathbb{C}$, which is a complex symplectic vector space of dimension $2(\kappa+1)$. The isometry $\gamma : H^{\oplus(\kappa+1)} \to H^3(X_q;\mathbb{Z})/\mathrm{Tor}$ extends to an isomorphism $\gamma : V \to H^3(X_q;\mathbb{C})$ of complex symplectic vector spaces. There is a period map

$$P_{\mathcal{T}} : \mathcal{T} \to \mathbb{P}(V) \cong \mathbb{P}^{2\kappa+1}, \quad (X_q, \gamma) \mapsto \gamma^{-1}(H^{3,0}(X_q)).$$

More explicitly,

$$P_{\mathcal{T}}(q) = [x^0(\xi), x^1(\xi), \dots, x^\kappa(\xi), p_0(\xi), p_1(\xi), \dots, p_\kappa(\xi)],$$

where $\xi \in \widetilde{\mathcal{T}}$ is any point in the fiber of $\widetilde{\mathcal{T}} \to \mathcal{T}$ over $q \in \mathcal{T}$.

Let $\mathcal{O}_{\mathbb{P}(V)}(-1)$ be the tautological line bundle over $\mathbb{P}(V)$. Then

$$P_{\mathcal{T}}^*\mathcal{O}_{\mathbb{P}(V)}(-1) = \pi^*\mathcal{L}^\vee.$$

Recall the Euler sequence:

$$0 \to \mathcal{O}_{\mathbb{P}(V)} \to V \otimes \mathcal{O}_{\mathbb{P}(V)}(1) \to T_{\mathbb{P}(V)} \to 0.$$

Pulling back the above sequence under $P_{\mathcal{T}}$, we obtain

$$0 \to \mathcal{O}_{\mathcal{T}} \to V \otimes \pi^*\mathcal{L} \to P_{\mathcal{T}}^* T_{\mathbb{P}(V)} \to 0.$$

Here, $\pi : \mathcal{T} \to \mathcal{M}_B$ is the projection from the Torelli space to the complex moduli as before, and

$$V \otimes \pi^*\mathcal{L} = \pi^*(\mathbb{H}_{\mathbb{C}} \otimes \mathcal{L}).$$

3.7. The Hodge metric. The symplectic form Q on $H^3(Y;\mathbb{R})$ extends to $V = H^3(Y;\mathbb{C})$. For $\alpha, \beta \in H^3(Y,\mathbb{C})$, define

$$(\alpha,\beta) := \sqrt{-1}Q(\alpha,\beta).$$

By the Hodge-Riemann bilinear relation,

- $(\alpha,\alpha) = 0$ if $\alpha \in H^{3,0}(Y) \oplus H^{2,1}(Y)$;
- $(\alpha,\bar{\alpha}) > 0$ if $\alpha \in H^{3,0}(Y)$ and $\alpha \neq 0$;
- $(\alpha,\bar{\alpha}) < 0$ if $\alpha \in H^{2,1}(Y)$ and $\alpha \neq 0$.

Define a Hermitian metric on the holomorphic line bundle $\mathcal{L}^\vee = \mathbb{F}^3$ by

$$\| \Omega \|^2 = (\Omega, \bar{\Omega}).$$

This is known as the Hodge metric on $\mathbb{F}^3$.

Let $\mathbb{H}^{2,1}$ be the complex vector bundle over $\mathcal{M}_B$ whose fiber over $q \in \mathcal{M}_B$ is $H^{2,1}(Y_q)$. Then $\mathbb{H}^{2,1}$ is a C^∞ complex subbundle of $\mathbb{H}_\mathbb{C}$ but not a holomorphic subbundle of $\mathbb{H}_\mathbb{C}$. Define the Hodge metric h on $\mathbb{H}^{2,1}$ by

$$h(s_1,s_2) = -(s_1,\bar{s}_2).$$

Then $(\mathbb{H}^{2,1},h)$ is a Hermitian vector bundle of rank κ over $\mathcal{M}_B$.

3.8. The Chern connection on $\mathcal{L}^\vee = \mathbb{F}^3$. Let

$$D : \Omega^0(\mathcal{M}_B,\mathcal{L}^\vee) \to \Omega^1(\mathcal{M}_B,\mathcal{L}^\vee) = \Omega^{1,0}(\mathcal{M}_B,\mathcal{L}^\vee) \oplus \Omega^{0,1}(\mathcal{M}_B,\mathcal{L}^\vee)$$

be the Chern connection determined by the holomorphic structure and the Hodge metric on $\mathcal{L}^\vee = \mathbb{F}^3$, and let D' and D'' be the $(1,0)$ and $(0,1)$ parts of D, so that

$$D = D' + D'',$$

where D'' depends on the holomorphic structure. Any C^∞ section s of $\mathcal{L}^\vee$ is also a C^∞ section of $\mathbb{H}_\mathbb{C}$, and we have

$$D''s = \nabla''s$$

since $\mathcal{L}^\vee$ is a holomorphic subbundle of $\mathbb{H}_\mathbb{C}$.

If (Y_q,Ω_q) is a local holomorphic frame of $\mathcal{L}^\vee$ over an open neighborhood U then $D''\Omega = 0$, and for any tangent vector $v \in T_q\mathcal{M}_B$, where $q \in U$, we have

$$D'_v\Omega = D_v\Omega = \frac{(\nabla_v\Omega,\bar{\Omega})}{(\Omega,\bar{\Omega})}\Omega. \tag{3.2}$$

The connection 1-form is

$$A = \partial\log(\Omega,\bar{\Omega}) \in \Omega^{1,0}(U).$$

The curvature form 2-form is

$$F = dA = -\partial\bar{\partial}\log\left(\| \Omega \|^2\right) \in \Omega^{1,1}(U).$$

The right-hand side is independent of the choice of the local holomorphic frame, so F is a global $(1,1)$ form on $\mathcal{M}_B$. More explicitly, we define

$$\mathbb{D} : \Omega^0(\mathcal{M}_B,\mathcal{L}^\vee) \to \Omega^{1,0}(\mathcal{M}_B,\mathbb{H}^{2,1})$$

by

$$\mathbb{D}s = \nabla's - D's = \nabla s - Ds.$$

Write

$$F = F_{i\bar{j}}dz^i \wedge d\bar{z}^j,$$

where $(z^1,\dots,z^\kappa)$ are the local holomorphic coordinates on $U\subset\mathcal{M}_B$. Define

$$\nabla_i := \nabla_{\frac{\partial}{\partial z^i}},\quad D_i := D_{\frac{\partial}{\partial z^i}},\quad \mathbb{D}_i = \nabla_i - D_i.$$

Then

$$\begin{aligned}
F_{i\bar{j}} &= -\frac{\partial^2}{\partial z^i\partial\bar{z}^j}\log(\Omega,\bar{\Omega})\\
&= -\frac{\partial}{\partial z^i}\frac{(\Omega,\overline{\nabla_j\Omega})}{(\Omega,\bar{\Omega})}\\
&= -\frac{(\nabla_i\Omega,\overline{\nabla_j\Omega})}{(\Omega,\bar{\Omega})}+\frac{(\nabla_i\Omega,\bar{\Omega})(\Omega,\overline{\nabla_j\Omega})}{(\Omega,\bar{\Omega})^2}\\
&= \frac{-(\nabla_i\Omega,\overline{\nabla_j\Omega})+(D_i\Omega,\overline{D_j\Omega})}{(\Omega,\bar{\Omega})}\\
&= \frac{-(\mathbb{D}_i\Omega,\overline{\mathbb{D}_j\Omega})}{(\Omega,\bar{\Omega})}
\end{aligned}$$

where the fourth equality follows from Equation (3.2), and the fifth (and last) equality follows from the identity $\nabla_k\Omega = D_k\Omega+\mathbb{D}_k\Omega$ where $D_k\Omega$ is a (3,0)-form and $\mathbb{D}_k\Omega$ is a (2,1)-form.

The above computation shows that the first Chern form

$$c_1(\mathcal{L}^\vee,D)=\frac{\sqrt{-1}}{2\pi}F$$

is a positive $(1,1)$-form.

3.9. The Weil-Petersson metric. The Weil-Petersson metric on $\mathcal{M}_B$ is defined by the Kähler form

$$\omega_{\mathrm{WP}} := 2\pi c_1(\mathcal{L}^\vee,D)=\sqrt{-1}F.$$

In local holomorphic coordinates, the Weil-Petersson metric is given by

$$G_{i\bar{j}} := \langle\frac{\partial}{\partial z_i},\frac{\partial}{\partial z_j}\rangle = \frac{-(\mathbb{D}_i\Omega,\overline{\mathbb{D}_j\Omega})}{(\Omega,\bar{\Omega})} = \frac{g_{i\bar{j}}}{g_{0\bar{0}}},$$

where $g_{i\bar{j}}$ is the Hodge metric on $\mathbb{H}^{2,1}$ and $g_{0\bar{0}}$ is the Hodge metric on $\mathcal{L}^\vee=\mathbb{F}^3$.

We have an isomorphism of Hermitian vector bundles:

$$T\mathcal{M}_B\otimes\mathcal{L}^\vee\cong\mathbb{H}^{2,1}, \tag{3.3}$$

where the tangent bundle $T\mathcal{M}_B$ is equipped with the Weil-Petersson metric $G_{i\bar{j}}$, while $\mathbb{H}^{2,1}$ is equipped with the Hodge metric $g_{i\bar{j}}$, and $\mathcal{L}^\vee$ is equipped with the Hodge metric $g_{0\bar{0}}$. The restriction of (3.3) to a point $q\in\mathcal{M}_B$ can be identified with

$$H^1(Y_q,T_Y)\otimes H^0(Y_q,\Omega^3_{Y_q})\cong H^1(Y_q,\Omega^2_{Y_q}).$$

4. B-model on compact Calabi-Yau threefold

4.1. Genus zero free energy.

$$x^0,x^1,\dots,x^\kappa,p_0,p_1,\dots,p_\kappa\in\Gamma(\mathcal{T},p^*\mathcal{L})$$

are holomorphic sections of the line bundle

$$p^*\mathcal{L}=\mathcal{P}_{\mathcal{T}}^*\mathcal{O}_{\mathbb{P}(V)}(1)$$

over the Torelli space $\mathcal{T}$. Note that

$$\mathcal{F}_0 := \frac{1}{2}\sum_{i=0}^{\kappa} x^i p_i$$

is a holomorphic section on $p^*\mathcal{L}^2 \to \mathcal{T}$ and is a multi-valued holomorphic section of $\mathcal{L}^2 \to \mathcal{M}_B$. Define the local holomorphic functions

$$z^i := \frac{x^i}{x^0}, \quad F_i := \frac{p_i}{x^0}, \quad g := \frac{p_0}{x^0}, \quad \check{F}_0 = (x^0)^{-2}\mathcal{F}_0, \tag{4.1}$$

where $i = 1, \ldots, \kappa$. In particular,

$$g = 2\check{F}_0 - \sum_{i=1}^{\kappa} z^i F_i.$$

The functions $(z^1, \ldots, z^\kappa)$ are known as special coordinates (or the B-model flat coordinates), which are local holomorphic coordinates on $\mathcal{T}$, defined in an open neighborhood of large complex structure with A^0 a vanishing cycle. The function $\check{F}_0$ is the B-model genus zero free energy.

We may write $\Omega = x^0\Omega_0$, where

$$\Omega_0 = \alpha_0 + \sum_{j=1}^{\kappa} z^j\alpha_j + \sum_{j=1}^{\kappa} F_j\beta^j + (2\check{F}_0 - \sum_{j=1}^{\kappa} z^j F_j)\beta^0.$$

For $i = 1, \ldots, \kappa$,

$$\nabla_i\Omega_0 = \alpha_i + \sum_{j=1}^{\kappa} \frac{\partial F_j}{\partial z^i}\beta^j + \left(2\frac{\partial \check{F}_0}{\partial z^i} - F_i - \sum z^j \frac{\partial F_j}{\partial z^i}\right)\beta^0.$$

Here, Ω_0 is a local holomorphic section of $\mathbb{F}^3$, and $\nabla_i\Omega_0$ is a local holomorphic section of $\mathbb{F}^2$ by Griffiths transversality, so Hodge-Riemann bilinear relations imply

$$0 = Q(\Omega_0, \nabla_i\Omega_0) = 2\frac{\partial \check{F}_0}{\partial z^i} - 2F_i.$$

To summarize:

- The B-model flat coordinates are given by

$$z^i = \frac{x^i}{x^0} = \frac{\int_{A^i}\Omega}{\int_{A^0}\Omega}, \quad i = 1, \ldots, \kappa.$$

- The B-model genus zero free energy $\check{F}_0$ is defined by

$$\check{F}_0 = \frac{1}{2}(x^0)^{-2}\cdot\sum_{i=0}^{\kappa} x^i p_i,$$

which satisfies

$$\frac{\partial \check{F}_0}{\partial z^i} = \frac{p_i}{x^0} = \frac{\int_{B_i}\Omega}{\int_{A^0}\Omega}, \quad i = 1, \ldots, \kappa.$$

4.2. Yukawa coupling.

$$\nabla_i\Omega_0 = \alpha_i + \sum_{j=1}^{\kappa} \frac{\partial^2 \check{F}_0}{\partial z^i \partial z^j}\beta^j \mod \beta^0.$$

$$\nabla_j\nabla_k\Omega_0 = \sum_{j,k,\ell=1}^{\kappa} \frac{\partial^3 \check{F}_0}{\partial z^j \partial z^k \partial z^\ell}\beta^\ell \mod \beta^0.$$

$$Q(\nabla_i\Omega_0, \nabla_j\nabla_k\Omega_0) = \frac{\partial^3 \check{F}_0}{\partial z^i \partial z^j \partial z^k}.$$

We also have

$$Q(\nabla_i\Omega_0, \nabla_j\nabla_k\Omega_0) = \frac{\partial}{\partial z^i}Q(\Omega_0, \nabla_j\nabla_k\Omega_0) - Q(\Omega_0, \nabla_i\nabla_j\nabla_k\Omega_0) = -Q(\Omega_0, \nabla_i\nabla_j\nabla_k\Omega_0).$$

So

$$\frac{\partial^3 \check{F}_0}{\partial z^i \partial z^j \partial z^k} = -Q(\Omega_0, \nabla_i\nabla_j\nabla_k\Omega_0) = -\int_Y \Omega_0 \wedge \nabla_i\nabla_j\nabla_k\Omega_0.$$

Define $C \in \Gamma((T^*M)^3 \otimes \mathcal{L}^2)$ by

$$\begin{aligned} C_{ijk} = C(\frac{\partial}{\partial z^i}, \frac{\partial}{\partial z^j}, \frac{\partial}{\partial z^k}) &:= -\int_Y \Omega \wedge \nabla_i\nabla_j\nabla_k\Omega \\ &= -(x^0)^2 \int_Y \Omega_0 \wedge \nabla_i\nabla_j\nabla_k\Omega_0 = (x^0)^2 \frac{\partial^3 \check{F}_0}{\partial z^i \partial z^j \partial z^k}. \end{aligned}$$

Then C_{ijk} is symmetric in i, j, k, so

$$C \in \Gamma(\mathrm{Sym}^3(T^*M) \otimes \mathcal{L}^2).$$

Define

$$C_{\bar{i}}^{jk} := \overline{C_{ilm}} g^{j\bar{l}} g^{k\bar{m}}.$$

Then

$$\mathcal{C} := C_{\bar{i}}^{jk} d\bar{z}^i \otimes \frac{\partial}{\partial z_j} \otimes \frac{\partial}{\partial z_k} \in \Omega^{0,1}(\mathcal{M}_B, (T\mathcal{M}_B)^{\otimes 2} \otimes \mathcal{L}^{-2}). \tag{4.2}$$

4.3. Genus one free energy. The genus one free energy $\check{F}_1$ is a linear combination of Ray-Singer torsions. A mathematical definition of $\check{F}_1$ is given in [**8**].

4.4. Genus $g \geq 2$ free energies and the holomorphic anomaly equations. The Weil-Petersson metric is Kähler, so the Chern connection on $T\mathcal{M}_B$ defined by the holomorphic structure and the Weil-Petersson metric is also torsion free. We equip $\mathcal{L}^\vee = \mathbb{F}^3$ with the connection D in Section 3.8. These two connections induce connections on tensor bundles

$$(T^*\mathcal{M}_B)^{\otimes m} \otimes \mathcal{L}^k$$

for any integers m, k. Let D be the covariant derivative defined by these connections.

The special homogeneous coordinates $x^0, x^1, \ldots, x^\kappa$ are local holomorphic sections of the vacuum line bundle $\mathcal{L}$. For $g \geq 2$,

$$\mathcal{F}_g(z, \bar{z}) \in \Gamma(\mathcal{M}_B, \mathcal{L}^{2-2g}).$$

The limit

$$\check{F}_g(z) = \lim_{\bar{z} \to \sqrt{-1}\infty} (x^0)^{2g-2} \mathcal{F}_g(z, \bar{z})$$

is a holomorphic function on U. The non-holomorphic section $\mathcal{F}_g(z,\bar{z})$ satisfies the following Holomorphic Anomaly Equation (BCOV [**4**]):

$$\bar{\partial}_i \mathcal{F}_g = \frac{1}{2} C_{\bar{i}}^{jk} \Big(D_j D_k \mathcal{F}_{g-1} + \sum_{\substack{g_1, g_2 > 0 \\ g_1 + g_2 = g}} D_j \mathcal{F}_{g_1} D_k \mathcal{F}_{g_2} \Big). \tag{4.3}$$

More precisely, we have

$$DD\mathcal{F}_{g-1} + \sum_{\substack{g_1, g_2 > 0 \\ g_1 + g_2 = g}} D\mathcal{F}_{g_1} \otimes D\mathcal{F}_{g_2} \in \Gamma(\mathcal{M}_B, (T^*\mathcal{M}_B)^{\otimes 2} \otimes \mathcal{L}^{4-2g}),$$

$$\mathcal{C} \in \Omega^{0,1}(\mathcal{M}_B, (T\mathcal{M}_B)^{\otimes 2} \otimes \mathcal{L}^{-2}),$$

where $\mathcal{C}$ is defined by Equation (4.2). Using the natural pairing between $T\mathcal{M}_B$ and $T^*\mathcal{M}_B$, we obtain

$$\mathcal{C} \diamond \Big(DD\mathcal{F}_{g-1} + \sum_{\substack{g_1, g_2 > 0 \\ g_1 + g_2 = g}} D\mathcal{F}_{g_1} \otimes D\mathcal{F}_{g_2} \Big) \in \Omega^{0,1}(\mathcal{M}_B, \mathcal{L}^{2-2g}).$$

With the above notation, the Holomorphic Anomaly Equation (4.3) can be rewritten in the following coordinate-free form:

$$D'' \mathcal{F}_g = \frac{1}{2} \mathcal{C} \diamond \Big(DD\mathcal{F}_{g-1} + \sum_{\substack{g_1, g_2 > 0 \\ g_1 + g_2 = g}} D\mathcal{F}_{g_1} \otimes D\mathcal{F}_{g_2} \Big). \tag{4.4}$$

4.5. B-model topological open string and the extended holomorphic anomaly equation. Let

$$\Delta_{ij} := D_j D_i \mathcal{F}_{(0,1)}$$

be the second covariant derivatives of the disk potential, and define

$$\Delta_{\bar{i}}^{j} := g^{j\bar{k}} \overline{\Delta_{ki}}.$$

Then

$$\Delta_{\bar{i}}^{j} d\bar{z}^i \otimes \frac{\partial}{\partial z_j} \in \Omega^{0,1}(\mathcal{M}_B, T\mathcal{M}_B \otimes \mathcal{L}^{-1}).$$

If $h > 0$ and $2g - 2 + h > 0$, then

$$\mathcal{F}_{(g,h)}(z,\bar{z}) \in \Gamma(\mathcal{L}^{2-2g-h}).$$

The limit

$$\check{F}_{(g,h)}(z) = \lim_{\bar{z} \to \infty} (x^0)^{2g-2+h} \mathcal{F}_{(g,h)}(z,\bar{z})$$

is a holomorphic function on U. The non-holomorphic section $\mathcal{F}_{(g,h)}(z,\bar{z})$ satisfies the following Extended Holomorphic Anomaly Equation (Walcher [**15**])

$$\bar{\partial}_i \mathcal{F}_{(g,h)} = \frac{1}{2} C_{\bar{i}}^{jk} \Big(D_j D_k \mathcal{F}_{g-1,h} + \sideset{}{'}\sum_{\substack{g_1 + g_2 = g \\ h_1 + h_2 = h}} D_j \mathcal{F}_{g_1,h_1} D_k \mathcal{F}_{g_2,h_2} \Big) - \Delta_{\bar{i}}^{j} D_j F_{g,h-1}, \tag{4.5}$$

where the sum $\sum'$ excludes the unstable case $(g_i, h_i) = (0,0), (0,1)$, and the last term on the RHS corresponds to the cases $(g_1, h_1) = (0,1)$ or $(g_2, h_2) = (0,1)$.

4.6. Mirror symmetry. Under the mirror map

$$(t^1,\dots,t^\kappa)=(z^1(q),\dots,z^\kappa(q))$$

we have the following mirror conjectures:

- For any $i,j,k\in\{1,\dots,\kappa\}$,
$$\frac{\partial^3\check{F}_0}{\partial z^i\partial z^j\partial z^k}(z)=\frac{\partial^3 F_0^X}{\partial t^i\partial t^j\partial t^k}(t).$$
- For $i=1,\dots,\kappa$,
$$\frac{\partial\check{F}_1}{\partial z^i}(z)=\frac{\partial F_1^X}{\partial t^i}(t),$$
or equivalently,
$$d\check{F}_1=dF_1^X.$$
- For $g\geq 2$,
$$\check{F}_g(z)=F_g^X(t).$$
- If $h>0$ and $2g-2+h>0$ then
$$\check{F}_{(g,h)}(z)=F^{X,L}_{(g,h)}(t).$$

Acknowledgments

This note is based on the author's talks at the 6th Workshop on Combinatorics of Moduli Spaces, Cluster Algebras, and Topological Recursion in Moscow on June 4–9, 2018, and the conference on Crossing the Walls in Enumerative Geometry in Snowbird, Utah on May 21–June 1, 2018. The author sincerely thanks the organizers of these events for the invitation to participate as a speaker. The author also wishes to thank Bohan Fang, Sheldon Katz, Zhengyu Zong, and the anonymous referee for their helpful comments on an earlier version of this note.

References

[1] Mina Aganagic, Vincent Bouchard, and Albrecht Klemm, *Topological strings and (almost) modular forms*, Comm. Math. Phys. **277** (2008), no. 3, 771–819, DOI 10.1007/s00220-007-0383-3. MR2365453

[2] Murad Alim and Jean Dominique Länge, *Polynomial structure of the (open) topological string partition function*, J. High Energy Phys. **10** (2007), 045, 13, DOI 10.1088/1126-6708/2007/10/045. MR2357933

[3] Murad Alim, Emanuel Scheidegger, Shing-Tung Yau, and Jie Zhou, *Special polynomial rings, quasi modular forms and duality of topological strings*, Adv. Theor. Math. Phys. **18** (2014), no. 2, 401–467. MR3273318

[4] M. Bershadsky, S. Cecotti, H. Ooguri, and C. Vafa, *Kodaira-Spencer theory of gravity and exact results for quantum string amplitudes*, Comm. Math. Phys. **165** (1994), no. 2, 311–427. MR1301851

[5] Giulio Bonelli and Alessandro Tanzini, *The holomorphic anomaly for open string moduli*, J. High Energy Phys. **10** (2007), 060, 17, DOI 10.1088/1126-6708/2007/10/060. MR2357918

[6] David A. Cox and Sheldon Katz, *Mirror symmetry and algebraic geometry*, Mathematical Surveys and Monographs, vol. 68, American Mathematical Society, Providence, RI, 1999. MR1677117

[7] Hao Fang and Zhiqin Lu, *Generalized Hodge metrics and BCOV torsion on Calabi-Yau moduli*, J. Reine Angew. Math. **588** (2005), 49–69, DOI 10.1515/crll.2005.2005.588.49. MR2196728

[8] Hao Fang, Zhiqin Lu, and Ken-Ichi Yoshikawa, *Analytic torsion for Calabi-Yau threefolds*, J. Differential Geom. **80** (2008), no. 2, 175–259. MR2454893

[9] Kentaro Hori, Sheldon Katz, Albrecht Klemm, Rahul Pandharipande, Richard Thomas, Cumrun Vafa, Ravi Vakil, and Eric Zaslow, *Mirror symmetry*, Clay Mathematics Monographs, vol. 1, American Mathematical Society, Providence, RI; Clay Mathematics Institute, Cambridge, MA, 2003. With a preface by Vafa. MR2003030

[10] Yukiko Konishi and Satoshi Minabe, *On solutions to Walcher's extended holomorphic anomaly equation*, Commun. Number Theory Phys. **1** (2007), no. 3, 579–603, DOI 10.4310/CNTP.2007.v1.n3.a5. MR2385247

[11] Atsushi Kanazawa and Jie Zhou, *Lectures on BCOV holomorphic anomaly equations*, Calabi-Yau varieties: arithmetic, geometry and physics, Fields Inst. Monogr., vol. 34, Fields Inst. Res. Math. Sci., Toronto, ON, 2015, pp. 445–473, DOI 10.1007/978-1-4939-2830-9_13. MR3409783

[12] R. Pandharipande, *Three questions in Gromov-Witten theory*, Proceedings of the International Congress of Mathematicians, Vol. II (Beijing, 2002), Higher Ed. Press, Beijing, 2002, pp. 503–512. MR1957060

[13] Andrew Strominger, *Special geometry*, Comm. Math. Phys. **133** (1990), no. 1, 163–180. MR1071240

[14] Gang Tian, *Smoothness of the universal deformation space of compact Calabi-Yau manifolds and its Petersson-Weil metric*, Mathematical aspects of string theory (San Diego, Calif., 1986), Adv. Ser. Math. Phys., vol. 1, World Sci. Publishing, Singapore, 1987, pp. 629–646. MR915841

[15] Johannes Walcher, *Extended holomorphic anomaly and loop amplitudes in open topological string*, Nuclear Phys. B **817** (2009), no. 3, 167–207, DOI 10.1016/j.nuclphysb.2009.02.006. MR2522663

[16] Satoshi Yamaguchi and Shing-Tung Yau, *Topological string partition functions as polynomials*, J. High Energy Phys. **7** (2004), 047, 20, DOI 10.1088/1126-6708/2004/07/047. MR2095047

Department of Mathematics, Columbia University, 2990 Broadway, New York, New York 10027

Email address: `ccliu@math.columbia.edu`

Selected Published Titles in This Series

763 **Tyler J. Jarvis and Nathan Priddis, Editors,** Singularities, Mirror Symmetry, and the Gauged Linear Sigma Model, 2021

760 **Olivier Collin, Stefan Friedl, Cameron Gordon, Stephan Tillmann, and Liam Watson, Editors,** Characters in Low-Dimensional Topology, 2020

759 **Omayra Ortega, Emille Davie Lawrence, and Edray Herber Goins, Editors,** The Golden Anniversary Celebration of the National Association of Mathematicians, 2020

758 **Jan Šťovíček and Jan Trlifaj, Editors,** Representation Theory and Beyond, 2020

757 **Kaïs Ammari and Stéphane Gerbi, Editors,** Identification and Control: Some New Challenges, 2020

756 **Joeri Van der Veken, Alfonso Carriazo, Ivko Dimitrić, Yun Myung Oh, Bogdan D. Suceavă, and Luc Vrancken, Editors,** Geometry of Submanifolds, 2020

755 **Marion Scheepers and Ondřej Zindulka, Editors,** Centenary of the Borel Conjecture, 2020

754 **Susanne C. Brenner, Igor Shparlinski, Chi-Wang Shu, and Daniel B. Szyld, Editors,** 75 Years of Mathematics of Computation, 2020

753 **Matthew Krauel, Michael Tuite, and Gaywalee Yamskulna, Editors,** Vertex Operator Algebras, Number Theory and Related Topics, 2020

752 **Samuel Coskey and Grigor Sargsyan, Editors,** Trends in Set Theory, 2020

751 **Ashish K. Srivastava, André Leroy, Ivo Herzog, and Pedro A. Guil Asensio, Editors,** Categorical, Homological and Combinatorial Methods in Algebra, 2020

750 **A. Bourhim, J. Mashreghi, L. Oubbi, and Z. Abdelali, Editors,** Linear and Multilinear Algebra and Function Spaces, 2020

749 **Guillermo Cortiñas and Charles A. Weibel, Editors,** K-theory in Algebra, Analysis and Topology, 2020

748 **Donatella Danielli and Irina Mitrea, Editors,** Advances in Harmonic Analysis and Partial Differential Equations, 2020

747 **Paul Bruillard, Carlos Ortiz Marrero, and Julia Plavnik, Editors,** Topological Phases of Matter and Quantum Computation, 2020

746 **Erica Flapan and Helen Wong, Editors,** Topology and Geometry of Biopolymers, 2020

745 **Federico Binda, Marc Levine, Manh Toan Nguyen, and Oliver Röndigs, Editors,** Motivic Homotopy Theory and Refined Enumerative Geometry, 2020

744 **Pieter Moree, Anke Pohl, L'ubomír Snoha, and Tom Ward, Editors,** Dynamics: Topology and Numbers, 2020

743 **H. Garth Dales, Dmitry Khavinson, and Javad Mashreghi, Editors,** Complex Analysis and Spectral Theory, 2020

742 **Francisco-Jesús Castro-Jiménez, David Bradley Massey, Bernard Teissier, and Meral Tosun, Editors,** A Panorama of Singularities, 2020

741 **Houssam Abdul-Rahman, Robert Sims, and Amanda Young, Editors,** Analytic Trends in Mathematical Physics, 2020

740 **Alina Bucur and David Zureick-Brown, Editors,** Analytic Methods in Arithmetic Geometry, 2019

739 **Yaiza Canzani, Linan Chen, and Dmitry Jakobson, Editors,** Probabilistic Methods in Geometry, Topology and Spectral Theory, 2019

738 **Shrikrishna G. Dani, Surender K. Jain, Jugal K. Verma, and Meenakshi P. Wasadikar, Editors,** Contributions in Algebra and Algebraic Geometry, 2019

737 **Fernanda Botelho, Editor,** Recent Trends in Operator Theory and Applications, 2019

For a complete list of titles in this series, visit the
AMS Bookstore at **www.ams.org/bookstore/conmseries/**.